STATA USER'S GUIDE
RELEASE 12

A Stata Press Publication
StataCorp LP
College Station, Texas

Version 12

Published by Stata Press, 4905 Lakeway Drive, College Station, Texas 77845
Typeset in TEX
Printed in the United States of America
10 9 8 7 6 5 4 3 2 1

ISBN-10: 1-59718-095-5
ISBN-13: 978-1-59718-095-5

For copyright information about the software, type `help copyright` within Stata.

The suggested citation for this software is

StataCorp. 2011. *Stata: Release 12*. Statistical Software. College Station, TX: StataCorp LP.

Table of contents

Stata basics

Elements of Stata

Advice

Stata basics

1 Read this—it will help

Contents

A Complete Stata Documentation Set contains more than 9,000 pages of information in the following manuals:

[GS]	*Getting Started with Stata* (Mac, Unix, or Windows)
[U]	*Stata User's Guide*
[R]	*Stata Base Reference Manual*
	Volume 1, A–F
	Volume 2, G–M
	Volume 3, N–R
	Volume 4, S–Z
[D]	*Stata Data-Management Reference Manual*
[G]	*Stata Graphics Reference Manual*
[XT]	*Stata Longitudinal-Data/Panel-Data Reference Manual*
[MI]	*Stata Multiple-Imputation Reference Manual*
[MV]	*Stata Multivariate Statistics Reference Manual*
[P]	*Stata Programming Reference Manual*
[SEM]	*Stata Structural Equation Modeling Reference Manual*
[SVY]	*Stata Survey Data Reference Manual*
[ST]	*Stata Survival Analysis and Epidemiological Tables Reference Manual*
[TS]	*Stata Time-Series Reference Manual*
[I]	*Stata Quick Reference and Index*
[M]	*Mata Reference Manual*
	Volume 1, [M-1]–[M-3]
	Volume 2, [M-4]–[M-6]

Detailed information about each of these manuals and the other Stata manuals may be found online at

http://www.stata-press.com/manuals/

In addition, installation instructions may be found in the *Installation Guide*, which comes in the DVD case.

1.1 Getting Started with Stata

There are three *Getting Started* manuals:

[GSM] *Getting Started with Stata for Mac*
[GSU] *Getting Started with Stata for Unix*
[GSW] *Getting Started with Stata for Windows*

1. Learn how to use Stata—read the *Getting Started* (GSM, GSU, or GSW) manual.
2. Now turn to the other manuals; see [U] **1.2 The User's Guide and the Reference manuals**.

1.2 The User's Guide and the Reference manuals

The *User's Guide* is divided into three sections: *Stata basics*, *Elements of Stata*, and *Advice*. At the beginning of each section is a list of the chapters found in that section. In addition to helping you explore the fundamentals of Stata—information that all users should know—this manual will guide you to other sources for Stata information.

The other manuals are the *Reference* manuals. The Stata *Reference* manuals are each arranged like an encyclopedia—alphabetically. Look at the *Base Reference Manual*. Look under the name of a command. If you do not find the command, look in the index. A few commands are so closely related that they are documented together, such as `ranksum` and `median`, which are both documented in [R] **ranksum**.

Not all the entries in the *Base Reference Manual* are Stata commands; some contain technical information, such as [R] **maximize**, which details Stata's iterative maximization process, or [R] **error messages**, which provides information on error messages and return codes.

Like an encyclopedia, the *Reference* manuals are not designed to be read from cover to cover. When you want to know what a command does, complete with all the details, qualifications, and pitfalls, or when a command produces an unexpected result, read its description. Each entry is written at the level of the command. The descriptions assume that you have little knowledge of Stata's features when they are explaining simple commands, such as those for using and saving data. For more complicated commands, they assume that you have a firm grasp of Stata's other features.

If a Stata command is not in the *Base Reference Manual*, you can find it in one of the other *Reference* manuals. The titles of the manuals indicate the types of commands that they contain. The *Programming Reference Manual*, however, contains commands not only for programming Stata but also for manipulating matrices (not to be confused with the matrix programming language described in the *Mata Reference Manual*).

1.2.1 PDF manuals

Every copy of Stata comes with a PDF version of Stata's complete documentation. Of course, printed manuals are also available.

The PDF documentation may be accessed from within Stata by selecting **Help** > **PDF Documentation**. Even more convenient, every help file in Stata links to the equivalent manual entry. If you are reading **help regress**, simply click on [R] **regress** in the `Title` section of the help file to go directly to the [R] **regress** manual entry.

We provide recommended settings for your PDF viewer to optimize it for Stata's documentation at http://www.stata.com/support/faqs/res/documentation.html.

1.2.2 Example datasets

Various examples in this manual use what is referred to as the automobile dataset, `auto.dta`. We have created a dataset on the prices, mileages, weights, and other characteristics of 74 automobiles and have saved it in a file called `auto.dta`. (These data originally came from the April 1979 issue of *Consumer Reports* and from the United States Government EPA statistics on fuel consumption; they were compiled and published by Chambers et al. [1983].)

In our examples, you will often see us type

```
. use http://www.stata-press.com/data/r12/auto
```

We include the auto.dta file with Stata. If you want to use it from your own computer rather than via the Internet, you can type

```
. sysuse auto
```

See [D] **sysuse**.

You can also access auto.dta by selecting **File > Example Datasets...**, clicking on *Example datasets installed with Stata*, and clicking on use beside the auto.dta filename.

There are many other example datasets that ship with Stata or are available over the web. Here is a partial list of the example datasets included with Stata:

auto.dta	1978 Automobile Data
autornd.dta	Subset of 1978 Automobile Data
bplong.dta	fictional blood pressure data
bpwide.dta	fictional blood pressure data
cancer.dta	Patient Survival in Drug Trial
census.dta	1980 Census data by state
citytemp.dta	City Temperature Data
citytemp4.dta	City Temperature Data
educ99gdp.dta	Education and GDP
gnp96.dta	U.S. GNP, 1967–2002
lifeexp.dta	Life expectancy, 1978
nlsw88.dta	U.S. National Longitudinal Study of Young Women (NLSW, 1988 extract)
pop2000.dta	U.S. Census, 2000, extract
sp500.dta	S&P 500
uslifeexp.dta	U.S. life expectancy, 1900–1999
uslifeexp2.dta	U.S. life expectancy, 1900–1940
voter.dta	1992 presidential voter data

All these datasets may be used or described from the **Example Datasets...** menu listing.

Even more example datasets, including most of the datasets used in the reference manuals, are available at the Stata Press website (http://www.stata-press.com/data/). You can download the datasets with your browser, or you can use them directly from the Stata command line:

```
. use http://www.stata-press.com/data/r12/nlswork
```

An alternative to the use command for these example datasets is webuse. For example, typing

```
. webuse nlswork
```

is equivalent to the above use command. For more information, see [D] **webuse**.

1.2.3 Cross-referencing

The *Getting Started* manual, the *User's Guide*, and the *Reference* manuals cross-reference each other.

[R] **regress**
[D] **reshape**
[XT] **xtmixed**

The first is a reference to the regress entry in the *Base Reference Manual*, the second is a reference to the reshape entry in the *Data-Management Reference Manual*, and the third is a reference to the xtmixed entry in the *Longitudinal-Data/Panel-Data Reference Manual*.

[GSW] **C Advanced Stata usage**
[GSM] **C Advanced Stata usage**
[GSU] **C Advanced Stata usage**

are instructions to see the appropriate section of the *Getting Started with Stata for Windows*, *Getting Started with Stata for Mac*, or *Getting Started with Stata for Unix* manual.

1.2.4 The index

At the end of each manual is an index for that manual. The index for the *Base Reference Manual* is found at the end of the fourth volume. The index for the *Mata Reference Manual* is found at the end of the second volume.

The *Quick Reference and Index* contains a combined index for all the manuals. It also contains quick reference information on subjects such as the estimation commands.

To find information and commands quickly, you can use Stata's `search` or `hsearch` commands; see [R] **search** and [R] **hsearch**. At the Stata command prompt, try typing `search geometric mean`. `search` searches Stata's keyword database and, optionally, will search the Internet, too. Try typing `search geometric mean, all`. Searching the Internet lets you find more commands and extensions for Stata written by Stata users. `hsearch`, on the other hand, searches the text of all the Stata help files on your computer, including help files that you have written and help files from user-written extensions that you have installed, and orders the outcome by relevancy. Try typing `hsearch geometric mean`. Between these two commands, if a capability exists, you should be able to find it.

1.2.5 The subject table of contents

A subject table of contents for the *User's Guide* and all the *Reference* manuals except the *Mata Reference Manual* is located in the *Quick Reference and Index*. This subject table of contents may also be accessed by clicking on **Contents** in the PDF bookmarks.

If you look under "Functions and expressions", you will see

[U] Chapter 13 .... Functions and expressions
[D] datetime .... Date and time (%t) values and variables
[D] egen .... Extensions to generate
[D] functions .... Functions

1.2.6 Typography

We mix the ordinary typeface that you are reading now with a typewriter-style typeface `that looks like this`. When something is printed in the typewriter-style typeface, it means that something is a command or an option—it is something that Stata understands and something that you might actually type into your computer. Differences in typeface are important. If a sentence reads, "You could list the result ...", it is just an English sentence—you *could* list the result, but the sentence provides no clue as to how you might actually do that. On the other hand, if the sentence reads, "You could `list` the result ...", it is telling you much more—you could list the result, and you could do that by using the `list` command.

We will occasionally lapse into periods of inordinate cuteness and write, "We `describe`d the data and then `list`ed the data." You get the idea. `describe` and `list` are Stata commands. We purposely began the previous sentence with a lowercase letter. Because `describe` is a Stata command, it must be typed in lowercase letters. The ordinary rules of capitalization are temporarily suspended in favor of preciseness.

We also mix in words printed in italic type, such as "To perform the rank-sum test, type `ranksum` *varname*, `by(`*groupvar*`)`". Italicized words are not supposed to be typed; instead, you are to substitute another word for them.

We would also like users to note our rule for punctuation of quotes. We follow a rule that is often used in mathematics books and British literature. The punctuation mark at the end of the quote is included in the quote only if it is a part of the quote. For instance, the pleased Stata user said she thought that Stata was a "very powerful program". Another user simply said, "I love Stata."

In this manual, however, there is little dialogue, and we follow this rule to precisely clarify what you are to type, as in, type "`cd c:`". The period is outside the quotation mark because you should not type the period. If we had wanted you to type the period, we would have included two periods at the end of the sentence: one inside the quotation and one outside, as in, type "the orthogonal polynomial operator, `p.`".

We have tried not to violate the other rules of English. If you find such violations, they were unintentional and resulted from our own ignorance or carelessness. We would appreciate hearing about them.

We have heard from Nicholas J. Cox of the Department of Geography at Durham University in the United Kingdom and express our appreciation. His efforts have gone far beyond dropping us a note, and there is no way with words that we can fully express our gratitude.

1.2.7 Vignettes

If you look, for example, at the entry [R] **brier**, you will see a brief biographical vignette of Glenn Wilson Brier (1913–1998), who did pioneering work on the measures described in that entry. A few such vignettes were added without fanfare in the Stata 8 manuals, just for interest, and many more were added in Stata 9, and even more have been added in Stata 10, Stata 11, and Stata 12. A vignette could often appropriately go in several entries. For example, George E. P. Box deserves to be mentioned in entries other than [TS] **arima**, such as [R] **boxcox**. However, to save space, each vignette is given once only, and an index of all vignettes is given in the *Quick Reference Manual and Index*.

All the vignettes were written by Nicholas J. Cox, Durham University, and were compiled using a wide range of reference books, articles in the literature, Internet sources, and information from individuals. Especially useful were the dictionaries of Upton and Cook (2006) and Everitt and Skrondal (2010) and the compilations of statistical biographies edited by Heyde and Seneta (2001) and Johnson and Kotz (1997). Of these, only the first provides information on people living at the time of publication.

1.3 What's new

This section is intended for users of the previous version of Stata. If you are new to Stata, you may as well skip to [U] **1.3.14 What's more**.

As always, Stata 12 is 100% compatible with the previous releases, but we remind programmers that it is important to put `version 11.2`, `version 11.1`, `version 11`, or `version 10`, etc., at the top of old do- and ado-files so that they continue to work as you expect. You were supposed to do that when you wrote them, but if you did not, go back and do it now.

We will list all the changes, item by item, but first, here are the highlights.

1.3.1 What's new (highlights)

Here are the highlights. There is more, and do not assume that because we mention a category, we have mentioned everything new in the category. Detailed sections follow the highlights.

1. **Automatic memory management**, which means that you no longer have to `set memory` and never again will you be told that there is no room because you set too little! Stata automatically adjusts its memory usage up and down according to current requirements.

   The memory manager is tunable. We recommend the default settings. See [D] **memory** if you are interested.

   Old do-files can still `set memory`. Stata merely responds, "`set memory` ignored".

2. **Structural equation modeling (SEM)**, via the new `sem` command, is itself the subject of the new *Stata Structural Equation Modeling Reference Manual*. SEM fits multivariate linear models that can include observed and latent variables. These models include confirmatory factor analysis, linear models, instrumental variables, 2SLS, 3SLS, multivariate regression, seemingly unrelated least squares, recursive systems, simultaneous systems, path analysis, latent variables, MIMIC, modeling of direct and indirect effects, and more. All the above can be estimated by maximum likelihood with or without missing values, GLS, or ADF (asymptotic distribution free, also known as GMM). Missing values are handled using FIML. Raw and standardized coefficients and effects are reported. All models may be fit across groups and include tests for group invariance. Modification indices and score tests are provided.

   Models may be specified and reported using commands or interactive path diagrams. See the new *Stata Structural Equation Modeling Reference Manual*.

3. **MI**, multiple imputation,

   a. **Chained equations**, which is to say, fully conditional specifications for imputing missing values given arbitrary patterns for continuous, binary, ordinal, cardinal, or count variables. See [MI] **mi impute chained**.

   b. **Four new imputation methods**. You can impute

      1) truncated data,

      2) interval-censored data,

      3) count data, and

      4) overdispersed count data.

   See [MI] **mi impute truncreg**, [MI] **mi impute intreg**, [MI] **mi impute poisson**, and [MI] **mi impute nbreg**.

   c. **Conditional imputation** is now supported by all univariate imputation methods, which is to say, you can impute values for variables with restrictions, such as the number of pregnancies being imputed only for females, even if female itself is imputed. See *Conditional imputation* in [MI] **mi impute** and new option `conditional()` in the univariate imputation entries such as [MI] **mi impute regress**.

   d. **Imputation by groups**, which is to say, imputations can be made separately for different groups of the data. See new option `by()` in [MI] **mi impute**.

   e. **Imputation by drawing posterior estimates from bootstrapped samples**. See new option `bootstrap` in the univariate imputation entries such as [MI] **mi impute regress**.

   f. **Panel-data and multilevel models** are now supported by `mi estimate`. Included are `xtcloglog`, `xtgee`, `xtlogit`, `xtmelogit`, `xtmepoisson`, `xtmixed`, `xtnbreg`, `xtpoisson`, `xtprobit`, `xtrc`, and `xtreg`. See [MI] **estimation**.

g. **Linear and nonlinear predictions after MI estimation** using new commands `mi predict` and `mi predictnl`. See [MI] **mi predict**.

h. **Monte Carlo jackknife error estimates** obtained by omitting one imputation at a time and reapplying the combination rules. See new option `mcerror` in [MI] **mi estimate**.

4. **Longitudinal/panel data**,

   a. **Survey feature support for xtmixed** including multilevel sampling weights and robust variance estimators. See [XT] **xtmixed**.

   b. **Documentation for xtmixed, xtmelogit, and xtmepoisson has been modified to adopt the standard "level" terminology from the literature on hierarchical models.** See the *Introduction* section of *Remarks* in both [XT] **xtmixed** and [XT] **xtmelogit**.

   c. **xtmixed now uses maximum likelihood (ML) as the default method of estimation**. See [XT] **xtmixed**.

5. **Contour plots**. Filled and outlined plots are available. See [G-2] **graph twoway contour** and [G-2] **graph twoway contourline**.

6. **Contrasts**, which is to say, tests of linear hypotheses involving factor variables and their interactions from the most recently fit model, and that model can be virtually any model that Stata can fit. Tests include ANOVA-style tests of main effects, simple effects, interactions, and nested effects. Effects can be decomposed into comparisons with reference categories, comparisons of adjacent levels, comparisons with the grand mean, and more. See [R] **contrast** and [R] **margins, contrast**.

7. **Pairwise comparisons** of means, estimated cell means, estimated marginal means, predictive margins of linear and nonlinear responses, intercepts, and slopes. In addition to ANOVA-style comparisons, comparisons can be made of population averages. See [R] **pwmean**, [R] **pwcompare**, and [R] **margins, pwcompare**.

8. **Graphs of margins, marginal effects, contrasts, and pairwise comparisons**. Margins and effects can be obtained from linear or nonlinear (for example, probability) responses. See [R] **marginsplot**.

9. **Time series**,

   a. **MGARCH**, which is to say, multivariate GARCH, which is to say, estimation of multivariate generalized autoregressive conditional heteroskedasticity models of volatility, and this includes constant, dynamic, and varying conditional correlations, also known as the CCC, DCC, and VCC models. Innovations in these models may follow multivariate normal or Student's t distributions. See [TS] **mgarch**.

   b. **UCM**, which is to say, unobserved components models, also known as structural time-series models that decompose a series into trend, seasonal, and cyclical components, and which were popularized by Harvey (1989). See [TS] **ucm**.

   c. **ARFIMA**, which is to say, autoregressive fractionally integrated moving-average models, useful for long-memory processes. See [TS] **arfima**.

   d. **Filters for extracting business and seasonal cycles**. Four popular time-series filters are provided: the Christiano–Fitzgerald and the Baxter–King band-pass filters, and the Hodrick–Prescott and the Butterworth high-pass filters. See [TS] **tsfilter**.

10. **Business dates** allow you to define your own calendars so that they display correctly and lags and leads work as they should. You could create file `lse.stbcal` that recorded the days the London Stock Exchange is open (or closed) and then Stata would understand format `%tblse` just as it understands the usual date format `%td`. Once you define a calendar, Stata deeply understands it. You can, for instance, easily convert between `%tblse` and `%td` values. See [D] **datetime business calendars**.

11. **Improved documentation for date and time variables**. Anyone who has ever been puzzled by Stata's date and time variables, which is to say, anyone who uses them, should see [D] **datetime**, [D] **datetime translation**, and [D] **datetime display formats**.

12. **ROC adjusted for covariates**, which is to say, you can model the ROC curve and obtain coefficients, standard errors, and graphs. Nonparametric and parametric estimation is supported. See [R] **rocreg** and [R] **rocregplot**.

13. **Survey SDR weights**, which is to say, successive difference replicate weights, which are supplied with many datasets from the U.S. Census Bureau. See [SVY] **svy sdr**.

14. **Bootstrap standard errors for survey data** using user-supplied bootstrap replicate weights. See [SVY] **svy bootstrap**.

15. **Importing and exporting**,

    a. **Excel files**. And the new import preview tool lets you see the data before you import them. See [D] **import excel**.

    b. **EBCDIC files, importing**. And you can convert between EBCDIC and ASCII formats; see [D] **infile (fixed format)** and [D] **filefilter**.

    c. **ODBC connection strings**. See [D] **odbc**.

    d. **PDF export for graphs and logs** lets you directly create PDFs from your Stata results. See [G-2] **graph export** and [R] **translate**.

16. **Renaming groups of variables** is now easy using `rename`'s new syntax that is 100% compatible with its old syntax. You can change names, swap names, renumber indices within variable names, and more. See [D] **rename group**.

17. **Stata interface**,

    a. **New layout** is wider and fits most screens better.

    b. **New Properties window** lets you manage the properties of your variables including names, labels, value labels, notes, display formats, and storage types. And you can manage the properties of your dataset.

    c. **Filtering of Review and Variables windows** lets you type text and see only the matches.

    d. **Searching in the Results window** lets you find results.

    e. **Expression Builder now accesses parameter estimates, returned results, macros, and more**, so you can build expressions for `nlcom` and `testnl`. It is worth a test drive.

    f. **Unified interface for Mac** means no more lost windows; all the Stata windows are tied together.

    g. **Gesture support for Mac** makes changing font sizes and moving forward and backward easy.

    h. **Tabbed graphs for Mac**.

    i. **File drag-and-drop for Windows**—Stata for Mac already had it—now Stata for Windows does, too.

18. **Data Editor**,

    a. **New tool for managing variables** lets you hide/show variables (and includes filtering!), sort variables, and reorder variables via drag and drop. And it includes Stata's new Properties tool, so you can manage your data more easily from the Data Editor. Try it. Click on the Variables tool in the toolbar.

    b. **New Clipboard Preview Tool** lets you see the data before you paste them into Stata and lets you control how the data will be pasted.

c. **Clipboard preserves variable properties** such as display formats and types when you copy-and-paste data within Stata.

19. **All-new Viewer**,

a. **Quick access to dialogs, sections, and "also see" references** via three pulldown menus at the top of the Viewer for quick navigation inside help files.

b. **Tabbed Viewer** lets you open multiple help files and documents and switch between them.

20. **Do-file Editor**,

a. **Tabbed for Mac and Unix**—Stata for Windows already had it—now Stata for Mac and Unix do, too.

b. **Syntax highlighting and bookmarks for Mac**—Stata for Windows already had it—now Stata for Mac does, too.

21. **Estimation output improved**,

a. **Baseline odds now shown**, which is to say, the exponentiated intercept is displayed by `logistic` and by `logit` with option `or`. In fact, all estimation commands show exponentiated intercepts when option `eform()` or its equivalent is specified. For example, `poisson` shows the baseline incidence rate when option `irr` is specified.

b. **Implied zero coefficients now shown**. When a coefficient is omitted, it is now shown as being zero and the reason it was omitted—collinearity, base, empty—is shown in the standard-error column. (The word "omitted" is shown if the coefficient was omitted because of collinearity.)

c. **You can set displayed precision for all values in coefficient tables** using `set cformat`, `set pformat`, and `set sformat`. Or you may use options `cformat()`, `pformat()`, and `sformat()` now allowed on all estimation commands. See [R] **set cformat** and [R] **estimation options**.

d. Estimation commands now respect the width of the Results window. This feature may be turned off by new display option `nolstretch`. See [R] **estimation options**.

e. **You can now set whether base levels, empty cells, and omitted are shown** using `set showbaselevels`, `set showemptycells`, and `set showomitted`. See [R] **set showbaselevels**.

22. **More MP speed ups**, meaning faster execution for those running Stata/MP.

a. **Improved MP support for factor variables used in estimation**. Execution is much faster when there are lots of levels.

b. **Faster maximum likelihood execution with large numbers of covariates**. Processors being assigned on the basis of variables rather than observations when there are 300 or more covariates results in improved performance.

c. **Improved performance on 16 or more cores** due to better tuning.

d. **Up to 64 cores now supported**, up from 32.

23. **Installation Qualification** is now provided by a new tool which you download for free. The tool produces a report for submission to regulatory agencies such as the FDA to establish that Stata is installed correctly. Visit http://www.stata.com/support/installation-qualification/.

1.3.2 What's new in the GUI and command interface

1. **Highlights,**

a. **New layout.**

b. **New Properties window.**

c. **Filtering of Review and Variable windows.**

d. **Searching in the Results window.**

e. **Expression Builder can access parameter estimates, ...**

f. **Unified interface for Mac.**

g. **Gesture support for Mac.**

h. **Tabbed Graphs for Mac.**

i. **File drag-and-drop for Windows.**

j. **Data Editor, new tool for managing variables.**

k. **Data Editor, new Clipboard Preview Tool.**

l. **Data Editor, Clipboard preserves variable properties.**

m. **New Viewer, quick access to dialogs, sections, ...**

n. **New Viewer, tabbed.**

o. **Do-file Editor, tabbed for Mac and Unix.**

p. **Do-file Editor, syntax highlighting and bookmarks for Mac.**

See [U] **1.3.1 What's new (highlights)**.

2. **Aero Snap functionality for Viewer in Windows 7.**
3. **Stata for Unix dialog boxes now have full varlist controls.**

1.3.3 What's new in data management

1. **Highlights**,

   a. **Automatic memory management.** See [D] **memory**.

   b. **Excel files, importing and exporting.** See [D] **import excel**.

   c. **EBCDIC files, importing.** See [D] **infile (fixed format)** and [D] **filefilter**.

   d. **ODBC connection strings, importing and exporting.** See [D] **odbc**.

   e. **PDF files, exporting of graphs and logs.** See [R] **translate**.

   f. **Business dates.** See [D] **datetime business calendars**.

   g. **Improved documentation for date and time variables.** See [D] **datetime**, [D] **datetime translation**, and [D] **datetime display formats**.

   h. **Renaming groups of variables.** See [D] **rename group**.

   See [U] **1.3.1 What's new (highlights)**.

2. **New functions**,

   a. **Tukey's Studentized range**, cumulative and inverse, `tukeyprob()` and `invtukeyprob()`.

   b. **Dunnett's multiple range**, cumulative and inverse, `dunnettprob()` and `invdunnettprob()`.

   c. **New date conversion functions** `dofb()` and `bofd()` convert between business dates and standard calendar dates. See [D] **datetime business calendars**.

   See [D] **functions**.

3. **ODBC support for Oracle Solaris.** See [D] **odbc**.

4. **New Stata commands getmata and putmata** make it easy to transfer your data into Mata, manipulate them, and then transfer them back to Stata. `getmata` and `putmata` are especially designed for interactive use. See [D] **putmata**.

5. **New Stata commands import sasxport, export sasxport, and import sasxport, describe** replace existing commands `fdause`, `fdasave`, and `fdadescribe`. `fdause`, `fdasave`, and `fdadescribe` are understood as synonyms. See [D] **import sasxport**.

6. **xshell** support for Mac. See [D] **shell**.

1.3.4 What's new in statistics (general)

1. **Highlights**,

   a. **Contrasts**. See [R] **contrast** and [R] **margins, contrast**.

   b. **Pairwise comparisons**. See [R] **pwmean**, [R] **pwcompare**, and [R] **margins, pwcompare**.

   c. **Graphs of margins, marginal effects, contrasts, ...** . See [R] **marginsplot**.

   d. **ROC adjusted for covariates.** See [R] **rocreg** and [R] **rocregplot**.

   e. **Estimation output improved**:
   —Baseline odds now shown.
   —Implied zero coefficients now shown.
   —You can set displayed precision. See [R] **set cformat** and [R] **estimation options**.
   —Estimation commands now respect the width of the Results window. See [R] **estimation options**.
   —You can now set whether base levels, empty cells, and omitted are shown. See [R] **set showbaselevels** and [R] **estimation options**.

   See [U] **1.3.1 What's new (highlights)**.

2. **test with coefficient names not using _b[] notation is now allowed**, even when the specified variables no longer exist in the current dataset. See [R] **test**.

3. **areg now faster.** `areg` is orders of magnitude faster when there are hundreds of absorption groups, even if you are not running Stata/MP. See [R] **areg**.

4. **misstable summarize will now create summary variable** recording the missing-values pattern. See new option `generate()` for `summarize` in [R] **misstable**.

5. **margins command supports contrasts.** See [R] **margins, contrast** and [R] **contrast**.

6. **sfrancia uses better algorithm.** `sfrancia` now uses an algorithm based on the log transformation for approximating the sampling distribution of the W' statistic for testing normality. The old algorithm, using the Box–Cox transformation, is available under version control or via the new `boxcox` option. Based on simulation, the new algorithm is more powerful for sample sizes greater than 1,000 and is comparable to the old algorithm for sample sizes less than 1,000. Also, similarly to `swilk`, `sfrancia` now allows you to suppress the treatment of ties when option `noties` is used. See [R] **swilk**.

7. **logistic now allows option noconstant.** See [R] **logistic**.

8. **Probability predictions now available.** `predict` after count-data models, such as `poisson` and `nbreg`, can now predict the probability of any count or count range. See [R] **nbreg postestimation**, [R] **poisson postestimation**, [R] **tnbreg postestimation**, [R] **tpoisson postestimation**, [R] **zinb postestimation**, and [R] **zip postestimation**.

9. **Truncated count-data models now available.** New estimation commands `tpoisson` and `tnbreg` fit models of count-data outcomes with any form of left truncation, including truncation that varies observation by observation. These new commands supersede `ztp` and `ztnb`. See [R] **tpoisson** and [R] **tnbreg**.

10. **cnsreg checks for collinear variables prior to estimation** and has new option `collinear`, which keeps the collinear variables instead of omitting them. The old behavior of always keeping collinear variables is preserved under version control. See [R] **cnsreg**.

11. **ml improved**,

    a. `ml` now distinguishes the Hessian matrix produced by `technique(nr)` from the other techniques that compute a substitute for the Hessian matrix. This means that `ml` will compute the real Hessian matrix of second derivatives to determine convergence when all other convergence tolerances are satisfied and `technique(bfgs)`, `technique(bhhh)`, or `technique(dfp)` is in effect.

       The old behavior was to use the `nrtolerance()` value with the *H* matrix associated with the `technique()` currently in effect to determine convergence; this behavior is preserved under version control.

    b. `ml` has new option `qtolerance()` that distinguishes itself from `nrtolerance()` when `technique(bfgs)`, `technique(bhhh)`, or `technique(dfp)` is specified. Option `qtolerance()` replaces `nrtolerance()` when `technique(bfgs)`, `technique(bhhh)`, or `technique(dfp)` is in effect.

    See [R] **ml** and [R] **maximize**.

12. **margins has new option estimtolerance() for setting tolerance** used to determine estimable functions. See [R] **margins**.

13. **Option addplot() now places added graphs above or below.** Commands that allow option `addplot()` can now place the added plots above or below the command's plots.

1.3.5 What's new in statistics (longitudinal/panel data)

1. **Highlights**,

    a. **MI support for panel-data and multilevel models** including `xtcloglog`, `xtgee`, `xtlogit`, `xtmelogit`, `xtmepoisson`, `xtmixed`, `xtnbreg`, `xtpoisson`, `xtprobit`, `xtrc`, and `xtreg`. See [MI] **estimation**.

    b. **Survey feature support for linear multilevel models**, `xtmixed`, including multilevel sampling weights and robust variance estimators. See [XT] **xtmixed**.

    c. **Documentation for xtmixed, xtmelogit, and xtmepoisson has been modified to adopt the standard "level" terminology from the literature on hierarchical models.** For example, what in previous Stata versions was considered a one-level model is now called a two-level model with the observations now being counted as "level one"; see the *Introduction* section of *Remarks* in both [XT] **xtmixed** and [XT] **xtmelogit** for more details.

    d. **Contrasts** available after most xt commands. See [R] **contrast** and [R] **margins, contrast**.

    e. **Pairwise comparisons** available after most xt estimation commands. See [R] **pwcompare** and [R] **margins, pwcompare**.

    f. **Graphs of margins, marginal effects, contrasts, and pairwise comparisons** available after all xt estimation commands. See [R] **marginsplot**.

g. **xtmixed now uses maximum likelihood (ML) as the default method of estimation**, where previously it used restricted maximum likelihood (REML). REML is still available with the `reml` option, and previous behavior is preserved under version control.

h. **Estimation output improved.**
—Baseline odds now shown.
—Implied zero coefficients now shown.
—You can set displayed precision. See [R] **set cformat** and [R] **estimation options**.
—Estimation commands now respect the width of the Results window. See [R] **estimation options**.
—You can now set whether base levels, empty cells, and omitted are shown. See [R] **set showbaselevels** and [R] **estimation options**.

See [U] **1.3.1 What's new (highlights)**.

2. **Robust and cluster–robust SEs after fixed-effects xtpoisson.** See [XT] **xtpoisson**.

3. **New residual covariance structures for multilevel models** include exponential, banded, and Toeplitz. See [XT] **xtmixed**.

4. **Probability predictions now available.** `predict` after random-effects and population-averaged count-data models, such as `xtpoisson` and `xtgee`, can now predict the probability of any count or count range. See [XT] **xtpoisson postestimation**, [XT] **xtgee postestimation**, and [XT] **xtnbreg postestimation**.

5. **Option addplot() now places added graphs above or below.** Commands that allow option `addplot()` can now place the added plots above or below the command's plots. Affected is the command `xtline`; see [XT] **xtline**.

1.3.6 What's new in statistics (time series)

1. **Highlights,**

a. **MGARCH.** See [TS] **mgarch**.

b. **UCM.** See [TS] **ucm**.

c. **ARFIMA.** See [TS] **arfima**.

d. **Filters for extracting business and seasonal cycles.** See [TS] **tsfilter**.

e. **Business dates.** See [D] **datetime business calendars**.

f. **Improved documentation for date and time variables.** See [D] **datetime**, [D] **datetime translation**, and [D] **datetime display formats**.

g. **Contrasts** available after many time-series estimation commands. See [R] **contrast** and [R] **margins, contrast**.

h. **Pairwise comparisons** available after many time-series estimation commands. See [R] **pwcompare** and [R] **margins, pwcompare**.

i. **Graphs of margins, marginal effects, contrasts, and pairwise comparisons** available after most time-series estimation commands. See [R] **marginsplot**.

j. **Estimation output improved.**
—Implied zero coefficients now shown.
—You can set displayed precision. See [R] **set cformat** and [R] **estimation options**.
—Estimation commands now respect the width of the Results window. See [R] **estimation options**.

—**You can now set whether base levels, empty cells, and omitted are shown.** See [R] **set showbaselevels** and [R] **estimation options**.

See [U] **1.3.1 What's new (highlights)**.

2. **Spectral densities from parametric models** via new postestimation command `psdensity` lets you estimate using `arfima`, `arima`, and `ucm` and then obtain the implied spectral density. See [TS] **psdensity**.
3. **dvech renamed mgarch dvech.** The command for fitting the diagonal VECH model is now named `mgarch dvech`, and innovations may follow multivariate normal or Student's t distributions. See [TS] **mgarch**.
4. **Loading data from Haver Analytics supported on all 64-bit Windows.** See [TS] **haver**.
5. **Option addplot() now places added graphs above or below.** Graph commands that allow option `addplot()` can now place the added plots above or below the command's plots. Affected by this are the commands `corrgram`, `cumsp`, `pergram`, `varstable`, `vecstable`, `wntestb`, and `xcorr`.

1.3.7 What's new in statistics (survey)

1. **Highlights**,
   a. **Contrasts** available after survey estimation. See [R] **contrast** and [R] **margins, contrast**.
   b. **Pairwise comparisons** available after survey estimation. See [R] **pwcompare** and [R] **pwcompare postestimation**.
   c. **Graphs of margins, marginal effects, contrasts, and pairwise comparisons** available after survey estimation. See [R] **marginsplot**.
   d. **Survey SDR weights.** See [SVY] **svy sdr**.
   e. **Bootstrap standard errors for survey data.** See [SVY] **svy bootstrap**.
   f. **Estimation output improved.**
      —**Baseline odds now shown.**
      —**Implied zero coefficients now shown.**
      —**You can set displayed precision.** See [R] **set cformat** and [R] **estimation options**.
      —**Estimation commands now respect the width of the Results window.** See [R] **estimation options**.
      —**You can now set whether base levels, empty cells, and omitted are shown.** See [R] **set showbaselevels** and [R] **estimation options**.

   See [U] **1.3.1 What's new (highlights)**.
2. **Survey estimation may be combined with new SEM** for structural equation modeling. See [SVY] **svy estimation** and the *Stata Structural Equation Modeling Reference Manual*.
3. **Survey goodness-of-fit** available after `logistic`, `logit`, and `probit` with new command `estat gof`. See [SVY] **estat**.
4. **Survey coefficient of variation (CV)** available with new command `estat cv`. See [SVY] **estat**.

1.3.8 What's new in statistics (survival analysis)

1. **Highlights**,

   a. **Contrasts** available after `stcox`, `stcrreg`, and `streg`. See [R] **contrast** and [R] **margins, contrast**.

   b. **Pairwise comparisons** available after `stcox`, `stcrreg`, and `streg`. See [R] **pwcompare** and [R] **margins, pwcompare**.

   c. **Graphs of margins, marginal effects, contrasts, and pairwise comparisons** available after `stcox`, `stcrreg`, and `streg`. See [R] **marginsplot**.

   d. **Estimation output improved.**
   —Implied zero coefficients now shown.
   —You can set displayed precision.
   —Estimation commands now respect the width of the Results window. See [R] **set cformat** and [R] **estimation options**.
   —You can now set whether base levels, empty cells, and omitted are shown. See [R] **set showbaselevels** and [R] **estimation options**.

   See [U] **1.3.1 What's new (highlights)**.

2. **Gönen and Heller's K concordance coefficient** available after Cox proportional hazards estimation. K is robust to censoring. See new option `gheller` for `estat concordance` in [ST] **stcox postestimation**.

3. **Option addplot() now places added graphs above or below.** Graph commands that allow option `addplot()` can now place the added plots above or below the command's plots. Affected by this are the commands `ltable`, `stci`, `stcoxkm`, `stcurve`, `stphplot`, `strate`, `sts graph`, and `tabodds`.

1.3.9 What's new in statistics (multivariate)

1. **Highlights**,

   a. **Structural equation modeling (SEM).** See the new *Stata Structural Equation Modeling Reference Manual.*

   b. **Contrasts.** See [R] **contrast** and [R] **margins, contrast**.

   c. **Pairwise comparisons.** See [R] **pwcompare** and [R] **margins, pwcompare**.

   d. **Graphs of margins, marginal effects, contrasts, and pairwise comparisons.** See [R] **marginsplot**.

   See [U] **1.3.1 What's new (highlights)**.

2. **Option addplot() now places added graphs above or below.** Graph commands that allow option `addplot()` can now place the added plots above or below the command's plots. Affected by this are the commands `screeplot` and `cluster dendrogram`; see [MV] **screeplot** and [MV] **cluster dendrogram**.

1.3.10 What's new in statistics (multiple imputation)

1. **Highlights**,

   a. **Chained equations.** See [MI] **mi impute chained**.

   b. **Four new imputation methods.** See [MI] **mi impute truncreg**, [MI] **mi impute intreg**, [MI] **mi impute poisson**, and [MI] **mi impute nbreg**.

   c. **Conditional imputation.** See *Conditional imputation* in [MI] **mi impute** and new option `conditional()` in the univariate imputation entries such as [MI] **mi impute regress**.

   d. **Imputation by groups.** See new option `by()` in [MI] **mi impute**.

   e. **Imputation by drawing posterior estimates from bootstrapped samples.** See new option `bootstrap` in the univariate imputation entries such as [MI] **mi impute regress**.

   f. **Panel-data and multilevel models are now supported.** Included are `xtcloglog`, `xtgee`, `xtlogit`, `xtmelogit`, `xtmepoisson`, `xtmixed`, `xtnbreg`, `xtpoisson`, `xtprobit`, `xtrc`, and `xtreg`. See [MI] **estimation**.

   g. **Linear and nonlinear predictions after MI estimation.** See [MI] **mi estimate postestimation**.

   h. **Monte Carlo jackknife error estimates.** See new option `mcerror` in [MI] **mi estimate**.

   i. **Estimation output improved.**
   —Baseline odds now shown.
   —Implied zero coefficients now shown.
   —You can set displayed precision. See [R] **set cformat** and [R] **estimation options**.
   —Estimation commands now respect the width of the Results window. See [R] **estimation options**.
   —You can now set whether base levels, empty cells, and omitted are shown. See [R] **set showbaselevels** and [R] **estimation options**.

   See [U] **1.3.1 What's new (highlights)**.

2. **Handling of perfect prediction** during imputation of categorical data using `logit`, `ologit`, and `mlogit`. See *The issue of perfect prediction during imputation of categorical data* in [MI] **mi impute** and see new option `augment` in [MI] **mi impute logit**, [MI] **mi impute ologit**, and [MI] **mi impute mlogit**.

3. **Faster imputation.** `mi impute` no longer secretly converts to `flongsep` and back again.

4. **mi estimate now supports total.** See [MI] **estimation**.

5. **misstable summarize will now create summary variables** recording the missing-values pattern. See new option `generate()` for `summarize` in [R] **misstable**. Note that `mi misstable` does not have this new option. The new option is useful before data are imputed.

1.3.11 What's new in graphics

1. **Highlights**,

   a. **Graphs of margins, marginal effects, contrasts, ...** See [R] **marginsplot**.

   b. **Contour plots.** See [G-2] **graph twoway contour** and [G-2] **graph twoway contourline**.

   c. **PDF export for graphs and logs** lets you directly create PDFs from your Stata graphs. See [G-2] **graph export** and [R] **translate**.

   See [U] **1.3.1 What's new (highlights)**.

2. **Time-series operators now supported** by `twoway lfit`, `twoway lfitci`, `twoway qfit`, and `twoway qfitci`. See [G-2] **graph twoway lfit**, [G-2] **graph twoway lfitci**, [G-2] **graph twoway qfit**, and [G-2] **graph twoway qfitci**.
3. **Graphs of marginal and covariate-specific ROC curves.** New command `rocregplot` plots the fitted ROC curve after `rocreg`. See [R] **rocregplot**.
4. **Option addplot() now places added graphs above or below.** Graph commands that allow option `addplot()` can now place the added plots above or below the command's plots.

1.3.12 What's new in programming

1. **Saved results r() and e() can be marked hidden or historical**, which means they do not show when the user types `return list` or `ereturn list` unless the user also specifies option `all`. See [P] **return**.
2. **Estimation commands now store in r() as well as e().** `r()` values are stored at estimation time and after replaying. Stored are
   a. `r(level)`, a scalar containing the confidence level for the CIs.
   b. `r(label#)`, a macro containing the label displayed with the #th coefficient, such as "(base)", "(omitted)", or "(empty)".
   c. `r(table)`, a matrix containing all the data displayed in the coefficient table. The matrix is the coefficient table, transposed; each column contains coefficients and associated statistics. To understand the matrix, do the following:

   ```
   . sysuse auto, clear
   . regress mpg weight displ
   . matrix list r(table)
   ```

   See [P] **ereturn**.
3. **ereturn display offers new options for controlling the look of the coefficient table.**
   a. Options `noomitted`, `vsquish`, `noemptycells`, `baselevels`, and `allbaselevels` control row spacing and display of omitted variables and base and empty cells.
   b. Formatting display options `cformat(%`*fmt*`)`, `pformat(%`*fmt*`)`, and `sformat(%`*fmt*`)` control the formats of numbers in the coefficient table.
   c. `ereturn display` now respects the width of the Results window. This feature may be turned off by new display option `nolstretch`.

   See [R] **estimation options**.
4. **Matrices can be in tables with equation names only** using new options `coleqonly` and `roweqonly`. See [P] **matlist**.
5. **matrix accum allows option absorb()** to accumulate deviations from the mean within groups. See [P] **matrix accum**.
6. **Version control for random-number generators** is now determined when the seed is set, not when the generator function is used; see [P] **version**. New `creturn` result `c(version_rng)` records the version number currently in effect for random-number generators; see [P] **creturn**.
7. **fvrevar has new option stub()**, which generates stub + index variables rather than temporary variables. See [R] **fvrevar**.
8. **mprobit now posts base outcome equation to e(b).** See [R] **mprobit**.

9. **Default time for network timeouts was reduced.** `timeout1` has been reduced from 120 seconds to 30, and `timeout2` has been reduced from 300 seconds to 180. See [R] **netio**.

1.3.13 What's new in Mata

1. **New Stata commands getmata and putmata** make it easy to transfer your data into Mata, manipulate them, and then transfer them back to Stata. `getmata` and `putmata` are especially designed for interactive use. See [D] **putmata**.
2. **New functions** imported from Stata,
   a. **Tukey's Studentized range**, cumulative and inverse, `tukeyprob()` and `invtukeyprob()`.
   b. **Dunnett's multiple range**, cumulative and inverse, `dunnettprob()` and `invdunnettprob()`.
   c. **New date conversion functions** `dofb()` and `bofd()` convert between business dates and standard calendar dates. See [D] **datetime business calendars**.

   See [D] **functions**, [M-5] **normal()**, and [M-5] **date()**.
3. **Support for hidden and historical saved results.** Existing Mata functions `st_global()`, `st_numscalar()`, and `st_matrix()` now allow an optional third argument specifying the hidden or historical status. Three new functions—`st_global_hcat()`, `st_numscalar_hcat()`, `st_matrix_hcat()`—allow you to determine the saved hidden or historical status. See [M-5] **st_global()**, [M-5] **st_numscalar()**, and [M-5] **st_matrix()**.
4. **Support for new ml features.** Stata's `ml` now distinguishes the Hessian matrix produced by `technique(nr)` from the other techniques that compute a substitute for the Hessian matrix. This means that `ml` will compute the real Hessian matrix of second derivatives to determine convergence when all other convergence tolerances are satisfied and `technique(bfgs)`, `technique(bhhh)`, or `technique(dfp)` is in effect.

   Mata's commands `optimize()` and `moptimize()` have been similarly changed. See [M-5] **optimize()** and [M-5] **moptimize()**.

1.3.14 What's more

We have not listed all the changes, but we have listed the important ones.

Stata is continually being updated, and those updates are available for free over the Internet. All you have to do is type

```
. update query
```

and follow the instructions.

To learn what has been added since this manual was printed, select **Help > What's New?** or type

```
. help whatsnew
```

We hope that you enjoy Stata 12.

1.4 References

Chambers, J. M., W. S. Cleveland, B. Kleiner, and P. A. Tukey. 1983. *Graphical Methods for Data Analysis*. Belmont, CA: Wadsworth.

Everitt, B. S., and A. Skrondal. 2010. *The Cambridge Dictionary of Statistics*. 4th ed. Cambridge: Cambridge University Press.

Freedman, L. S. 1982. Tables of the number of patients required in clinical trials using the logrank test. *Statistics in Medicine* 1: 121–129.

Harvey, A. C. 1989. *Forecasting, Structural Time Series Models and the Kalman Filter*. Cambridge: Cambridge University Press.

Heyde, C. C., and E. Seneta, ed. 2001. *Statisticians of the Centuries*. New York: Springer.

Johnson, N. L., and S. Kotz, ed. 1997. *Leading Personalities in Statistical Sciences: From the Seventeenth Century to the Present*. New York: Wiley.

Lachin, J. M., and M. A. Foulkes. 1986. Evaluation of sample size and power for analyses of survival with allowance for nonuniform patient entry, losses to follow-up, noncompliance, and stratification. *Biometrics* 42: 507–519.

Rubinstein, L. V., M. H. Gail, and T. J. Santner. 1981. Planning the duration of a comparative clinical trial with loss to follow-up and a period of continued observation. *Journal of Chronic Diseases* 34: 469–479.

Schoenfeld, D. A. 1981. The asymptotic properties of nonparametric tests for comparing survival distributions. *Biometrika* 68: 316–319.

Tarone, R. E. 1985. On heterogeneity tests based on efficient scores. *Biometrika* 72: 91–95.

Upton, G., and I. Cook. 2006. *A Dictionary of Statistics*. 2nd ed. Oxford: Oxford University Press.

2 A brief description of Stata

Stata is a statistical package for managing, analyzing, and graphing data.

Stata is available for a variety of platforms. Stata may be used either as a point-and-click application or as a command-driven package.

Stata's GUI provides an easy interface for those new to Stata and for experienced Stata users who wish to execute a command that they seldom use.

The command language provides a fast way to communicate with Stata and to communicate more complex ideas.

Here is an extract of a Stata session using the GUI:

(Throughout the Stata manuals, we will refer to various datasets. These datasets are all available from http://www.stata-press.com/data/r12/. For easy access to them within Stata, type `webuse` *dataset_name*, or select **File > Example Datasets...** and click on *Stata 12 manual datasets*.)

```
. webuse lbw
(Hosmer & Lemeshow data)
```

We select **Data > Describe data > Summary statistics** and choose to summarize variables `low`, `age`, and `smoke`, whose names we obtained from the Variables window. We click on **OK**.

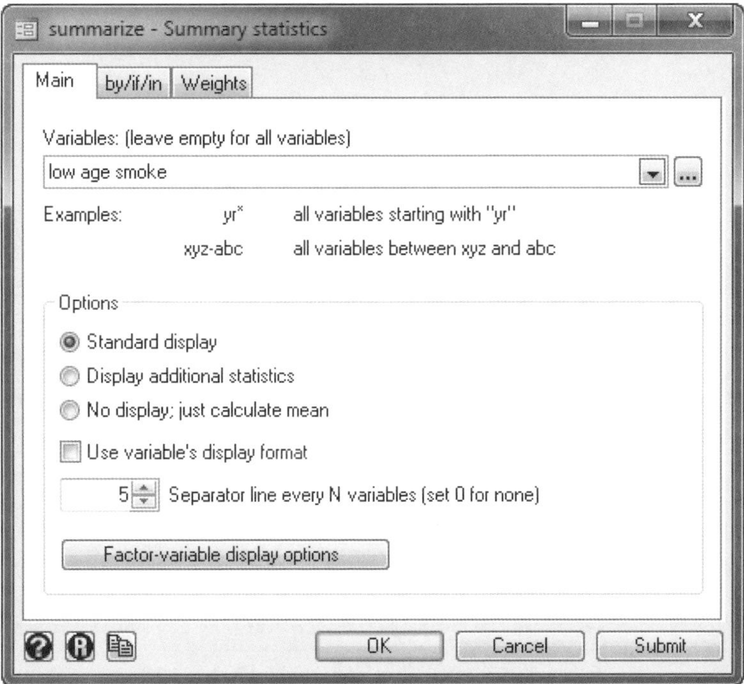

```
. summarize low age smoke

    Variable |       Obs        Mean    Std. Dev.       Min        Max
-------------+--------------------------------------------------------
         low |       189    .3121693    .4646093          0          1
         age |       189     23.2381    5.298678         14         45
       smoke |       189    .3915344    .4893898          0          1
```

Stata shows us the command that we could have typed in command mode—`summarize low age smoke`—before displaying the results of our request.

Next we fit a logistic regression model of `low` on `age` and `smoke`. We select **Statistics > Binary outcomes > Logistic regression (reporting odds ratios)**, fill in the fields, and click on **OK**.

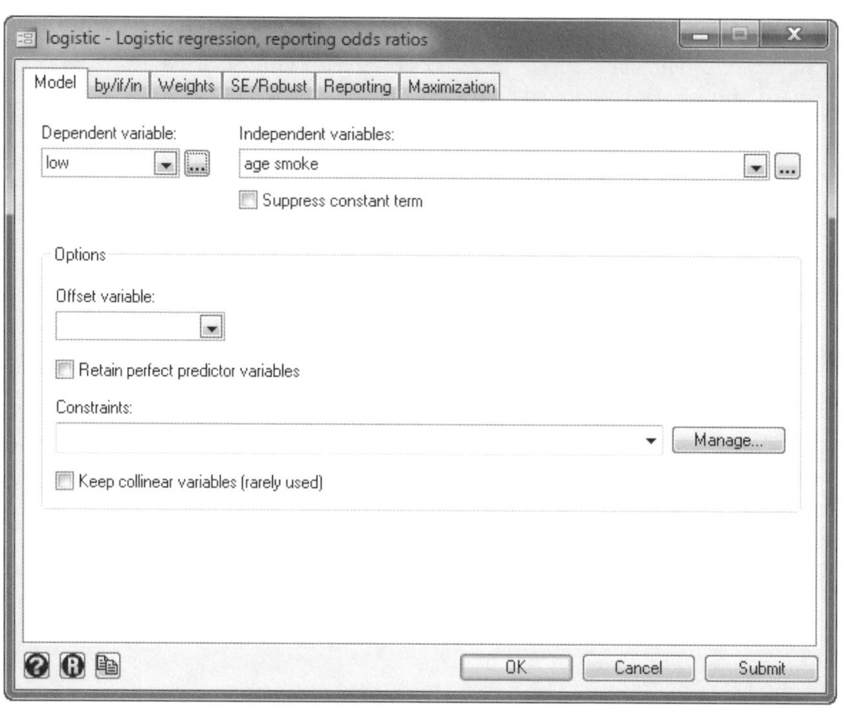


```
. logistic low age smoke

Logistic regression                               Number of obs   =        189
                                                  LR chi2(2)      =       7.40
                                                  Prob > chi2     =     0.0248
Log likelihood = -113.63815                       Pseudo R2       =     0.0315

------------------------------------------------------------------------------
         low | Odds Ratio   Std. Err.      z    P>|z|     [95% Conf. Interval]
-------------+----------------------------------------------------------------
         age |   .9514394   .0304194    -1.56   0.119     .8936482    1.012968
       smoke |   1.997405    .642777     2.15   0.032     1.063027    3.753081
       _cons |   1.062798   .8048781     0.08   0.936     .2408901    4.689025
------------------------------------------------------------------------------
```

Here is an extract of a Stata session using the command language:

```
. use http://www.stata-press.com/data/r12/auto
(1978 Automobile Data)

. summarize mpg weight

    Variable |       Obs        Mean    Std. Dev.       Min        Max
-------------+--------------------------------------------------------
         mpg |        74     21.2973    5.785503         12         41
      weight |        74    3019.459    777.1936       1760       4840
```

The user typed `summarize mpg weight` and Stata responded with a table of summary statistics. Other commands would produce different results:

```
. generate gp100m = 100/mpg

. label var gp100m "Gallons per 100 miles"

. format gp100m %5.2f

. correlate gp100m weight
(obs=74)

             |   gp100m   weight
-------------+------------------
      gp100m |   1.0000
      weight |   0.8544   1.0000

. regress gp100m weight gear_ratio

      Source |       SS       df       MS              Number of obs =      74
-------------+------------------------------           F(  2,    71) =   96.65
       Model |  87.4543721     2  43.7271861           Prob > F      =  0.0000
    Residual |  32.1218886    71  .452420967           R-squared     =  0.7314
-------------+------------------------------           Adj R-squared =  0.7238
       Total |  119.576261    73  1.63803097           Root MSE      =  .67262

------------------------------------------------------------------------------
      gp100m |      Coef.   Std. Err.      t    P>|t|     [95% Conf. Interval]
-------------+----------------------------------------------------------------
      weight |   .0014769   .0001556     9.49   0.000     .0011665    .0017872
  gear_ratio |   .1566091   .2651131     0.59   0.557    -.3720115    .6852297
       _cons |   .0878243   1.198434     0.07   0.942    -2.301786    2.477435
------------------------------------------------------------------------------

. scatter gp100m weight, by(foreign)
```

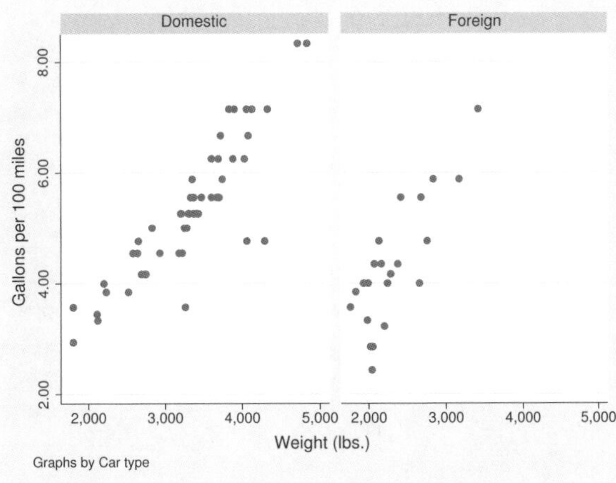

The user-interface model is type a little, get a little, etc., so that the user is always in control.

Stata's model for a dataset is that of a table—the rows are the observations and the columns are the variables:

```
. list mpg weight gp100m in 1/10

     ┌──────────────────────────┐
     │ mpg   weight    gp100m   │
     ├──────────────────────────┤
  1. │  22    2,930      4.55   │
  2. │  17    3,350      5.88   │
  3. │  22    2,640      4.55   │
  4. │  20    3,250      5.00   │
  5. │  15    4,080      6.67   │
     ├──────────────────────────┤
  6. │  18    3,670      5.56   │
  7. │  26    2,230      3.85   │
  8. │  20    3,280      5.00   │
  9. │  16    3,880      6.25   │
 10. │  19    3,400      5.26   │
     └──────────────────────────┘
```

Observations are numbered; variables are named.

Stata is fast. That speed is due partly to careful programming, and partly because Stata keeps the data in memory. Stata's file model is that of a word processor: a dataset may exist on disk, but the dataset in memory is a copy. Datasets are loaded into memory, where they are worked on, analyzed, changed, and then perhaps stored back on disk.

Working on a copy of the data in memory makes Stata safe for interactive use. The only way to harm the permanent copy of your data on disk is if you explicitly save over it.

Having the data in memory means that the dataset size is limited by the amount of computer memory. Stata stores the data in memory in an efficient format—you will be surprised how much data can fit. Nevertheless, if you work with extremely large datasets, you may run into memory constraints. You will want to learn how to store your data as efficiently as possible; see [D] **compress**.

3 Resources for learning and using Stata

Contents

3.1 Overview

The *Getting Started* manual, *User's Guide*, and *Reference* manuals are the primary tools for learning about Stata; however, there are many other sources of information. A few are listed below.

- Stata itself. Stata has a subject table of contents online with links to the help system and dialog boxes that make it easy to find and to execute a Stata command. See [U] **4 Stata's help and search facilities**.
- The Stata website. Visit http://www.stata.com. Much of the site is dedicated to user support; see [U] **3.2.1 The Stata website (www.stata.com)**.
- The Stata Blog, Twitter, and Facebook. Visit http://blog.stata.com/, http://twitter.com/stata, and http://facebook.com/statacorp. See [U] **3.2.2 The Stata Blog—Not Elsewhere Classified** and [U] **3.2.3 Stata on Twitter and Facebook**.
- The Stata Press website. Visit http://www.stata-press.com. This site contains the datasets used throughout the Stata manuals; see [U] **3.3 Stata Press**.
- The Stata listserver. An active group of Stata users communicate over an Internet listserver, which you can join for free; see [U] **3.4 The Stata listserver**.

- The *Stata Journal* and the *Stata Technical Bulletin*. The *Stata Journal* contains reviewed papers, regular columns, book reviews, and other material of interest to researchers applying statistics in a variety of disciplines. The *Stata Technical Bulletin*, the predecessor to the *Stata Journal*, contains articles and user-written commands. See [U] **3.5 The Stata Journal**.
- The Stata software distribution site and other user-provided software distribution sites. Stata itself can download and install updates and additions. We provide official updates to Stata—type `update query` or select **Help > Check for Updates**. We also provide user-written additions to Stata and links to other user-provided sites—type `net` or select **Help > SJ and User-written Programs**; see [U] **3.6 Updating and adding features from the web**.
- NetCourses. We offer training via the Internet. Details are in [U] **3.7.2 NetCourses**.
- Public training courses. We offer in-depth training courses at third-party sites around the country. Details are in [U] **3.7.3 Public training courses**.
- On-site training courses. We can come to your institution to provide customized training. Details are in [U] **3.7.4 On-site training courses**.
- Books and support materials. Supplementary Stata materials are available; see [U] **3.8 Books and other support materials**.
- Technical support. We provide technical support by email, telephone, and fax; see [U] **3.9 Technical support**.

3.2 Stata on the Internet (www.stata.com and other resources)

3.2.1 The Stata website (www.stata.com)

Point your browser to http://www.stata.com and click on **Support**. More than half our website is dedicated to providing support to users.

- The website provides answers to FAQs (frequently asked questions) on Windows, Mac, Unix, statistics, programming, Internet capabilities, graphics, and data management. These FAQs run the gamut from "I cannot save/open files" to "What does 'completely determined' mean in my logistic regression output?" Most users will find something of interest.
- Visiting the website is one way that you can subscribe to the Stata listserver; see [U] **3.4 The Stata listserver**.
- The website provides detailed information about NetCourses, along with the current schedule; see [U] **3.7.2 NetCourses**.
- The website provides information about Stata courses and meetings, both in the United States and elsewhere.
- The website provides an online bookstore for Stata-related books and other supplementary materials; see [U] **3.8 Books and other support materials**.
- The website provides links to information about statistics: other statistical software providers, book publishers, statistical journals, statistical organizations, and statistical listservers.
- The website provides links to resources for learning Stata at http://www.stata.com/links/resources.html. Be sure to look at these materials, as many outstanding resources about Stata are listed here.

In short, the website provides up-to-date information on all support materials and, where possible, provides the materials themselves. Visit http://www.stata.com if you can.

3.2.2 The Stata Blog—Not Elsewhere Classified

Stata's official blog can be found at http://blog.stata.com/ and contains news and advice related to the use of Stata. The articles appearing in the blog are individually signed and are written by the same people who develop, support, and sell Stata.

The Stata Blog also has links to other blogs about Stata, written by Stata users around the world.

3.2.3 Stata on Twitter and Facebook

StataCorp has an official presence on Twitter and Facebook. You can follow us on Twitter at http://twitter.com/stata and find us on Facebook at http://facebook.com/statacorp. These are both good ways to stay up-to-the-minute with the latest Stata information.

3.2.4 Other Internet resources on Stata

Many other people have published information on the Internet about Stata such as tutorials, examples, and datasets. Visit http://www.stata.com/links/ to explore other Stata and statistics resources on the Internet.

3.3 Stata Press

Stata Press is the publishing arm of StataCorp LP and publishes books, manuals, and journals about Stata statistical software and about general statistics topics for professional researchers of all disciplines.

Point your browser to http://www.stata-press.com. This site is devoted to the publications and activities of Stata Press.

- Datasets that are used in the Stata *Reference* manuals and other books published by Stata Press may be downloaded. Visit http://www.stata-press.com/data/. These datasets can be used in Stata by simply typing `use http://www.stata-press.com/data/r12/`*dataset_name*; for example, type `use http://www.stata-press.com/data/r12/auto`. You could also type `webuse auto`; see [D] **webuse**.
- An online catalog of all our books and multimedia products is at http://www.stata-press.com/catalog.html. We have tried to include enough information, such as table of contents and preface material, so that you may tell whether the book is appropriate for you.
- Information about forthcoming publications is posted at http://www.stata-press.com/forthcoming.html.

3.4 The Stata listserver

The Stata listserver (Statalist) is an independently operated, real-time list of Stata users on the Internet. Anyone may join. Instructions for doing so can be found at http://www.stata.com/statalist/.

Many knowledgeable users are active on the list, as are the StataCorp technical staff. We recommend that new users subscribe, observe the exchanges, and, if it turns out not to be useful, unsubscribe.

A searchable archive of Statalist postings from 2002 through the present can be found at http://www.stata.com/statalist/archive/.

3.5 The Stata Journal

The *Stata Journal* (SJ) is a printed and electronic journal, published quarterly, containing articles about statistics, data analysis, teaching methods, and effective use of Stata's language. The *Journal* publishes reviewed papers together with shorter notes and comments, regular columns, tips, book reviews, and other material of interest to researchers applying statistics in a variety of disciplines. The *Journal* is a publication for all Stata users, both novice and experienced, with different levels of expertise in statistics, research design, data management, graphics, reporting of results, and in Stata, in particular.

Tables of contents for past issues and abstracts of the articles are available at http://www.stata-journal.com/archives.html. PDF copies of articles published at least three years ago are available for free from the *Stata Journal* website.

We recommend that all users subscribe to the SJ. Visit http://www.stata-journal.com to learn more about the *Stata Journal* and to order your subscription.

To obtain any programs associated with articles in the SJ, type

```
. net from http://www.stata-journal.com/software
```

or

- Select **Help > SJ and User-written Programs**
- Click on **Stata Journal**

The Stata Technical Bulletin

For 10 years, the *Stata Technical Bulletin* (STB) served as the means of distributing new commands and Stata upgrades, both user-written and "official". After 10 years of continual publication, the STB evolved into the *Stata Journal*. The Internet provided an alternative delivery mechanism for user-written programs, so the emphasis shifted from user-written programs to more expository articles. Although the STB is no longer published, many of the programs and articles that appeared in it are still valuable today. PDF copies of all issues of the STB are available for free at http://www.stata.com/bookstore/stbj.html. To obtain the programs that were published in the STB, type

```
. net from http://www.stata.com
. net cd stb
```

3.6 Updating and adding features from the web

Stata itself can open files on the Internet.

First, try this:

```
. use http://www.stata.com/manual/oddeven, clear
```

That will load an uninteresting dataset into your computer from our website. If you have a home page, you can use this feature to share datasets with coworkers. Save a dataset on your home page, and researchers worldwide can use it. See [R] **net**.

3.6.1 Official updates

Although we follow no formal schedule for the release of updates, we typically provide updates to Stata approximately once a month. Installing the updates is easy. Type

```
. update query
```

or select **Help > Check for Updates**. Do not be concerned; nothing will be installed unless and until you say so. Once you have installed the update, you can type

```
. help whatsnew
```

or select **Help > What's New?** to find out what has changed. We distribute official updates to fix bugs and to add new features.

3.6.2 Unofficial updates

There are also "unofficial" updates—additions to Stata written by Stata users, which includes members of the StataCorp technical staff. Stata is programmable, and even if you never write a Stata program, you may find these additions useful, some of them spectacularly so. Start by typing

```
. net from http://www.stata.com
```

or select **Help > SJ and User-written Programs**.

Be sure to visit the Statistical Software Components (SSC) Archive, which hosts a large collection of free additions to Stata. The `ssc` command makes it easy for you to find, install, and uninstall packages from the SSC. Type

```
. ssc whatsnew
```

to find out what's new at the site. If you find something that interests you, type

```
. ssc describe pkgname
```

for more information. If you have already installed a package, you can check for and optionally install updates by typing

```
. adoupdate pkgname
```

To check for and optionally install updates to all the packages you have previously installed, type

```
. adoupdate all
```

Periodically, you can type

```
. news
```

or select **Help > News** to display a short message from our website telling you what is newly available.

See [U] **28 Using the Internet to keep up to date**.

3.7 Conferences and training

3.7.1 Conferences and users group meetings

StataCorp organizes the annual Stata Conference in the United States. Other conferences and users group meetings are held in several countries around the world each year.

These meetings provide in-depth presentations from experienced Stata users and experts from StataCorp. They also provide you with the opportunity to interact directly with the people who develop Stata and to share your thoughts and ideas with them.

Visit http://www.stata.com/meeting/ for a list of upcoming conferences and meetings.

3.7.2 NetCourses

We offer courses on Stata at both introductory and advanced levels. Courses on software are typically expensive and time consuming. They are expensive because, in addition to the direct costs of the course, participants must travel to the course site. Courses over the Internet save everyone time and money.

We offer courses over the Internet and call them Stata NetCourses™.

- **What is a NetCourse?**
  A NetCourse is a course offered through the Stata website that varies in length from 7 to 8 weeks. Everyone with an email address and a web browser can participate.
- **How does it work?**
  Every Friday a lecture is posted on a password-protected website. After reading the lecture over the weekend or perhaps on Monday, participants then post questions and comments on a message board. Course leaders typically respond to the questions and comments on the same day they are posted. Other participants are encouraged to amplify or otherwise respond to the questions or comments as well. The next lecture is then posted on Friday, and the process repeats.
- **How much of my time does it take?**
  It depends on the course, but the introductory courses are designed to take roughly 3 hours per week.
- **There are three of us here—can just one of us enroll and then redistribute the NetCourse materials ourselves?**
  We ask that you not. NetCourses are priced to cover the substantial time input of the course leaders. Moreover, enrollment is typically limited to prevent the discussion from becoming unmanageable. The value of a NetCourse, just like a real course, is the interaction of the participants, both with each other and with the course leaders.
- **I've never taken a course by Internet before. I can see that it might work, but then again, it might not. How do I know I will benefit?**
  All Stata NetCourses come with a 30-day satisfaction guarantee. The 30 days begins after the conclusion of the final lecture.

You can learn more about the current NetCourse offerings by visiting http://www.stata.com/netcourse.

NetCourseNow

A NetCourseNow offers the same material as NetCourses but it allows you to choose the time and pace of the course, and you have a personal NetCourse instructor.

- **What is a NetCourseNow?**
A NetCourseNow offers the same material as a NetCourse, but allows you to move at your own pace and to specify a starting date. With a NetCourseNow, you also have the added benefit of a personal NetCourse instructor whom you can email directly with questions about lectures and exercises. You must have an email address and a web browser to participate.
- **How does it work?**
All course lectures and exercises are posted at once, and you are free to study at your own pace. You will be provided with the email address of your personal NetCourse instructor to contact when you have questions.
- **How much of my time does it take?**
A NetCourseNow allows you to set your own pace. How long the course takes and how much time you spend per week is up to you.

3.7.3 Public training courses

Public training courses are intensive, in-depth courses taught by StataCorp at third-party sites around the country.

- **How is a public training course taught?**
These are interactive, hands-on sessions. Participants work along with the instructor so that they can see firsthand how to use Stata. Questions are encouraged.
- **Do I need my own computer?**
Because the sessions are in computer labs running the latest version of Stata, there is no need to bring your own computer. Of course, you may bring your own computer if you have a registered copy of Stata you can use.
- **Do I get any notes?**
You get a complete set of notes for each class, which includes not only the materials from the lessons but also all the output from the example commands.

See http://www.stata.com/training/public.html for all course offerings.

3.7.4 On-site training courses

On-site training courses are courses that are tailored to the needs of an institution. StataCorp personnel can come to your site to teach what you need, whether it be to teach new users or to show how to use a specialized tool in Stata.

- **How is an on-site training course taught?**
These are interactive, hands-on sessions, just like our public-training courses. You will need a computer for each participant.
- **What topics are available?**
We offer training in anything and everything related to Stata. You work with us to put together a curriculum that matches your needs.
- **How does licensing work?**
We will supply you with the licenses you need for the training session, whether the training is in a lab or for individuals working on laptops. We will ship the licensing and installation instructions so that you can have everything up and running before the session starts.

See http://www.stata.com/training/onsite.html for all the details.

3.8 Books and other support materials

3.8.1 For readers

There are books published about Stata, both by us and by others. Visit the Stata bookstore at http://www.stata.com/bookstore. For the books that we carry, we include the table of contents and comments written by a member of our technical staff, explaining why we think this book might interest you.

3.8.2 For authors

If you have written a book related to Stata and would like us to consider carrying it in our bookstore, email bookstore@stata.com.

If you are writing a book, join our free Author Support Program. Stata professionals are available to review your Stata code to ensure that it is efficient and reflects modern usage, production specialists are available to help format Stata output, and editors and statisticians are available to ensure the accuracy of Stata-related content. Visit http://www.stata.com/authorsupport.

If you are thinking about writing a Stata-related book, consider publishing it with Stata Press. Email submissions@statapress.com.

3.9 Technical support

We are committed to providing superior technical support for Stata software. To assist you as efficiently as possible, please follow the procedures listed below.

3.9.1 Register your software

You must register your software to be eligible for technical support, updates, special offers, and other benefits. By registering, you will receive the *Stata News*, and you may access our support staff for free with any question that you encounter. You may register your software either electronically or by mail.

Electronic registration: After installing Stata and successfully entering your License and Authorization Key, your default web browser will open to the online registration form at the Stata website. You may also manually point your web browser to http://www.stata.com/register/ if you wish to register your copy of Stata at a later time.

Mail-in registration: Fill in the registration card that came with Stata and mail it to StataCorp.

3.9.2 Before contacting technical support

Before you spend the time gathering the information our technical support department needs, make sure that the answer does not already exist in the help files. You can use the `help` and `search` commands to find all the entries in Stata that address a given subject. Be sure to try selecting **Help > Contents**. Check the manual for a particular command. There are often examples that address questions and concerns. Another good source of information is our website. You should keep a bookmark to our frequently asked questions page (http://www.stata.com/support/faqs/) and check it occasionally for new information.

If you do need to contact technical support, visit http://www.stata.com/support/tech-support/ for more information.

3.9.3 Technical support by email

This is the preferred method of asking a technical support question. It has the following advantages:

- You will receive a prompt response from us saying that we have received your question and that it has been forwarded to *Technical Services* to answer.
- We can route your question to a specialist for your particular question.
- Questions submitted via email may be answered after normal business hours, or even on weekends or holidays. Although we cannot promise that this will happen, it may, and your email inquiry is bound to receive a faster response than leaving a message on Stata's voicemail.
- If you are receiving an error message or an unexpected result, it is easy to include a log file that demonstrates the problem.

Please see visit http://www.stata.com/support/tech-support/ for information about contacting technical support.

3.9.4 Technical support by phone or fax

Our installation support telephone number is 979-696-4600. Please have your serial number handy. It is also best if you are at your computer when you call. Telephone support is reserved for installation questions. If your question does not involve installation, the question should be submitted via email or fax.

Send fax requests to 979-696-4601. If possible, collect the relevant information in a log file and include the file in your fax.

Please see visit http://www.stata.com/support/tech-support/ for information about contacting technical support.

3.9.5 Comments and suggestions for our technical staff

By all means, send in your comments and suggestions. Your input is what determines the changes that occur in Stata between releases, so if we do not hear from you, we may not include your most desired new feature! Email is preferred, as this provides us with a permanent copy of your request. When requesting new features, please include any references that you would like us to review should we develop those new features. Email your suggestions to service@stata.com.

4 Stata's help and search facilities

Contents

4.1 Introduction

To access Stata's help, you will either

1. select **Help** from the menus, or
2. use the `help`, `search`, `findit`, and `hsearch` commands.

Regardless of the method you use, results will be shown in the Viewer or Results windows. Blue text indicates a hypertext link, so you can click to go to related entries.

4.2 Getting started

The first time you use help, try one of the following:

1. Select **Help > Advice** form the menu bar, or
2. Type `help advice`.

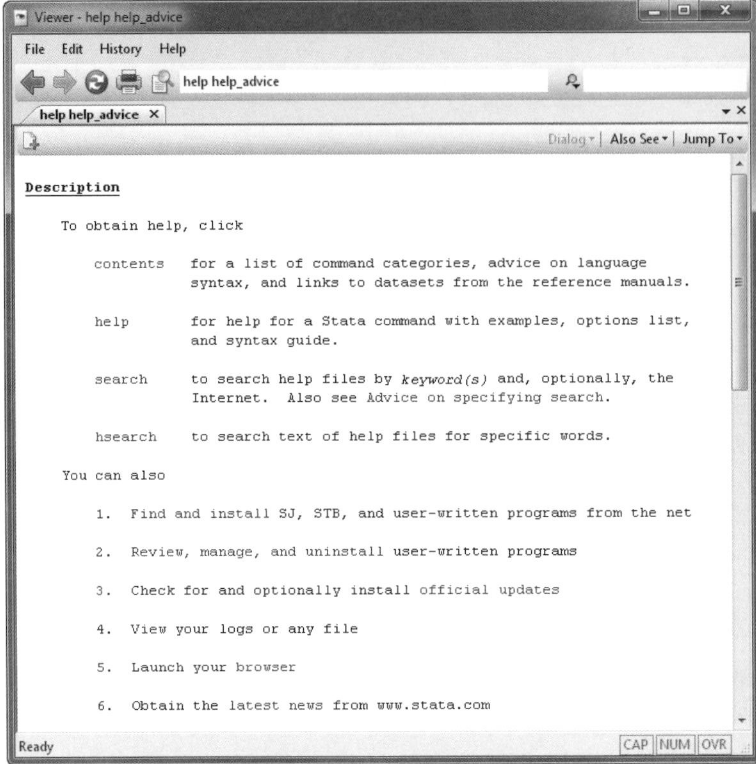

Advice presents you with four alternatives:

1. `contents`: Click on that and you will be taken to a system organized by topic: Basics, Data management, Statistics, Graphics, and Programming and matrices.
2. `help`: Click on that and you can type the name of a Stata command and go directly to its online help.
3. `search`: Click on that and you can perform a keyword search.
4. `hsearch`: Click on that and you can do a search of the contents of the help files much like search engines on the Internet search the contents of the web.

The above are the four main features of Stata's help system, and you can access each in various ways:

1. Topical help:

   Select **Help > Advice**, click on `contents`, or

   Select **Help > Contents**, or

   Type `help contents`

2. Help on a command or topic:

   Select **Help > Advice**, click on `help`, or

   Select **Help > Stata Command...**, or

   Type `help` *commandname*

3. Keyword search:

Select **Help > Advice**, click on `search`, or

Select **Help > Search...**, or

Type `search` *keyword(s)*

4. Text search:

Select **Help > Advice**, click on `hsearch`, or

Type `hsearch` *whatever*

4.3 help: Stata's help system

When you

1. Select **Help > Stata command...**
   Type a command name in the Command edit field
   Click on OK, or
2. Type `help` followed by a command name

you access Stata's online help files. These files provide shortened versions of what is in the printed manuals. Let's access the online help for Stata's `ttest` command. Do one of the following:

1. Select **Help > Stata command...**
   Type `ttest` in the Command edit field
   Click on OK, or
2. Type `help ttest`

Regardless of which you do, the result will be

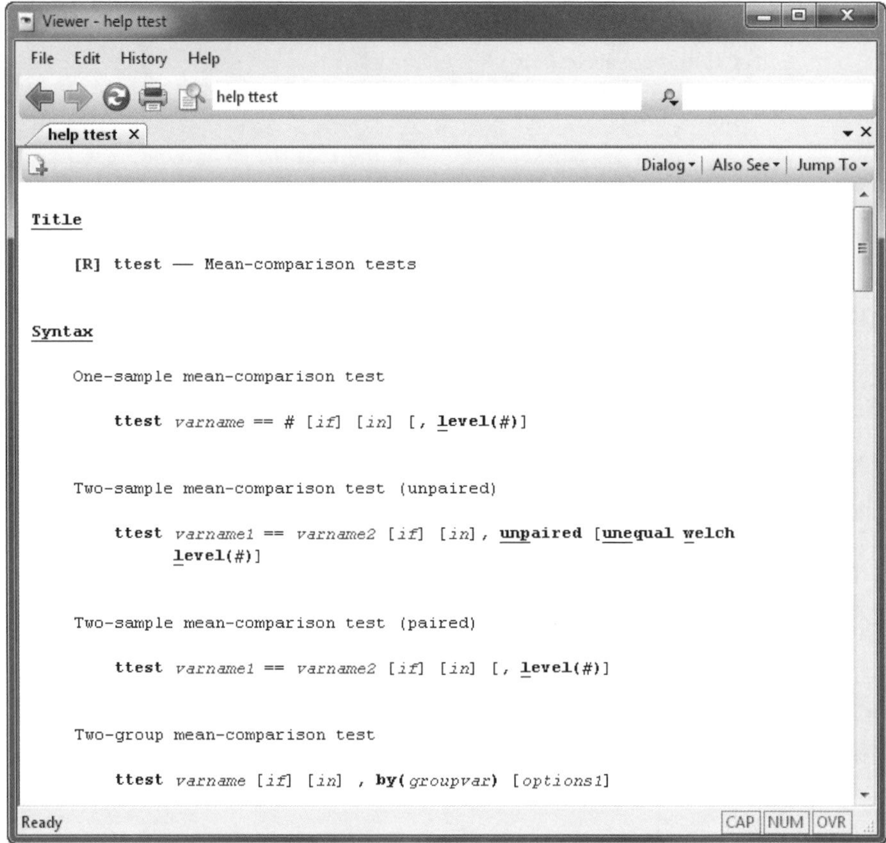

The trick is in already knowing that Stata's command for testing equality of means is `ttest` and not, say `meanstest`. There are two solutions to that problem, topical help and searching. Topical help you access by selecting **Help > Contents** or by typing `help contents`.

4.4 Accessing PDF manuals from help entries

Every help file in Stata links to the equivalent manual entry. If you are reading `help ttest`, simply click on `[R] ttest` in the **Title** section of the help file to go directly to the [R] **ttest** manual entry.

We provide recommended settings for your PDF viewer to optimize it for Stata's documentation at http://www.stata.com/support/faqs/res/documentation.html.

4.5 Searching

If you do not know the name of the Stata command you are looking for, you can search for it by keyword,

1. Select **Help > Search...**
   Type keywords in the edit field
   Click on OK

2. Type search followed by the keywords

or you can search the text of Stata's help files,

3. Type hsearch followed by text

search and hsearch each have their own advantages.

search is available by both menu and command; hsearch is available by command only (and from **Help > Advice**).

search matches the keywords you specify to a database and returns matches found in Stata commands, FAQs at www.stata.com, and articles that have appeared in the *Stata Journal*. Optionally, it can also find user-written additions to Stata available over the web. To do that, from the Search dialog box select *Search all* or, on the search command, specify the all option.

hsearch searches the text of the Stata help files on your computer. That does not include FAQs at www.stata.com and articles that have appeared in the *Stata Journal*, but it does include any help files you have written and the help files associated with any user-written additions you may have installed.

search does a better job when what you want is based on terms commonly used or when what you are looking for might not already be installed on your computer.

hsearch does a better job when the search terms are odder or when what you want might already be installed.

It is sometimes a good idea to use both.

4.6 More on search

However you access search—command or menu—it does the same thing. You tell search what you want information about, and it searches for relevant entries. If you want search to look for the topic across all sources, including the online help, the FAQs at the Stata website, the *Stata Journal*, and all Stata-related Internet sources including user-written additions, specify "Search all" from the Search dialog or, on the search command, specify the all option. (Command findit is a synonym for search, all, so many users type findit ... rather than search ..., all.)

search can be used broadly or narrowly. For instance, if you want to perform the Kolmogorov–Smirnov test for equality of distributions, you could type

```
. search Kolmogorov-Smirnov test of equality of distributions
[R]     ksmirnov . . . . . . Kolmogorov-Smirnov equality of distributions test
        (help ksmirnov)
```

In fact, we did not have to be nearly so complete—typing search Kolmogorov-Smirnov would have been adequate. Had we specified our request more broadly—looking up equality of distributions—we would have obtained a longer list that included ksmirnov.

Here are guidelines for using search.

- Capitalization does not matter. Look up Kolmogorov-Smirnov or kolmogorov-smirnov.
- Punctuation does not matter. Look up kolmogorov smirnov.
- Order of words does not matter. Look up smirnov kolmogorov.
- You may abbreviate, but how much depends. Break at syllables. Look up kol smir. search tends to tolerate a lot of abbreviation; it is better to abbreviate than to misspell.
- The prepositions for, into, of, on, to, and with are ignored. Use them—look up equality of distributions—or omit them—look up equality distributions—it makes no difference.

- `search` tolerates plurals, especially when they can be formed by adding an *s*. Even so, it is better to look up the singular. Look up `normal distribution`, not `normal distributions`.
- Specify the search criterion in English, not in computer jargon.
- Use American spellings. Look up `color`, not `colour`.
- Use nouns. Do not use -ing words or other verbs. Look up `median tests`, not `testing medians`.
- Use few words. Every word specified further restricts the search. Look up `distribution`, and you get one list; look up `normal distribution`, and the list is a sublist of that.
- Sometimes words have more than one context. The following words can be used to restrict the context:

  a. `data`, meaning in the context of data management. Order could refer to the order of data or to order statistics. Look up `order data` to restrict order to its data-management sense.

  b. `statistics` (abbreviation `stat`), meaning in the context of statistics. Look up `order statistics` to restrict order to the statistical sense.

  c. `graph` or `graphs`, meaning in the context of statistical graphics. Look up `median graphs` to restrict the list to commands for graphing medians.

  d. `utility` (abbreviation `util`), meaning in the context of utility commands. The `search` command itself is not data management, not statistics, and not graphics; it is a utility.

  e. `programs` or `programming` (abbreviation `prog`), to mean in the context of programming. Look up `programming scalar` to obtain a sublist of scalars in programming.

`search` has other features, as well; see [U] **4.9 search: All the details**.

4.7 More on help

Both `help` and `search` are understanding of some mistakes. For instance, you may abbreviate a command name. If you type either `help regres` or `help regress`, you will bring up the help file for `regress`.

When `help` cannot find the command you are looking for, try `search`. In this case, typing `search regres` would find the command, but that would be because 'regres' is an abbreviation of the word regression, which probably appears in the keyword database.

Stata can run into some problems with abbreviations. For instance, Stata has a command with the inelegant name `ksmirnov`. You forget and think the command is called `ksmir`:

```
. help ksmir
help for ksmir not found
try help contents or search ksmir
```

This is a case where `help` gives bad advice because typing `search ksmir` will do you no good. You should type `search` followed by what you are really looking for: `search kolmogorov smirnov`.

4.8 help contents: Table of contents for Stata's help system

Typing `help contents` or selecting **Help > Contents** provides another way of locating entries in the documentation and online help. Whichever you use, you will be presented with a long table of contents, organized topically.

```
. help contents
```

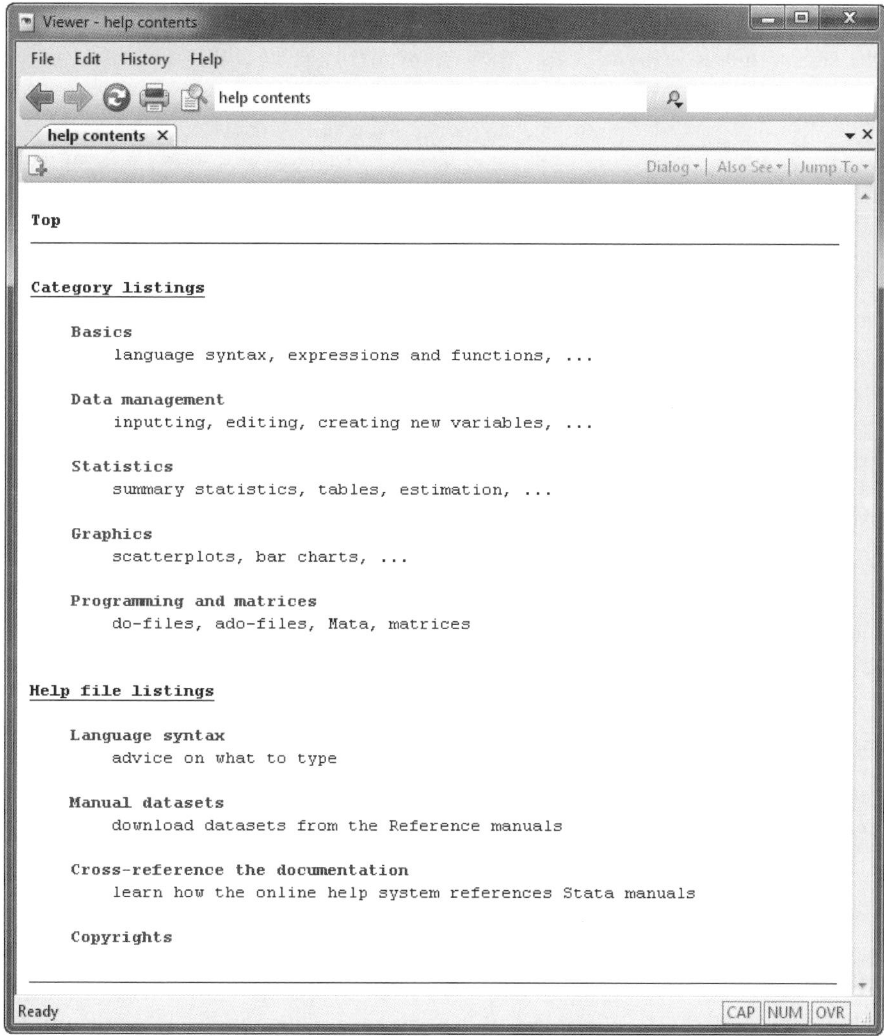

4.9 search: All the details

The `search` command actually provides a few features that are not available from the **Help** menu. The full syntax of the `search` command is

`search` *word* [*word* ...] [, [`local` | `net` | `all`] `author` `entry` `exact` `faq` `historical` `or` `manual` `sj`]

where underlining indicates the minimum allowable abbreviation and [`brackets`] indicate optional.

`local`, the default (unless changed by `set searchdefault`), specifies that the search be performed using only Stata's keyword database.

`net` specifies that the search be performed across the materials available via Stata's `net` command. Using `search word` [*word* ...], `net` is equivalent to typing `net search word` [*word* ...] (without options); see [R] **net**.

`all` specifies that the search be performed across both the local keyword database and the net materials.

`author` specifies that the search be performed on the basis of author's name rather than keywords.

`entry` specifies that the search be performed on the basis of entry IDs rather than keywords.

`exact` prevents matching on abbreviations.

`faq` limits the search to entries found in the FAQs at http://www.stata.com.

`historical` adds to the search entries that are of historical interest only. By default, such entries are not listed. Past entries are classified as historical if they discuss a feature that later became an official part of Stata. Updates to historical entries will always be found, even if `historical` is not specified.

`or` specifies that an entry be listed if any of the words typed after `search` are associated with the entry. The default is to list the entry only if all the words specified are associated with the entry.

`manual` limits the search to entries in the *User's Guide* and all the *Reference* manuals.

`sj` limits the search to entries in the *Stata Journal* and the *Stata Technical Bulletin*.

4.9.1 How search works

`search` has a database—files—containing the titles, etc., of every entry in the *User's Guide*, *Reference* manuals, undocumented help files, NetCourses, Stata Press books, FAQs posted on the Stata website, selected articles on StataCorp's official blog, selected user-written FAQs and examples, and the articles in the *Stata Journal* and in the *Stata Technical Bulletin*. In this file is a list of words associated with each entry, called keywords.

When you type `search` *xyz*, `search` reads this file and compares the list of keywords with *xyz*. If it finds *xyz* in the list or a keyword that allows an abbreviation of *xyz*, it displays the entry.

When you type `search` *xyz abc*, `search` does the same thing but displays an entry only if it contains both keywords. The order does not matter, so you can `search linear regression` or `search regression linear`.

How many entries `search` finds depends on how the search database was constructed. We have included a plethora of keywords under the theory that, for a given request, it is better to list too much rather than risk listing nothing at all. Still, you are in the position of guessing the keywords. Do you look up normality test, normality tests, or tests of normality? Normality test would be best, but all would work. In general, use the singular, and strike the unnecessary words. We provide guidelines for specifying keywords in [U] **4.6 More on search** above.

4.9.2 Author searches

search ordinarily compares the words following search with the keywords for the entry. If you specify the author option, however, it compares the words with the author's name. In the search database, we have filled in author names for *Stata Journal* and STB articles and for FAQs.

For instance, in [R] **kdensity**, you will discover that Isaías H. Salgado-Ugarte wrote the first version of Stata's kdensity command and published it in the STB. Assume that you have read his original and find the discussion useful. You might now wonder what else he has written in the STB. To find out, type

```
. search Salgado-Ugarte, author
  (output omitted)
```

Names like Salgado-Ugarte are confusing to some people. search does not require you specify the entire name; what you type is compared with each "word" of the name, and, if any part matches, the entry is listed. The hyphen is a special character, and you can omit it. Thus you can obtain the same list by looking up Salgado, Ugarte, or Salgado Ugarte without the hyphen.

Actually, to find all entries written by Salgado-Ugarte, you need to type

```
. search Salgado-Ugarte, author historical
  (output omitted)
```

Prior inserts in the STB that provide a feature that later was superseded by a built-in feature of Stata are marked as historical in the search database and, by default, are not listed. The historical option ensures that all entries are listed.

4.9.3 Entry ID searches

If you specify the entry option, search compares what you have typed with the entry ID. The entry ID is not the title—it is the reference listed to the left of the title that tells you where to look. For instance, in

```
[R]      regress . . . . . . . . . . . . . . . . . . . . . . . . Linear regression
         (help regress)
```

"[R] regress" is the entry ID. In

```
GS       . . . . . . . . . . . . . . . . . . . . . . . . . . . Getting Started manual
```

"GS" is the entry ID. In

```
SJ-6-4   st0113  . Testing for cross-sectional dependence in panel-data models
         (help xtcsd if installed) . . . . . . R. E. De Hoyos and V. Sarafidis
         Q4/06   SJ 6(4): 482--496
         tests for the presence of cross-sectional dependence in
         panels with many cross-sectional units and few time-series
         observations
```

"SJ-6-4 st0113" is the entry ID.

search with the entry option searches these entry IDs.

Thus, you could generate a table of contents for the *Reference* manuals by typing

```
. search [R], entry
  (output omitted)
```

You could generate a table of contents for the 16th issue of the STB by typing

```
. search STB-16, entry historical
  (output omitted)
```

The `historical` option here is possibly important. STB-16 was published in November 1993, and perhaps some of its inserts have been marked as historical.

You could obtain a list of all inserts associated with *dm36* by typing

```
. search dm36, entry historical
  (output omitted)
```

Again, we include the `historical` option if any of the relevant inserts have been marked historical.

4.9.4 FAQ searches

To search across the FAQs, specify the `faq` option:

```
. search logistic regression, faq
  (output omitted)
```

4.9.5 Return codes

In addition to indexing the entries in the *User's Guide* and all the *Stata Reference* manuals, `search` also can be used to look up return codes.

To see information about return code 131, type

```
. search rc 131
[R]     error messages  . . . . . . . . . . . . . . . . . . .  Return code 131
        not possible with test;
        You requested a test of a hypothesis that is nonlinear in the
        variables.  test tests only linear hypotheses.  Use testnl.
```

To get a list of all Stata return codes, type

```
. search rc
  (output omitted)
```

4.10 net search: Searching net resources

When you select **Help > Search**, there are two types of searches to choose. The first, which has been discussed in the previous sections, is to **Search documentation and FAQs**. The second is to **Search net resources**. This feature of Stata searches resources over the Internet.

When you choose **Search net resources** in the search dialog box and enter *keywords* in the field, Stata searches all user-written programs on the Internet, including user-written additions published in the *Stata Journal* and the STB. The results are displayed in the Viewer, and you can click to go to any of the matches found.

Equivalently, you can type `net search` *keywords* on the Stata command line to display the results in the Results window. For the full syntax for using the `net search` command, see [R] **net search**.

4.11 hsearch: An alternative to search

`hsearch` is available from command mode, and by clicking on `hsearch` from **Help > Advice**. Just like you can type `search` *anything*, you can type `hsearch` *anything*. Whereas `search` bases its results on a file of keywords for official Stata commands, official FAQs, and the *Stata Journal*, `hsearch` searches the text of the help files on your computer. `hsearch` searches all the help files, whether official or not, that is, whether the help files are part of Stata as supplied by StataCorp or from user-written additions you have installed or from help files you wrote yourself.

`hsearch` shows what is installed on your computer.

`search` shows what is official and includes FAQs and *Stata Journal* contributions.

5 Flavors of Stata

Contents

5.1 Platforms

Stata is available for a variety of computers, including

Stata for Windows, 64-bit x86-64
Stata for Windows, 32-bit x86

Stata for Mac, 64-bit Intel
Stata for Mac, 32-bit Intel
Stata for Mac, PowerPC

Stata for Linux, 64-bit x86-64
Stata for Linux, 32-bit x86
Stata for Solaris, 64-bit SPARC
Stata for Solaris, 64-bit x86-64

Which version of Stata you run does not matter—Stata is Stata. You instruct Stata in the same way and Stata produces the same results, right down to the random-number generator. Even files can be shared. A dataset created on one computer can be used on any other computer, and the same goes for graphs, programs, or any file Stata uses or produces. Moving files across platforms is simply a matter of copying them; no translation is required.

Some computers, however, are faster than others. Some computers have more memory than others. Computers with more memory, and faster computers, are better.

The list above includes both 64- and 32-bit computers. 64-bit Stata runs faster than 32-bit Stata and 64-bit Stata will allow processing data in excess of 2 gigabytes, assuming you have enough memory. 32-bit Stata will run on 64-bit hardware.

When you purchase Stata, you may install it on any of the above platforms. Stata licenses are not locked to a single operating system.

5.2 Stata/MP, Stata/SE, Stata/IC, and Small Stata

Stata is available in four flavors, although perhaps sizes would be a better word. The flavors are, from largest to smallest, Stata/MP, Stata/SE, Stata/IC, and Small Stata.

Stata/MP is the multiprocessor version of Stata. It runs on multiple CPUs or on multiple cores, from 2 to 64. Stata/MP uses however many cores you tell it to use (even one), up to the number of cores for which you are licensed. Stata/MP is the fastest version of Stata. Even so, all the details of parallelization are handled internally and you use Stata/MP just like you use any other flavor of Stata. You can read about how Stata/MP works and see how its speed increases with more cores in the Stata/MP performance report at http://www.stata.com/statamp/report.pdf.

Stata/SE is like Stata/MP, but for single CPUs. Stata/SE will run on multiple CPUs or multiple-core computers, but it will use only one CPU or core. SE stands for special edition.

Both Stata/MP and Stata/SE have the same limits and the same capabilities and are intended for those who work with large datasets. You may have up to 32,767 variables with either. Statistical models may have up to 11,000 variables.

Stata/IC is standard Stata. IC stands for intercooled, a name used before the introduction of Stata/SE and Stata/MP to indicate that it was at the time the top version of Stata. Up to 2,047 variables are allowed. Statistical models may have up to 800 variables.

Stata/MP, Stata/SE, and Stata/IC all allow up to 2,147,583,647 observations, assuming you have enough memory.

Small Stata is intended for students and limited to 99 variables and 1,200 observations.

5.2.1 Determining which version you own

Check your License and Authorization Key. Included with every copy of Stata is a paper License and Authorization Key that contains codes that you will input during installation. This determines which flavor of Stata you have and for which platform.

Contact us or your distributor if you want to upgrade from one flavor to another. Usually, all you need is an upgraded paper License and Authorization Key with the appropriate codes. All flavors of Stata are on the same DVD.

If you purchased one flavor of Stata and want to use a lesser version, you may. You might want to do this if you had a large computer at work and a smaller one at home. Please remember, however, that you have only one license (or however many licenses you purchased). You may, both legally and ethically, install Stata on both computers and then use one or the other, but you should not use them both simultaneously.

5.2.2 Determining which version is installed

If Stata is already installed, you can find out which Stata you are using by entering Stata as you normally do and typing `about`:

```
. about
Stata/MP 12.0 for Windows (64-bit x86-64)
Revision date
Copyright 1985-2011 StataCorp LP
Total physical memory:     8388608 KB
Available physical memory:  937932 KB
10-user 32-core Stata network perpetual license:
       Serial number:  5012041234
         Licensed to:  Alan R. Riley
                       StataCorp
```

5.3 Size limits of Stata/MP, SE, IC, and Small Stata

Here are some of the different size limits for Stata/MP, Stata/SE, Stata/IC, and Small Stata. Type `help limits` for a longer list.

Maximum size limits for Stata/MP, Stata/SE, Stata/IC, and Small Stata

	Stata/MP and SE	Stata/IC	Small Stata
Number of observations	limited only by memory	limited only by memory	fixed at 1,200
Number of variables	32,767	2,047	fixed at 99
Width of a dataset	393,192	24,564	800
Maximum # of right-hand-side variables	10,998	798	99
Number of characters in a macro	1,081,511	165,200	8,681
Number of characters in a command	1,081,527	165,216	8,697

Stata/MP and Stata/SE allow more variables, larger models, longer macros, and a longer command line than Stata/IC. The longer command line and macro length are required because of the greater number of variables allowed. Larger models means that Stata/MP and Stata/SE can fit statistical models with more independent variables.

Small Stata is limited. It is intended for student use and often used in undergraduate labs.

5.4 Speed comparison of Stata/MP, SE, IC, and Small Stata

We have written a white paper comparing the performance of Stata/MP with Stata/SE; see http://www.stata.com/statamp/report.pdf. The white paper includes command-by-command performance measurements.

In summary, on a 2-CPU or dual-core computer, Stata/MP will run commands in 71% of the time required by Stata/SE. There is variation; some commands run in half the time and others are not sped up at all. Statistical estimation commands run in 59% of the time. Numbers quoted are medians. Average performance gains are higher because commands that take longer to execute are generally sped up more.

Stata/MP running on four CPUs runs in 50% (all commands) and 35% (estimation commands) of the time required by Stata/SE. Both numbers are median measures.

Stata/MP supports up to 64 cores.

Stata/IC is slower than Stata/SE, but those differences emerge only when processing datasets that are pushing the limits of Stata/IC. Stata/SE has a larger memory footprint and uses that extra memory for larger look-aside tables to more efficiently process large datasets. The real benefits of the larger tables become apparent only after exceeding the limits of Stata/IC. Stata/SE was designed for processing large datasets.

Small Stata is, by comparison to all the above, slow, but given its limits, no one notices. Small Stata was designed to have a minimal memory footprint, and to achieve that, different logic is sometimes used. For instance, in Stata's `test` command, it must compute the matrix calculation $\mathbf{RZR}'$ (where $\mathbf{Z} = (\mathbf{X}'\mathbf{X})^{-1}$). Stata/MP, Stata/SE, and Stata/IC make the calculation in a straightforward way, which is to form $\mathbf{T} = \mathbf{RZ}$ and then calculate $\mathbf{TR}'$. This requires temporarily storing the matrix $\mathbf{T}$. Small Stata, on the other hand, goes into more complicated code to form the result directly—code that requires temporary storage of only one scalar. This code, in effect, recalculates intermediate results over and over again, and so it is slower.

The differences are all technical and internal. From the user's point of view, Stata/MP, Stata/SE, Stata/IC, and Small Stata work the same way.

5.5 Feature comparison of Stata/MP, SE, and IC

The features of all flavors of Stata on all platforms are the same. The differences are in speed and in limits as discussed above. To learn more, type `help stata/mp`, `help stata/se`, `help stata/ic`, or `help small stata`.

6 Managing memory

Contents

6.1 Memory-size considerations

Stata works with a copy of data that it loads into memory. Memory allocation is automatic.

Stata automatically sizes itself up and down as your session progresses. Stata obtains memory from the operating system and draws no distinction between real and virtual memory. Virtual memory is memory that resides on disk that operating systems supply when physical memory runs short. Virtual memory is slow but adequate in cases when you have a dataset that is too large to load into real memory. If you wish to limit the maximum amount of memory Stata can use, you can set `max_memory`; see [D] **memory**. If you use the Linux operating system, we strongly suggest you set `max_memory`; see *Serious bug in Linux OS* in [D] **memory**.

6.2 Compressing data

Stata stores data in memory. The `compress` command reduces the amount of memory required to store the data without loss of precision or any other disadvantages; see [D] **compress**. Typing `compress` every so often is a good idea.

`compress` works by examining the values you have stored and changing the data types of variables when that can be done without loss of precision. For instance, you may have a variable stored as `float` but that records only integer values between -127 and 100. `compress` would change the storage type of that variable to `byte` and save 3 bytes per observation. If you had 100 variables like that, the savings would be 300 bytes per observation, and if you had 3,000,000 observations, the total savings would be nearly 900 megabytes.

6.3 Setting maxvar

If you get the error message “no room to add more variables”, r(901), do not jump to the conclusion that you have exceeded Stata’s capacity.

`maxvar` specifies the maximum number of variables you can use. The default setting depends on whether you are using Stata/MP, Stata/SE, Stata/IC, or Small Stata. To determine the current setting, type `query memory` at the Stata prompt.

If you use Stata/MP or Stata/SE, you can reset this maximum number all the way up to 32,767. Set `maxvar` to more than you need—at least 20 more than you need but not too much more than you need. Figure that each 10,000 variables consumes roughly 0.5 megabytes of memory.

You reset maxvar using the `set maxvar` command,

`set maxvar # [, permanently]`

where 2,048 $\leq$ # $\leq$ 32,767. You can reset `maxvar` repeatedly during a session. If you specify the `permanently` option, you change `maxvar` not only for this session but also for future sessions. See [D] **memory**.

6.4 Setting matsize

You may issue an estimation command and obtain the error message "matsize too small", r(908). Stata uses matrices in making many calculations. `matsize` specifies the maximum size of those matrices in terms of (roughly speaking) the number of estimated coefficients. The default value of `matsize` is 400. `matsize` can be set to any value between 10 and 11,000, inclusive. The command is

`set matsize # [, permanently]`

where 10 $\leq$ # $\leq$ 11,000.

Increasing `matsize` increases Stata's memory consumption:

matsize	memory use
400	1.254M
800	4.950M
1,600	19.666M
3,200	78.394M
6,400	313.037M
11,000	924.080M

The table above understates the amount of memory Stata will use. The table was derived under the assumption of one matrix and eleven vectors. If two matrices are required, the numbers above would be nearly doubled.

If you use a 32-bit computer, you likely will be unable to set `matsize` to 11,000. A value of 11,000 would require nearly 1 gigabyte per matrix. The total memory consumption most 32-bit operating systems will grant to Stata is 2 gigabytes, so if you had two matrices, there would be no memory left for data or for Stata's code!

You should not `set matsize` larger than is necessary. Doing so will at best waste memory and at worst slow Stata down or prevent Stata from having enough memory for other tasks. If you receive the error message "matsize too small", increase `matsize` only as much as is necessary to eliminate the error message.

6.5 The memory command

The `memory` command will show you the major components of Stata's memory footprint.

You may use

```
. use http://www.stata-press.com/data/r12/regsmpl
(NLS Women 14-26 in 1968)

. memory

  Memory usage
                                          used            allocated
    ---------------------------------------------------------------
    data (incl. buffers)               913,120           67,108,864
    var. names, %fmts, ...               1,780               25,492
    overhead                         1,081,352            1,082,144

    Stata matrices                           0                    0
    ado-files                           44,763               44,763
    saved results                        1,142                1,142

    Mata matrices                       45,448               45,448
    Mata functions                      17,328               17,328

    set maxvar usage                 1,366,792            1,366,792

    other                                5,457                5,457
    ---------------------------------------------------------------
    total                            3,426,882           69,697,430
```

See [D] **memory**.

7 –more– conditions

Contents

7.1 Description

When you see —`more`— at the bottom of the screen,

Press ...	and Stata...
letter *l* or *Enter*	displays the next line
letter *q*	acts as if you pressed *Break*
Spacebar or any other key	displays the next screen

Also, you can press the *clear –more– condition* button, the button labeled **Go** with a circle around it.

—`more`— is Stata's way of telling you that it has something more to show you, but showing you that something more will cause the information on the screen to scroll off.

7.2 set more off

If you type `set more off`, —`more`— conditions will never arise and Stata's output will scroll by at full speed.

If you type `set more on`, —`more`— conditions will be restored at the appropriate places.

Programmers: Do-file writers sometimes include `set more off` in their do-files because they do not care to interactively watch the output. They want Stata to proceed at full speed because they plan on making a log of the output that they will review later. Do-filers need not bother to `set more on` at the conclusion of their do-file. Stata automatically restores the previous `set more` when the do-file (or program) concludes.

7.3 The more programming command

Ado-file programmers need take no special action to have —`more`— conditions arise when the screen is full. Stata handles that automatically.

If, however, you wish to force a —`more`— condition early, you can include the `more` command in your program. The syntax of `more` is

```
more
```

`more` takes no arguments.

For more information, see [P] **more**.

8 Error messages and return codes

Contents

8.1 Making mistakes

When an error occurs, Stata produces an error message and a *return code*. For instance,

```
. list myvar
no variables defined
r(111);
```

We ask Stata to list the variable named `myvar`. Because we have no data in memory, Stata responds with the message "no variables defined" and a line that reads "`r(111)`".

The "no variables defined" is called the error message.

The 111 is called the return code. You can click on blue return codes to get a detailed explanation of the error.

8.1.1 Mistakes are forgiven

After "no variables defined" and `r(111)`, all is forgiven; it is as if the error never occurred.

Typically, the message will be enough to guide you to a solution, but if it is not, the numeric return codes are documented in [P] **error**.

8.1.2 Mistakes stop user-written programs and do-files

Whenever an error occurs in a user-written program or do-file, the program or do-file immediately stops execution and the error message and return code are displayed.

For instance, consider the following do-file:

begin myfile.do

```
use http://www.stata-press.com/data/r12/auto
decribe
list
```

end myfile.do

Note the second line—you meant to type `describe` but typed `decribe`. Here is what happens when you execute this do-file by typing `do myfile`:

```
. do myfile

. use http://www.stata-press.com/data/r12/auto
(1978 Automobile Data)
```

```
. decribe
unrecognized command:  decribe
r(199);
end of do-file
r(199);
. _
```

The first error message and return code were caused by the illegal `decribe`. This then caused the do-file itself to be aborted; the valid `list` command was never executed.

8.1.3 Advanced programming to tolerate errors

Errors are not only of the typographical kind; some are substantive. A command that is valid in one dataset might not be valid in another. Moreover, in advanced programming, errors are sometimes anticipated: use one dataset if it is there, but use another if you must.

Programmers can access the return code to determine whether an error occurred, which they can then ignore, or, by examining the return code, code their programs to take the appropriate action. This is discussed in [P] **capture**.

You can also prevent do-files from stopping when errors occur by using the `do` command's `nostop` option.

```
. do myfile, nostop
```

8.2 The return message for obtaining command timings

In addition to error messages and return codes, there is something called a return message, which you normally do not see. Normally, if you typed `summarize tempjan`, you would see

```
. use http://www.stata-press.com/data/r12/citytemp
(City Temperature Data)
. summarize tempjan
    Variable |       Obs        Mean    Std. Dev.       Min        Max
-------------+--------------------------------------------------------
     tempjan |       954    35.74895    14.18813        2.2       72.6
```

If you were to type

```
. set rmsg on
r; t=0.00 10:21:22
```

sometime during your session, Stata would display return messages:

```
. summarize tempjan
    Variable |       Obs        Mean    Std. Dev.       Min        Max
-------------+--------------------------------------------------------
     tempjan |       954    35.74895    14.18813        2.2       72.6
r; t=0.01 10:21:26
```

The line that reads `r; t=0.01 10:21:26` is called the return message.

The `r;` indicates that Stata successfully completed the command.

The `t=0.01` shows the amount of time, in seconds, it took Stata to perform the command (timed from the point you pressed *Enter* to the time Stata typed the message). This command took a hundredth of a second. Stata also shows the time of day with a 24-hour clock. This command completed at 10:21 a.m.

Stata can run commands stored in files (called do-files) and can log output. Some users find the detailed return message helpful with do-files. They construct a long program and let it run overnight, logging the output. They come back the next morning, look at the output, and discover a mistake in some portion of the job. They can look at the return messages to determine how long it will take to rerun that portion of the program.

You may `set rmsg on` whenever you wish.

When you want Stata to stop displaying the detailed return message, type `set rmsg off`.

9 The Break key

Contents

9.1 Making Stata stop what it is doing

When you want to make Stata stop what it is doing and return to the Stata dot prompt, you click on *Break*:

Stata for Windows:	click on the **Break** button (it is the button with the big red X), or press *Ctrl+Pause/Break*
Stata for Mac:	click on the **Break** button or press *Command+.* (period)
Stata for Unix(GUI):	click on the **Break** button or press *Ctrl+k*
Stata for Unix(console):	press *Ctrl+c* or press *q*

Elsewhere in this manual, we describe this action as simply clicking on *Break*. Break tells Stata to cancel what it is doing and return control to you as soon as possible.

If you click on *Break* in response to the input prompt or while you are typing a line, Stata ignores it, because you are already in control.

If you click on *Break* while Stata is doing something—creating a new variable, sorting a dataset, making a graph, etc.—Stata stops what it is doing, undoes it, and issues an input prompt. The state of the system is the same as if you had never issued the command.

▷ Example 1

You are fitting a logit model, type the command, and, as Stata is working on the problem, realize that you omitted an important variable:

```
. logit foreign mpg weight
Iteration 0:   log likelihood =  -45.03321
Iteration 1:   log likelihood = -29.898968
—Break—
r(1);

. _
```

When you clicked on *Break*, Stata responded by typing `—Break—` and then typing `r(1);`. Clicking on *Break* always results in a return code of 1—that is why return codes are called return codes and not error codes. The 1 does not indicate an error, but it does indicate that the command did not complete its task.

◁

9.2 Side effects of clicking on Break

In general, there are no side effects of clicking on Break. We said above that Stata undoes what it is doing so that the state of the system is the same as if you had never issued the command. There are two exceptions to that statement.

If you are reading data from disk by using `insheet`, `infile`, or `infix`, whatever data have already been read will be left behind in memory, the theory being that perhaps you stopped the process so you could verify that you were reading the right data correctly before sitting through the whole process. If not, you can always `clear`.

```
. infile v1-v9 using workdata
(eof not at end of obs)
(4 observations read)
—Break—
r(1);
```

The other exception is `sort`. You have a large dataset in memory, decide to sort it, and then change your mind.

```
. sort price
—Break—
r(1);
```

If the dataset was previously sorted by, say, the variable `prodid`, it is no longer. When you click on *Break* in the middle of a `sort`, Stata marks the data as unsorted.

9.3 Programming considerations

There are basically no programming considerations for handling Break because Stata handles it all automatically. If you write a program or do-file, execute it, and then click on *Break*, Stata stops execution just as it would with an internal command.

Advanced programmers may be concerned about cleaning up after themselves; perhaps they have generated a temporary variable they intended to drop later or a temporary file they intended to erase later. If a Stata user clicks on *Break*, how can you ensure that these temporary variables and files will be erased?

If you obtain names for such temporary items from Stata's `tempname`, `tempvar`, and `tempfile` commands, Stata will automatically erase the temporary items; see [U] **18.7 Temporary objects**.

There are instances, however, when a program must commit to executing a group of commands without interruption, or the user's data would be left in an intermediate or undefined state. In these instances, Stata provides a

```
nobreak {
        ...
}
```

construct; see [P] **break**. Also see [M-5] **setbreakintr()** to read about Break-key processing in Mata.

10 Keyboard use

Contents

10.1 Description

The keyboard should operate much the way you would expect, with a few additions:

- There are some unexpected keys you can press to obtain previous commands you have typed. Also, you can click once on a command in the Review window to reload it, or click on it twice to reload and execute; this feature is discussed in the *Getting Started* manuals.
- There are a host of command-editing features for Stata for Unix(console) users because their user interface does not offer such features.
- Regardless of operating system or user interface, if there are *F*-keys on your keyboard, they have special meaning and you can change the definitions of the keys.

10.2 F-keys

Windows users: *F10* is reserved internally by Windows; you cannot program this key.

By default, Stata defines the *F*-keys to mean

F-key	Definition
F1	`help advice;`
F2	`describe;`
F7	`save`
F8	`use`

The semicolons at the end of some entries indicate an implied *Enter*.

Stata provides several methods for obtaining help. To learn about these methods, select **Help** >**Advice**. Or you can just press *F1*.

`describe` is the Stata command to report the contents of data loaded into memory. It is explained in [D] **describe**. Normally, you type `describe` and press *Enter*. You can also press *F2*.

`save` is the command to save the data in memory into a file, and `use` is the command to load data; see [D] **use** and [D] **save**. The syntax of each is the same: `save` or `use` followed by a filename. You can type the commands or you can press *F7* or *F8* followed by the filename.

You can change the definitions of the *F*-keys. For instance, the command to list data is `list`; you can read about it in [D] **list**. The syntax is `list` to list all the data, or `list` followed by the names of some variables to list just those variables (there are other possibilities).

If you wanted *F3* to mean `list`, you could type

```
. global F3 "list "
```

In the above, `F3` refers to the letter *F* followed by *3*, not the *F3* key. Note the capitalization and spacing of the command.

You type `global` in lowercase, type `F3`, and then type `"list "`. The space at the end of `list` is important. In the future, rather than typing `list mpg weight`, you want to be able to press the *F3* key and then type only `mpg weight`. You put a space in the definition of `F3` so that you would not have to type a space in front of the first variable name after pressing *F3*.

Now say you wanted *F5* to mean list all the data—`list` followed by *Enter*. You could define

```
. global F5 "list;"
```

Now you would have two ways of listing all the data: press *F3*, and then press *Enter*, or press *F5*. The semicolon at the end of the definition of *F5* will press *Enter* for you.

If you really want to change the definitions of *F3* and *F5*, you will probably want to change the definition every time you invoke Stata. One way would be to type the two `global` commands every time you invoke Stata. Another way would be to type the two commands into a text file named `profile.do`. Stata executes the commands in `profile.do` every time it is launched if `profile.do` is placed in the appropriate directory:

Windows:	see [GSW] **C.3 Executing commands every time Stata is started**
Mac:	see [GSM] **C.1 Executing commands every time Stata is started**
Unix:	see [GSU] **C.1 Executing commands every time Stata is started**

You can use the *F*-keys any way you desire: they contain a string of characters, and pressing the *F*-key is equivalent to typing those characters.

❑ Technical note

[*Stata for Unix(console) users.*] Sometimes Unix assigns a special meaning to the *F*-keys, and if it does, those meanings supersede our meanings. Stata provides a second way to get to the *F*-keys. Press *Ctrl+F*, release the keys, and then press a number from `0` through `9`. Stata interprets *Ctrl+F* plus 1 as equivalent to the *F1* key, *Ctrl+F* plus 2 as *F2*, and so on. *Ctrl+F* plus 0 means *F10*. These keys will work only if they are properly mapped in your `termcap` or `terminfo` entry.

❑

❑ Technical note

On some international keyboards, the left single quote is used as an accent character. In this case, we recommend mapping this character to one of your function keys. In fact, you might find it convenient to map both the left single quote (‘) and right single quote (’) characters so that they are next to each other.

Within Stata, open the Do-file Editor. Type the following two lines in the Do-file Editor:

```
global F4 ‘
global F5 ’
```

Save the file as `profile.do` into your Stata directory. If you already have a `profile.do` file, append the two lines to your existing `profile.do` file.

Exit Stata and restart it. You should see the startup message

```
running C:\Program Files\Stata12\profile.do ...
```

or some variant of it depending on where your Stata is installed. Press *F4* and *F5* to verify that they work.

If you did not see the startup message, you did not save the `profile.do` in your home folder.

You can, of course, map to any other function keys, but *F1*, *F2*, *F7*, and *F8* are already used.

❑

10.3 Editing keys in Stata

Users have available to them the standard editing keys for their operating system. So, Stata should just edit what you type in the natural way—the Stata Command window is a standard edit window.

Also, you can fetch commands from the Review window into the Command window. Click on a command in the Review window, and it is loaded into the Command window, where you can edit it. Alternatively, if you double-click on a line in the Review window, it is loaded and executed.

Another way to get lines from the Review window into the Command window is with the *PgUp* and *PgDn* keys. Press *PgUp* and Stata loads the last command you typed into the Command window. Press it again and Stata loads the line before that, and so on. *PgDn* goes in the opposite direction.

Another editing key that interests users is *Esc*. This key clears the Command window.

In summary,

Press	Result
PgUp	Steps back through commands and moves command from Review window to Command window
PgDn	Steps forward through commands and moves command from Review window to Command window
Esc	Clears Command window

10.4 Editing keys in Stata for Unix(console)

Certain keys allow you to edit the line that you are typing. Because Stata supports a variety of computers and keyboards, the location and the names of the editing keys are not the same for all Stata users.

Every keyboard has the standard alphabet keys (*QWERTY* and so on), and every keyboard has a *Ctrl* key. Some keyboards have extra keys located to the right, above, or left, with names like *PgUp* and *PgDn*.

Throughout this manual we will refer to Stata's editing keys using names that appear on nobody's keyboard. For instance, PrevLine is one of the Stata editing keys—it retrieves a previous line. Hunt all you want, but you will not find it on your keyboard. So, where is PrevLine? We have tried to put it where you would naturally expect it. On keyboards with a key labeled *PgUp*, *PgUp* is the PrevLine key, but on everybody's keyboard, no matter which version of Unix, brand of keyboard, or anything else, *Ctrl+R* also means PrevLine.

When we say press PrevLine, now you know what we mean: press *PgUp* or *Ctrl+R*. The editing keys are the following:

Name for editing key	Editing key	Function
Kill	*Esc* on PCs and *Ctrl+U*	Deletes the line and lets you start over.
Dbs	*Backspace* on PCs and *Backspace* or *Delete* on other computers	Backs up and deletes one character.
Lft	←, *4* on the numeric keypad for PCs, and *Ctrl+H*	Moves the cursor left one character without deleting any characters.
Rgt	→, *6* on the numeric keypad for PCs, and *Ctrl+L*	Moves the cursor forward one character.
Up	↑, *8* on the numeric keypad for PCs, and *Ctrl+O*	Moves the cursor up one physical line on a line that takes more than one physical line. Also see PrevLine.
Dn	↓, *2* on the numeric keypad for PCs, and *Ctrl+N*	Moves the cursor down one physical line on a line that takes more than one physical line. Also see NextLine.
PrevLine	*PgUp* and *Ctrl+R*	Retrieves a previously typed line. You may press PrevLine multiple times to step back through previous commands.
NextLine	*PgDn* and *Ctrl+B*	The inverse of PrevLine.
Seek	*Ctrl+Home* on PCs and *Ctrl+W*	Goes to the line number specified. Before pressing Seek, type the line number. For instance, typing *3* and then pressing Seek is the same as pressing PrevLine three times.
Ins	*Ins* and *Ctrl+E*	Toggles insert mode. In insert mode, characters typed are inserted at the position of the cursor.
Del	*Del* and *Ctrl+D*	Deletes the character at the position of the cursor.
Home	*Home* and *Ctrl+K*	Moves the cursor to the start of the line.
End	*End* and *Ctrl+P*	Moves the cursor to the end of the line.
Hack	*Ctrl+End* on PCs, and *Ctrl+X*	Hacks off the line at the cursor.
Tab	⇥ on PCs, *Tab*, and *Ctrl+I*	Expand variable name.
Btab	⇤ on PCs, and *Ctrl+G*	The inverse of Tab.

▷ Example 1

It is difficult to demonstrate the use of editing keys on paper. You should try each of them. Nevertheless, here is an example:

```
. summarize price waht
```

You typed `summarize price waht` and then pressed the *Lft* key (← key or *Ctrl+H*) three times to maneuver the cursor back to the `a` of `waht`. If you were to press *Enter* right now, Stata would see the command `summarize price waht`, so where the cursor is does not matter when you press *Enter*. If you wanted to execute the command `summarize price`, you could back up one more character and then press the Hack key. We will assume, however, that you meant to type `weight`.

If you were now to press the letter *e* on the keyboard, an `e` would appear on the screen to replace the `a`, and the cursor would move under the character `h`. We now have `weht`. You press *Ins*, putting Stata into insert mode, and press *i* and *g*. The line now says `summarize price weight`, which is correct, so you press *Enter*. We did not have to press *Ins* before every character we wanted to insert. The *Ins* key is a toggle: If we press it again, Stata turns off insert mode, and what we type replaces what was there. When we press *Enter*, Stata forgets all about insert mode, so we do not have to remember from one command to the next whether we are in insert mode.

◁

❑ Technical note

Stata performs its editing magic from the information about your terminal recorded in `/etc/termcap`(5) or, under System V, `/usr/lib/terminfo`(4). If some feature does not appear to work, the entry for your terminal in the `termcap` file or `terminfo` directory is probably incorrect. Contact your system administrator.

❑

10.5 Editing previous lines in Stata

In addition to what is said below, remember that the Review window also shows the contents of the review buffer.

One way to retrieve lines is with the PrevLine and NextLine keys. Remember, PrevLine and NextLine are the names we attach to these keys—there are no such keys on your keyboard. You have to look back at the previous section to find out which keys correspond to PrevLine and NextLine on your computer. To save you the effort this time, PrevLine probably corresponds to *PgUp* and NextLine probably corresponds to *PgDn*.

Suppose you wanted to reissue the third line back. You could press PrevLine three times and then press *Enter*. If you made a mistake and pressed PrevLine four times, you could press NextLine to go forward in the buffer. You do not have to count lines because, each time you press PrevLine or NextLine, the current line is displayed on your monitor. Simply press the key until you find the line you want.

Another method for reviewing previous lines, `#review`, is convenient for Unix(console) users.

▷ Example 2

Typing `#review` by itself causes Stata to list the last five commands you typed. For instance,

```
. #review
5 list make mpg weight if abs(res)>6
4 list make mpg weight if abs(res)>5
3 tabulate foreign if abs(res)>5
2 regress mpg weight weight2
1 test weight2=0
. _
```

We can see from the listing that the last command typed by the user was `test weight2=0`. Or, you may just look at the Review window to see the history of commands you typed.

◁

▷ Example 3

Perhaps the command you are looking for is not among the last five commands you typed. You can tell Stata to go back any number of lines. For instance, typing **#review 15** tells Stata to show you the last 15 lines you typed:

```
. #review 15
15 replace resmpg=mpg-pred
14 summarize resmpg, detail
13 drop predmpg
12 describe
11 sort foreign
10 by foreign: summarize mpg weight
9 * lines that start with a * are comments.
8 * they go into the review buffer too.
7 summarize resmpg, detail
6 list make mpg weight
5 list make mpg weight if abs(res)>6
4 list make mpg weight if abs(res)>5
3 tabulate foreign if abs(res)>5
2 regress mpg weight weight2
1 test weight2=0
. _
```

If you wanted to resubmit the 10th previous line, you could type `10` and press Seek, or you could press PrevLine 10 times. No matter which of the above methods you prefer for retrieving lines, you may edit previous lines by using the editing keys.

◁

10.6 Tab expansion of variable names

Another way to quickly enter a variable name is to take advantage of Stata's variable name completion feature. Simply type the first few letters of the variable name in the Command window and press the *Tab* key. Stata will automatically type the rest of the variable name for you. If more than one variable name matches the letters you have typed, Stata will complete as much as it can and beep at you to let you know that you have typed a nonunique variable abbreviation.

Elements of Stata

11 Language syntax

Contents

11.1 Overview

With few exceptions, the basic Stata language syntax is

[`by` *varlist*`:`] *command* [*varlist*] [`=`*exp*] [`if` *exp*] [`in` *range*] [*weight*] [`,` *options*]

where square brackets distinguish optional qualifiers and options from required ones. In this diagram, *varlist* denotes a list of variable names, *command* denotes a Stata command, *exp* denotes an algebraic expression, *range* denotes an observation range, *weight* denotes a weighting expression, and *options* denotes a list of options.

11.1.1 varlist

Most commands that take a subsequent *varlist* do not require that you explicitly type one. If no *varlist* appears, these commands assume a *varlist* of `_all`, the Stata shorthand for indicating all the variables in the dataset. In commands that alter or destroy data, Stata requires that the *varlist* be specified explicitly. See [U] **11.4 varlists** for a complete description.

Some commands take a *varname*, rather than a *varlist*. A *varname* refers to exactly one variable. The `tabulate` command requires a *varname*; see [R] **tabulate oneway**.

▷ Example 1

The `summarize` command lists the mean, standard deviation, and range of the specified variables. In [R] **summarize**, we see that the syntax diagram for summarize is

`summarize` [*varlist*] [*if*] [*in*] [*weight*] [, *options*]

Farther down on the manual page is a table summarizing `options`, but let's focus on the syntax diagram itself first. Because everything except the word `summarize` is enclosed in square brackets, the simplest form of the command is "`summarize`". Typing `summarize` without arguments is equivalent to typing `summarize _all`; all the variables in the dataset are summarized. Underlining denotes the shortest allowed abbreviation, so we could have typed just `su`; see [U] **11.2 Abbreviation rules**.

The table that defines *options* looks like this:

options	Description
Main	
`detail`	display additional statistics
`meanonly`	suppress the display; calculate only the mean; programmer's option
`format`	use variable's display format
`separator(#)`	draw separator line after every # variables; default is `separator(5)`

Thus we learn we could also type, for instance, `summarize, detail` or `summarize, detail format`.

As another example, the `drop` command eliminates variables or observations from a dataset. When dropping variables, its syntax is

`drop` *varlist*

`drop` has no option table because it has no options.

In fact, nothing is optional. Typing `drop` by itself would result in the error message "varlist or in range required". To drop all the variables in the dataset, we must type `drop _all`.

Even before looking at the syntax diagram, we could have predicted that *varlist* would be required—`drop` is destructive, so Stata requires us to spell out our intent. The syntax diagram informs us that *varlist* is required because *varlist* is not enclosed in square brackets. Because `drop` is not underlined, it cannot be abbreviated.

◁

11.1.2 by varlist:

The by *varlist*: prefix causes Stata to repeat a command for each subset of the data for which the values of the variables in *varlist* are equal. When prefixed with by *varlist*:, the result of the command will be the same as if you had formed separate datasets for each group of observations, saved them, and then gave the command on each dataset separately. The data must already be sorted by *varlist*, although by has a sort option; see [U] **11.5 by varlist: construct** for more information.

▷ Example 2

Typing `summarize marriage_rate divorce_rate` produces a table of the mean, standard deviation, and range of `marriage_rate` and `divorce_rate`, using all the observations in the data:

```
. use http://www.stata-press.com/data/r12/census12
(1980 Census data by state)

. summarize marriage_rate divorce_rate

    Variable |       Obs        Mean    Std. Dev.       Min        Max
-------------+--------------------------------------------------------
marriage_r~e |        50    .0133221    .0188122   .0074654   .1428282
divorce_rate |        50    .0056641    .0022473   .0029436   .0172918
```

Typing `by region: summarize marriage_rate divorce_rate` produces one table for each region of the country:

```
. sort region

. by region: summarize marriage_rate divorce_rate

----------------------------------------------------------------------------
-> region = N Cntrl

    Variable |       Obs        Mean    Std. Dev.       Min        Max
-------------+--------------------------------------------------------
marriage_r~e |        12    .0099121    .0011326   .0087363   .0127394
divorce_rate |        12    .0046974    .0011315   .0032817   .0072868

----------------------------------------------------------------------------
-> region = NE

    Variable |       Obs        Mean    Std. Dev.       Min        Max
-------------+--------------------------------------------------------
marriage_r~e |         9    .0087811     .001191   .0075757   .0107055
divorce_rate |         9     .004207    .0010264   .0029436   .0057071

----------------------------------------------------------------------------
-> region = South

    Variable |       Obs        Mean    Std. Dev.       Min        Max
-------------+--------------------------------------------------------
marriage_r~e |        16    .0114654    .0025721   .0074654   .0172704
divorce_rate |        16     .005633    .0013355   .0038917   .0080078

----------------------------------------------------------------------------
-> region = West

    Variable |       Obs        Mean    Std. Dev.       Min        Max
-------------+--------------------------------------------------------
marriage_r~e |        13    .0218987    .0363775   .0087365   .1428282
divorce_rate |        13    .0076037    .0031486   .0046004   .0172918
```

The dataset must be sorted on the by variables:

```
. use http://www.stata-press.com/data/r12/census12
(1980 Census data by state)
. by region: summarize marriage_rate divorce_rate
not sorted
r(5);
. sort region
. by region: summarize marriage_rate divorce_rate
(output appears)
```

We could also have asked that `by` sort the data:

```
. by region, sort: summarize marriage_rate divorce_rate
(output appears)
```

`by` *varlist*`:` can be used with most Stata commands; we can tell which ones by looking at their syntax diagrams. For instance, we could obtain the correlations by `region`, between `marriage_rate` and `divorce_rate`, by typing `by region: correlate marriage_rate divorce_rate`.

◁

❑ Technical note

The *varlist* in `by` *varlist*`:` may contain up to 32,767 variables with Stata/MP and Stata/SE or 2,047 variables with Stata/IC; these are the maximum allowed in the dataset. For instance, if we had data on automobiles and wished to obtain means according to market category (`market`) broken down by manufacturer (`origin`), we could type `by market origin: summarize`. That *varlist* contains two variables: `market` and `origin`. If the data were not already sorted on `market` and `origin`, we would first type `sort market origin`.

❑

❑ Technical note

The *varlist* in `by` *varlist*`:` may contain string variables, numeric variables, or both. In the example above, `region` is a string variable, in particular, a `str7`. The example would have worked, however, if `region` were a numeric variable with values 1, 2, 3, and 4, or even 12.2, 16.78, 32.417, and 152.13.

❑

11.1.3 if exp

The `if` *exp* qualifier restricts the scope of a command to those observations for which the value of the expression is *true* (which is equivalent to the expression being nonzero; see [U] **13 Functions and expressions**).

▷ Example 3

Typing `summarize marriage_rate divorce_rate if region=="West"` produces a table for the western region of the country:

```
. summarize marriage_rate divorce_rate if region == "West"

    Variable |       Obs        Mean    Std. Dev.       Min        Max
-------------+--------------------------------------------------------
marriage_r~e |        13    .0218987    .0363775   .0087365   .1428282
divorce_rate |        13    .0076037    .0031486   .0046004   .0172918
```

The double equal sign in `region=="West"` is not an error. Stata uses a *double* equal sign to denote equality testing and one equal sign to denote assignment; see [U] **13 Functions and expressions**.

A command may have at most one `if` qualifier. If you want the summary for the West restricted to observations with values of `marriage_rate` in excess of 0.015, do *not* type `summarize marriage_rate divorce_rate if region=="West" if marriage_rate>.015`. Instead type

```
. summarize marriage_rate divorce_rate if region == "West" & marriage_rate >.015

    Variable |       Obs        Mean    Std. Dev.       Min        Max
-------------+--------------------------------------------------------
marriage_r~e |         1    .1428282           .   .1428282   .1428282
divorce_rate |         1    .0172918           .   .0172918   .0172918
```

You may not use the word *and* in place of the symbol "&" to join conditions. To select observations that meet one condition *or* another, use the "|" symbol. For instance, `summarize marriage_rate divorce_rate if region=="West" | marriage_rate>.015` summarizes all observations for which `region` is West *or* `marriage_rate` is greater than 0.015.

◁

▷ Example 4

`if` may be combined with `by`. Typing `by region: summarize marriage_rate divorce_rate if marriage_rate>.015` produces a set of tables, one for each region, reflecting summary statistics on `marriage_rate` and `divorce_rate` among observations for which `marriage_rate` exceeds 0.015:

```
. by region: summarize marriage_rate divorce_rate if marriage_rate >.015

------------------------------------------------------------------------------
-> region = N Cntrl

    Variable |       Obs        Mean    Std. Dev.       Min        Max
-------------+--------------------------------------------------------
marriage_r~e |         0
divorce_rate |         0

------------------------------------------------------------------------------
-> region = NE

    Variable |       Obs        Mean    Std. Dev.       Min        Max
-------------+--------------------------------------------------------
marriage_r~e |         0
divorce_rate |         0

------------------------------------------------------------------------------
-> region = South

    Variable |       Obs        Mean    Std. Dev.       Min        Max
-------------+--------------------------------------------------------
marriage_r~e |         2    .0163219    .0013414   .0153734   .0172704
divorce_rate |         2    .0061813    .0025831   .0043548   .0080078

------------------------------------------------------------------------------
-> region = West

    Variable |       Obs        Mean    Std. Dev.       Min        Max
-------------+--------------------------------------------------------
marriage_r~e |         1    .1428282           .   .1428282   .1428282
divorce_rate |         1    .0172918           .   .0172918   .0172918
```

The results indicate that there are no states in the Northeast and North Central regions for which `marriage_rate` exceeds 0.015, whereas there are two such states in the South and one state in the West.

◁

11.1.4 in range

The `in` *range* qualifier restricts the scope of the command to a specific observation range. A range specification takes the form $\#_1\left[/\#_2\right]$, where $\#_1$ and $\#_2$ are positive or negative integers. Negative integers are understood to mean "from the end of the data", with -1 referring to the last observation. The implied first observation must be less than or equal to the implied last observation.

The first and last observations in the dataset may be denoted by `f` and `l` (lowercase letter), respectively. `F` is allowed as a synonym for `f`, and `L` is allowed as a synonym for `l`. A range specifies absolute observation numbers within a dataset. As a result, the `in` qualifier may not be used when the command is preceded by the `by` *varlist*`:` prefix; see [U] **11.5 by varlist: construct**.

▷ Example 5

Typing `summarize marriage_rate divorce_rate in 5/25` produces a table based on the values of `marriage_rate` and `divorce_rate` in observations 5–25:

```
. summarize marriage_rate divorce_rate in 5/25

    Variable |       Obs        Mean    Std. Dev.       Min        Max
-------------+--------------------------------------------------------
marriage_r~e |        21    .0096001    .0013263   .0075757   .0125884
divorce_rate |        21     .004726    .0012025   .0029436   .0072868
```

This is, admittedly, a rather odd thing to want to do. It would not be odd, however, if we substituted `list` for `summarize`. If we wanted to see the states with the 10 lowest values of `marriage_rate`, we could type `sort marriage_rate` followed by `list marriage_rate in 1/10`.

Typing `summarize marriage_rate divorce_rate in f/l` is equivalent to typing `summarize marriage_rate divorce_rate`—all observations are summarized.

◁

▷ Example 6

Typing `summarize marriage_rate divorce_rate in 5/25 if region=="South"` produces a table based on the values of the two variables in observations 5–25 for which the value of `region` is South:

```
. summarize marriage_rate divorce_rate in 5/25 if region=="South"

    Variable |       Obs        Mean    Std. Dev.       Min        Max
-------------+--------------------------------------------------------
marriage_r~e |         4    .0108886    .0015061   .0089201   .0125884
divorce_rate |         4    .0054448    .0011166   .0041485   .0068685
```

The ordering of the `in` and `if` qualifiers is not significant. The command could also have been specified as `summarize marriage_rate divorce_rate if region=="South" in 5/25`.

◁

▷ Example 7

Negative in ranges can be useful with sort. For instance, we have data on automobiles and wish to list the five with the highest mileage ratings:

```
. use http://www.stata-press.com/data/r12/auto
(1978 Automobile Data)

. sort mpg

. list make mpg in -5/l

     +----------------------+
     | make             mpg |
     |----------------------|
 70. | Toyota Corolla    31 |
 71. | Plym. Champ       34 |
 72. | Subaru            35 |
 73. | Datsun 210        35 |
 74. | VW Diesel         41 |
     +----------------------+
```

◁

11.1.5 =exp

=exp specifies the value to be assigned to a variable and is most often used with generate and replace. See [U] **13 Functions and expressions** for details on expressions and [D] **generate** for details on the generate and replace commands.

▷ Example 8

Expression	Meaning
generate newvar=oldvar+2	creates a new variable named newvar equal to oldvar+2
replace oldvar=oldvar+2	changes the contents of the existing variable oldvar
egen newvar=rank(oldvar)	creates newvar containing the ranks of oldvar (see [D] **egen**)

◁

11.1.6 weight

weight indicates the weight to be attached to each observation. The syntax of *weight* is

[*weightword*=*exp*]

where you actually type the square brackets and where *weightword* is one of

weightword	Meaning
weight	default treatment of weights
fweight *or* frequency	frequency weights
pweight	sampling weights
aweight *or* cellsize	analytic weights
iweight	importance weights

The underlining indicates the minimum acceptable abbreviation. Thus weight may be abbreviated w or we, etc.

▷ Example 9

Before explaining what the different types of weights mean, let's obtain the population-weighted mean of a variable called median_age from data containing observations on all 50 states of the United States. The dataset also contains a variable named pop, which is the total population of each state.

```
. use http://www.stata-press.com/data/r12/census12
(1980 Census data by state)
. summarize median_age [weight=pop]
(analytic weights assumed)
    Variable |     Obs      Weight        Mean   Std. Dev.       Min        Max
-------------+------------------------------------------------------------------
  median_age |      50   225907472    30.11047    1.66933       24.2       34.7
```

In addition to telling us that our dataset contains 50 observations, Stata informs us that the sum of the weight is 225,907,472, which was the number of people living in the United States as of the 1980 census. The weighted mean is 30.11. We were also informed that Stata assumed that we wanted "analytic" weights. ◁

weight is each command's idea of what the "natural" weights are and is one of fweight, pweight, aweight, or iweight. When you specify the vague weight, the command informs you which kind it assumes. Not every command supports every kind of weight. A note below the syntax diagram for a command will tell you which weights the command supports.

Stata understands four kinds of weights:

1. fweights, or frequency weights, indicate duplicated observations. fweights are always integers. If the fweight associated with an observation is 5, that means there are really 5 such observations, each identical.
2. pweights, or sampling weights, denote the inverse of the probability that this observation is included in the sample because of the sampling design. A pweight of 100, for instance, indicates that this observation is representative of 100 subjects in the underlying population. The scale of these weights does not matter in terms of estimated parameters and standard errors, except when estimating totals and computing finite-population corrections with the svy commands; see [SVY] **survey**.
3. aweights, or analytic weights, are inversely proportional to the variance of an observation; that is, the variance of the jth observation is assumed to be σ^2/w_j, where w_j are the weights. Typically, the observations represent averages, and the weights are the number of elements that gave rise to the average. For most Stata commands, the recorded scale of aweights is irrelevant; Stata internally rescales them to sum to N, the number of observations in your data, when it uses them.
4. iweights, or importance weights, indicate the relative "importance" of the observation. They have no formal statistical definition; this is a catch-all category. Any command that supports iweights will define how they are treated. They are usually intended for use by programmers who want to produce a certain computation.

See [U] **20.22 Weighted estimation** for a thorough discussion of weights and their meaning.

❑ Technical note

When you do not specify a weight, the result is equivalent to specifying [fweight=1]. ❑

11.1.7 options

Many commands take command-specific options. These are described along with each command in the *Reference* manuals. Options are indicated by typing a comma at the end of the command, followed by the options you want to use.

▷ Example 10

Typing `summarize marriage_rate` produces a table of the mean, standard deviation, minimum, and maximum of the variable `marriage_rate`:

```
. summarize marriage_rate
    Variable |       Obs        Mean    Std. Dev.       Min        Max
-------------+--------------------------------------------------------
marriage_r~e |        50    .0133221    .0188122   .0074654   .1428282
```

The syntax diagram for `summarize` is

`summarize` [*varlist*] [*if*] [*in*] [*weight*] [, *options*]

followed by the option table

options	Description
Main	
`detail`	display additional statistics
`meanonly`	suppress the display; calculate only the mean; programmer's option
`format`	use variable's display format
`separator(#)`	draw separator line after every # variables; default is `separator(5)`

Thus the options allowed by `summarize` are `detail` or `meanonly`, `format`, and `separator()`. The shortest allowed abbreviations for these options are `d` for `detail`, `mean` for `meanonly`, `f` for `format`, and `sep()` for `separator()`; see [U] **11.2 Abbreviation rules**.

Typing `summarize marriage_rate, detail` produces a table that also includes selected percentiles, the four largest and four smallest values, the skewness, and the kurtosis.

```
. summarize marriage_rate, detail
                        marriage_rate
-------------------------------------------------------------
      Percentiles      Smallest
 1%     .0074654       .0074654
 5%     .0078956       .0075757
10%     .0080043       .0078956       Obs                  50
25%     .0089399       .0079079       Sum of Wgt.          50

50%     .0105669                      Mean           .0133221
                        Largest       Std. Dev.      .0188122
75%     .0122899       .0146266
90%     .0137832       .0153734       Variance       .0003539
95%     .0153734       .0172704       Skewness       6.718494
99%     .1428282       .1428282       Kurtosis       46.77306
```

◁

Some commands have options that are required. For instance, the `ranksum` command requires the `by(`*groupvar*`)` option, which identifies the grouping variable. A *groupvar* is a specific kind of *varname*. It identifies to which group each observation belongs.

❑ Technical note

Once you have typed the *varlist* for the command, you can place options anywhere in the command. You can type `summarize marriage_rate divorce_rate if region=="West", detail`, or you can type `summarize marriage_rate divorce_rate, detail, if region=="West"`. You use a second comma to indicate a return to the command line as opposed to the option list. Leaving out the comma after the word `detail` would cause an error because Stata would attempt to interpret the phrase `if region=="West"` as an option rather than as part of the command.

You may not type an option in the middle of a *varlist*. Typing `summarize marriage_rate, detail, divorce_rate` will result in an error.

Options need not be specified contiguously. You may type `summarize marriage_rate divorce_rate, detail, if region=="South", noformat`. Both `detail` and `noformat` are options.

❑

❑ Technical note

Most options are toggles—they indicate that something either is or is not to be done. Sometimes it is difficult to remember which is the default. The following rule applies to all options: if *option* is an option, then no*option* is an option as well, and vice versa. Thus if we could not remember whether `detail` or `nodetail` were the default for `summarize` but we knew that we did not want the detail, we could type `summarize, nodetail`. Typing the `nodetail` option is unnecessary, but Stata will not complain.

Some options take *arguments*. The Stata `kdensity` command has an `n(`*#*`)` option that indicates the number of points at which the density estimate is to be evaluated. When an option takes an argument, the argument is enclosed in parentheses.

Some options take more than one argument. In such cases, arguments should be separated from one another by commas. For instance, you might see in a syntax diagram

`saving(`*filename*[`, replace`]`)`

Here `replace` is the (optional) second argument. *Lists*, such as lists of variables (varlists) and lists of numbers (numlists), are considered to be one argument. If a syntax diagram reported

`powers(`*numlist*`)`

the list of numbers would be one argument, so the elements would not be separated by commas. You would type, for instance, `powers(1 2 3 4)`. In fact, Stata will tolerate commas here, so you could type `powers(1,2,3,4)`.

Some options take string arguments. `regress` has an `eform()` option that works this way—for instance, `eform("Exp Beta")`. To play it safe, you should type the quotes surrounding the string, although it is not required. If you do not type the quotes, any sequence of two or more consecutive blanks will be interpreted as one blank. Thus `eform(Exp  beta)` would be interpreted the same as `eform(Exp beta)`.

❑

11.1.8 numlist

A *numlist* is a list of numbers. Stata allows certain shorthands to indicate ranges:

Numlist	Meaning
`2`	just one number
`1 2 3`	three numbers
`3 2 1`	three numbers in reversed order
`.5 1 1.5`	three different numbers
`1 3 -2.17 5.12`	four numbers in jumbled order
`1/3`	three numbers: 1, 2, 3
`3/1`	the same three numbers in reverse order
`5/8`	four numbers: 5, 6, 7, 8
`-8/-5`	four numbers: −8, −7, −6, −5
`-5/-8`	four numbers: −5, −6, −7, −8
`-1/2`	four numbers: −1, 0, 1, 2
`1 2 to 4`	four numbers: 1, 2, 3, 4
`4 3 to 1`	four numbers: 4, 3, 2, 1
`10 15 to 30`	five numbers: 10, 15, 20, 25, 30
`1 2:4`	same as 1 2 to 4
`4 3:1`	same as 4 3 to 1
`10 15:30`	same as 10 15 to 30
`1(1)3`	three numbers: 1, 2, 3
`1(2)9`	five numbers: 1, 3, 5, 7, 9
`1(2)10`	the same five numbers, 1, 3, 5, 7, 9
`9(-2)1`	five numbers: 9, 7, 5, 3, and 1
`-1(.5)2.5`	the numbers −1, −.5, 0, .5, 1, 1.5, 2, 2.5
`1[1]3`	same as 1(1)3
`1[2]9`	same as 1(2)9
`1[2]10`	same as 1(2)10
`9[-2]1`	same as 9(−2)1
`-1[.5]2.5`	same as −1(.5)2.5
`1 2 3/5 8(2)12`	eight numbers: 1, 2, 3, 4, 5, 8, 10, 12
`1,2,3/5,8(2)12`	the same eight numbers
`1 2 3/5 8 10 to 12`	the same eight numbers
`1,2,3/5,8,10 to 12`	the same eight numbers
`1 2 3/5 8 10:12`	the same eight numbers

`poisson`'s `constraints()` option has syntax `constraints(`*numlist*`)`. Thus you could type `constraints(2 4 to 8)`, `constraints(2(2)8)`, etc.

11.1.9 datelist

A *datelist* is a list of dates or times and is often used with graph options when the variable being graphed has a date format. For a description of how dates and times are stored and manipulated in Stata, see [U] **24 Working with dates and times**. Calendar dates, also known as `%td` dates, are recorded in Stata as the number of days since 01jan1960, so 0 means 01jan1960, 1 means 02jan1960, and 16,541 means 15apr2005. Similarly, −1 means 31dec1959, −2 means 30dec1959, and −16,541 means 18sep1914. In such a case, a datelist is either a list of dates, as in

```
15apr1973 17apr1973 20apr1973 23apr1973
```

or it is a first and last date with an increment between, as in

```
17apr1973(3)23apr1973
```

or it is a combination:

```
15apr1973 17apr1973(3)23apr1973
```

Dates specified with spaces, slashes, or commas must be bound in parentheses, as in

```
(15 apr 1973) (april 17, 1973)(3)(april 23, 1973)
```

Evenly spaced calendar dates are not especially useful, but with other time units, even spacing can be useful, such as

```
1999q1(1)2005q1
```

when `%tq` dates are being used. `1999q1(1)2005q1` means every quarter between 1999q1 and 2005q1. `1999q1(4)2005q1` would mean every first quarter.

To interpret a datelist, Stata first looks at the format of the related variable and then uses the corresponding date-to-numeric translation function. For instance, if the variable has a `%td` format, the `td()` function is used to translate the date; if the variable has a `%tq` format, the `tq()` function is used; and so on. See *Conveniently typing SIF values.*

11.1.10 Prefix commands

Stata has a handful of commands that are used to prefix other Stata commands. `by varlist:`, discussed in section [U] **11.1.2 by varlist:**, is in fact an example of a prefix command. In that section, we demonstrated `by` using

```
by region: summarize marriage_rate divorce_rate
```

and later,

```
by region, sort: summarize marriage_rate divorce_rate
```

and although we did not, we could also have demonstrated

```
by region, sort: summarize marriage_rate divorce_rate, detail
```

Each of the above runs the `summarize` command separately on the data for each region.

`by` itself follows standard Stata syntax:

`by` *varlist*[`,` *options*]`:` ...

In `by region, sort: summarize marriage_rate divorce_rate, detail`, `region` is `by`'s varlist and `sort` is `by`'s option, just as `marriage_rate` and `divorce_rate` are `summarize`'s varlist and `detail` is `summarize`'s option.

by is not the only prefix command, and the full list of such commands is

Prefix command	Description
by	run command on subsets of data
statsby	same as by, but collect statistics from each run
rolling	run command on moving subsets and collect statistics
bootstrap	run command on bootstrap samples
jackknife	run command on jackknife subsets of data
permute	run command on random permutations
simulate	run command on manufactured data
svy	run command and adjust results for survey sampling
mi estimate	run command on multiply imputed data and adjust results for multiple imputation (MI)
nestreg	run command with accumulated blocks of regressors, and report nested model comparison tests
stepwise	run command with stepwise variable inclusion/exclusion
xi	run command after expanding factor variables and interactions; for most commands, using factor variables is preferred to using xi (see [U] **11.4.3 Factor variables**)
fracpoly	run command with fractional polynomials of one regressor
mfp	run command with multiple fractional polynomial regressors
capture	run command and capture its return code
noisily	run command and show the output
quietly	run command and suppress the output
version	run command under specified version

The last group—capture, noisily, quietly, and version—have to do with programming Stata and, for historical reasons, capture, noisily, and quietly allow you to omit the colon, so one programmer might code

```
quietly regress ...
```

and another

```
quietly: regress ...
```

All the other prefix commands require the colon. In addition to the corresponding reference manual entries, you may want to consult Baum (2009) for a richer discussion of prefix commands.

11.2 Abbreviation rules

Stata allows abbreviations. In this manual, we usually avoid abbreviating commands, variable names, and options to ensure readability:

```
. summarize myvar, detail
```

Experienced Stata users, on the other hand, tend to abbreviate the same command as

```
. sum myv, d
```

As a general rule, command, option, and variable names may be abbreviated to the shortest string of characters that uniquely identifies them.

This rule is violated if the command or option does something that cannot easily be undone; the command must then be spelled out in its entirety.

Also, a few common commands and options are allowed to have even shorter abbreviations than the general rule would allow.

The general rule is applied, without exception, to variable names.

11.2.1 Command abbreviation

The shortest allowed abbreviation for a command or option can be determined by looking at the command's syntax diagram. This minimal abbreviation is shown by underlining:

```
regress
rename
replace
rotate
run
```

If there is no underlining, no abbreviation is allowed. For example, `replace` may not be abbreviated, the underlying reason being that `replace` changes the data.

`regress` can be abbreviated `reg`, `regr`, `regre`, or `regres`, or it can be spelled out in its entirety.

Sometimes short abbreviations are also allowed. Commands that begin with the letter *d* include `decode`, `describe`, `destring`, `dir`, `discard`, `display`, `do`, and `drop`, which suggests that the shortest allowable abbreviation for `describe` is `desc`. However, because `describe` is such a commonly used command, you may abbreviate it with the single letter `d`. You may also abbreviate the `list` command with the single letter `l`.

The other exception to the general abbreviation rule is that commands that alter or destroy data must be spelled out completely. Two commands that begin with the letter *d*, `discard` and `drop`, are destructive in the sense that, once you give one of these commands, there is no way to undo the result. Therefore, both must be spelled out.

The final exceptions to the general rule are commands implemented as ado-files. Such commands may not be abbreviated. Ado-file commands are external, and their names correspond to the names of disk files.

11.2.2 Option abbreviation

Option abbreviation follows the same logic as command abbreviation: you determine the minimum acceptable abbreviation by examining the command's syntax diagram. The syntax diagram for `summarize` reads, in part,

```
summarize ..., detail format
```

The `detail` option may be abbreviated `d`, `de`, `det`, ..., `detail`. Similarly, option `format` may be abbreviated `f`, `fo`, ..., `format`.

The `clear` and `replace` options occur with many commands. The `clear` option indicates that even though completing this command will result in the loss of all data in memory, and even though the data in memory have changed since the data were last saved on disk, you want to continue. `clear` must be spelled out, as in `use newdata, clear`.

The `replace` option indicates that it is okay to save over an existing dataset. If you type `save mydata` and the file `mydata.dta` already exists, you will receive the message "file mydata.dta already exists", and Stata will refuse to overwrite it. To allow Stata to overwrite the dataset, you would type `save mydata, replace`. `replace` may not be abbreviated.

❑ Technical note

`replace` is a stronger modifier than `clear` and is one you should think about before using. With a mistaken `clear`, you can lose hours of work, but with a mistaken `replace`, you can lose days of work.

❑

11.2.3 Variable-name abbreviation

- Variable names may be abbreviated to the shortest string of characters that uniquely identifies them given the data currently loaded in memory.

  If your dataset contained four variables, `state`, `mrgrate`, `dvcrate`, and `dthrate`, you could refer to the variable `dvcrate` as `dvcrat`, `dvcra`, `dvcr`, `dvc`, or `dv`. You might type `list dv` to list the data on `dvcrate`. You could not refer to the variable `dvcrate` as `d`, however, because that abbreviation does not distinguish `dvcrate` from `dthrate`. If you were to type `list d`, Stata would respond with the message "ambiguous abbreviation". (If you wanted to refer to *all* variables that started with the letter *d*, you could type `list d*`; see [U] **11.4 varlists**.)

- The character `~` may be used to mean that "zero or more characters go here". For instance, `r~8` might refer to the variable `rep78`, or `rep1978`, or `repair1978`, or just `r8`. (The `~` character is similar to the `*` character in [U] **11.4 varlists**, except that it adds the restriction "and only one variable matches this specification".)

  Above, we said that you could abbreviate variables. You could type `dvcr` to refer to `dvcrate`, but, if there were more than one variable that started with the letters `dvcr`, you would receive an error. Typing `dvcr` is the same as typing `dvcr~`.

11.3 Naming conventions

A name is a sequence of one to 32 letters (`A–Z` and `a–z`), digits (`0–9`), and underscores (`_`).

Programmers: Local macro names can have no more than 31 characters in the name; see [U] **18.3.1 Local macros**.

Stata reserves the following names:

```
_all        double      long        _rc
_b          float       _n          _skip
byte        if          _N          str#
_coef       in          _pi         using
_cons       int         _pred       with
```

You may not use these reserved names for your variables.

The first character of a name must be a letter or an underscore. We recommend, however, that you not begin your variable names with an underscore. All of Stata's built-in variables begin with an underscore, and we reserve the right to incorporate new *_variables* freely.

Stata respects case; that is, `myvar`, `Myvar`, and `MYVAR` are three distinct names.

All objects in Stata—not just variables—follow this naming convention.

11.4 varlists

A *varlist* is a list of variable names. The variable names in a *varlist* refer either exclusively to new (not yet created) variables or exclusively to existing variables. A *newvarlist* always refers exclusively to new (not yet created) variables. Similarly, a *varname* refers to one variable, either existing or not yet created. A *newvar* always refers to one new variable.

Sometimes a command will refer to a varname in another way, such as "*groupvar*". This is still a varname. The different name for it is used to give you an extra hint about the purpose of that variable. For example, a *groupvar* is the name of a variable that defines groups within your data.

11.4.1 Lists of existing variables

In lists of existing variable names, variable names may be repeated.

▷ Example 11

If you type `list state mrgrate dvcrate state`, the variable `state` will be listed twice, once in the leftmost column and again in the rightmost column of the list.

◁

Existing variable names may be abbreviated as described in [U] **11.2 Abbreviation rules**. You may also use "`*`" to indicate that "zero or more characters go here". For instance, if you suffix `*` to a partial variable name (for example, `sta*`), you are referring to all variable names that start with that letter combination. If you prefix `*` to a letter combination (for example, `*rate`), you are referring to all variables that end in that letter combination. If you put `*` in the middle (for example, `m*rate`), you are referring to all variables that begin and end with the specified letters. You may put more than one `*` in an abbreviation.

▷ Example 12

If the variables `poplt5`, `pop5to17`, and `pop18p` are in our dataset, we may type `pop*` as a shorthand way to refer to all three variables. For instance, `list state pop*` lists the variables `state`, `poplt5`, `pop5to17`, and `pop18p`.

If we had a dataset with variables `inc1990`, `inc1991`, ..., `inc1999` along with variables `incfarm1990`, ..., `incfarm1999`; `pop1990`, ..., `pop1999`; and `ms1990`, ..., `ms1999`, then `*1995` would be a shorthand way of referring to `inc1995`, `incfarm1995`, `pop1995`, and `ms1995`. We could type, for instance, `list *1995`.

In that same dataset, typing `list i*95` would be a shorthand way of listing `inc1995` and `incfarm1995`.

Typing `list i*f*95` would be a shorthand way of listing to `incfarm1995`.

◁

`~` is an alternative to `*`, and really, it means the same thing. The difference is that `~` indicates that if more than one variable matches the specified pattern, Stata will complain rather than substituting all the variables that match the specification.

▷ Example 13

In the previous example, we could have typed `list i~f~95` to list `incfarm1995`. If, however, our dataset also included variable `infant1995`, then `list i*f*95` would list both variables and `list i~f~95` would complain that `i~f~95` is an ambiguous abbreviation.

◁

You may use `?` to specify that one character goes here. Remember, `*` means zero or more characters go here, so `?*` can be used to mean one or more characters goes here, `??*` can be used to mean two or more characters go here, and so on.

▷ Example 14

In a dataset containing variables `rep1`, `rep2`, ..., `rep78`, `rep?` would refer to `rep1`, `rep2`, ..., `rep9`, and `rep??` would refer to `rep10`, `rep11`, ..., `rep78`.

◁

You may place a dash (`-`) between two variable names to specify all the variables stored between the two listed variables, inclusive. You can determine storage order by using `describe`; it lists variables in the order in which they are stored.

▷ Example 15

If the dataset contains the variables `state`, `mrgrate`, `dvcrate`, and `dthrate`, in that order, typing `list state-dvcrate` is equivalent to typing `list state mrgrate dvcrate`. In both cases, three variables are listed.

◁

11.4.2 Lists of new variables

In lists of *new variables*, no variable names may be repeated or abbreviated.

You may specify a dash (`-`) between two variable names that have the same letter prefix and that end in numbers. This form of the dash notation indicates a range of variable names in ascending numerical order.

For example, typing `input v1-v4` is equivalent to typing `input v1 v2 v3 v4`. Typing `infile state v1-v3 ssn using rawdata` is equivalent to typing `infile state v1 v2 v3 ssn using rawdata`.

You may specify the storage type before the variable name to force a storage type other than the default. The numeric storage types are `byte`, `int`, `long`, `float` (the default), and `double`. The string storage types are `str`*#*, where *#* is replaced with an integer between 1 and 244, inclusive, representing the maximum length of the string. See [U] **12 Data**.

For instance, the list `var1 str8 var2 var3` specifies that `var1` and `var3` be given the default storage type and that `var2` be stored as a `str8`—a string whose maximum length is eight characters.

The list `var1 int var2 var3` specifies that `var2` be stored as an `int`. You may use parentheses to bind a list of variable names. The list `var1 int(var2 var3)` specifies that both `var2` and `var3` be stored as `int`s. Similarly, the list `var1 str20(var2 var3)` specifies that both `var2` and `var3` be stored as `str20`s. The different storage types are listed in [U] **12.2.2 Numeric storage types** and [U] **12.4.4 String storage types**.

▷ Example 16

Typing `infile str2 state str10 region v1-v5 using mydata` reads the `state` and `region` strings from the file `mydata.raw` and stores them as `str2` and `str10`, respectively, along with the variables `v1` through `v5`, which are stored as the default storage type `float` (unless we have specified a different default with the `set type` command).

Typing `infile str10(state region) v1-v5 using mydata` would achieve almost the same result, except that the `state` and `region` values recorded in the data would both be assigned to `str10` variables. (We could then use the `compress` command to shorten the strings. See [D] **compress**; it is well worth reading.)

◁

❑ Technical note

You may append a colon and a *value label name* to numeric variables. (See [U] **12.6 Dataset, variable, and value labels** for a description of value labels.) For instance, `var1 var2:myfmt` specifies that the variable `var2` be associated with the value label stored under the name `myfmt`. This has the same effect as typing the list `var1 var2` and then subsequently giving the command `label values var2 myfmt`.

The advantage of specifying the value label association with the colon notation is that value labels can then be assigned by the current command; see [D] **input** and [D] **infile (free format)**.

❑

▷ Example 17

Typing `infile int(state:stfmt region:regfmt) v1-v5 using mydata, automatic` reads the state and region data from the file `mydata.raw` and stores them as `int`s, along with the variables `v1` through `v5`, which are stored as the default storage type.

In our previous example, both state and region were strings, so how can strings be stored in a numeric variable? See [U] **12.6 Dataset, variable, and value labels** for the complete answer. The colon notation specifies the name of the value label, and the `automatic` option tells Stata to assign unique numeric codes to all character strings. The numeric code for state, which Stata will make up on the fly, will be stored in the `state` variable. The mapping from numeric codes to words will be stored in the value label named `stfmt`. Similarly, regions will be assigned numeric codes, which are stored in `region`, and the mapping will be stored in `regfmt`.

If we were to `list` the data, the `state` and `region` variables would look like strings. `state`, for instance, would appear to contain things like `AL`, `CA`, and `WA`, but actually it would contain only numbers like 1, 2, 3, and 4.

◁

11.4.3 Factor variables

Factor variables are extensions of varlists of existing variables. When a command allows factor variables, in addition to typing variable names from your data, you can type factor variables, which might look like i.*varname*

i.*varname*#i.*varname*

i.*varname*#i.*varname*#i.*varname*

i.*varname*##i.*varname*

i.*varname*##i.*varname*##i.*varname*

Factor variables create indicator variables from categorical variables and are allowed with most estimation and postestimation commands, along with a few other commands.

Consider a variable named group that takes on the values 1, 2, and 3. Stata command list allows factor variables, so we can see how factor variables are expanded by typing

```
. list group i.group in 1/5
```

	group	1b. group	2. group	3. group
1.	1	0	0	0
2.	1	0	0	0
3.	2	0	1	0
4.	2	0	1	0
5.	3	0	0	1

There are no variables named 1b.group, 2.group, and 3.group in our data; there is only the variable named group. When we type i.group, however, Stata acts as if the variables 1b.group, 2.group, and 3.group exist. 1b.group, 2.group, and 3.group are called virtual variables.

Start at the right of the listing. 3.group is the virtual variable that equals 1 when group = 3, and 0 otherwise. 2.group is the virtual variable that equals 1 when group = 2, and 0 otherwise. 1b.group is different. The b is a marker indicating base value. 1b.group is a virtual variable equal to 0. If the i.group collection was included in a linear regression, virtual variable 1b.group would drop out from the estimation because it does not vary, and thus the coefficients on 2.group and 3.group would measure the change from group = 1. Hence the term base value.

The categorical variable to which factor-variable operators are applied must contain nonnegative integers.

❑ Technical note

We said above that 3.group equals 1 when group = 3 and equals 0 otherwise. We should have added that 3.group equals missing when group contains missing. To be precise, 3.group equals 1 when group = 3, equals system missing (.) when group ≥ ., and equals 0 otherwise.

❑

❑ Technical note

We said above that when we typed i.group, Stata acts as if the variables 1b.group, 2.group, and 3.group exist, and that might suggest that the act of typing i.group somehow created the virtual variables. That is not true; the virtual variables always exist.

In fact, i.group is an abbreviation for 1b.group, 2.group, and 3.group. In any command that allows factor variables, you can specify virtual variables. Thus the listing above could equally well have been produced by typing

```
. list group 1b.group 2.group 3.group in 1/5
```

#.varname is defined as equal to 1 when *varname* = #, equal to system missing (.) when *varname* $\geq$., and equal to 0 otherwise. Thus 4.group is defined even when group takes on only the values 1, 2, and 3. 4.group would be equal to 0 in all observations. Referring to 4.group would not produce an error such as "virtual variable not found".

❑

11.4.3.1 Factor-variable operators

i.group is called a factor variable, although more correctly, we should say that group is a categorical variable to which factor-variable operators have been applied. There are four factor-variable operators:

Operator	Description
i.	unary operator to specify indicators
c.	unary operator to treat as continuous
#	binary operator to specify interactions
##	binary operator to specify full-factorial interactions

When you type i.group, it forms the indicators for the unique values of group. We will usually say this more briefly as i.group forms indicators for the levels of group, and sometimes we will abbreviate the statement even more and say i.group forms indicators for group.

The c. operator means continuous. We will get to that below.

and ##, pronounced cross and factorial cross, are operators for use with pairs of variables.

i.group#i.sex means to form indicators for each combination of the levels of group and sex.

group#sex means the same thing, which is to say that use of # implies the i. prefix.

group#c.age (or i.group#c.age) means the interaction of the levels of group with the continuous variable age. This amounts to forming i.group and then multiplying each level by age. We already know that i.group expands to the virtual variables 1b.group, 2.group, and 3.group, so group#c.age results in the collection of variables equal to 1b.group*age, 2.group*age, and 3.group*age. 1b.group*age will just be zero because 1b.group is zero. 2.group*age will be age when group = 2, and 0 otherwise. 3.group*age will be age when group = 3, and 0 otherwise. In a linear regression of y on age and group#c.age, 1b.group*age will be omitted, 2.group*age will measure the change in the age coefficient for group = 2 relative to the base group, and 3.group*age will measure the change for group = 3 relative to the base.

Here are some more examples of use of the operators:

Factor specification	Result
i.group	indicators for levels of group
i.group#i.sex	indicators for each combination of levels of group and sex, a two-way interaction
group#sex	same as i.group#i.sex
group#sex#arm	indicators for each combination of levels of group, sex, and arm, a three-way interaction
group##sex	same as i.group i.sex group#sex
group##sex##arm	same as i.group i.sex i.arm group#sex group#arm sex#arm group#sex#arm
sex#c.age	two variables—age for males and 0 elsewhere, and age for females and 0 elsewhere; if age is also in the model, one of the two virtual variables will be treated as a base
sex##c.age	same as i.sex age sex#c.age
c.age	same as age
c.age#c.age	age squared
c.age#c.age#c.age	age cubed

Several factor-variable terms are often specified in the same varlist, such as

```
. regress y  i.sex i.group sex#group age sex#c.age
```

or, equivalently,

```
. regress y  sex##group sex##c.age
```

11.4.3.2 Base levels

When we typed i.group, group = 1 became the base level. When we do not specify otherwise, the smallest level becomes the base level.

You can specify the base level of a factor variable by using the ib. operator. The syntax is

Base operator[a]	Description
ib#.	use # as base, # = value of variable
ib(##).	use the #th ordered value as base[b]
ib(first).	use smallest value as base (default)
ib(last).	use largest value as base
ib(freq).	use most frequent value as base
ibn.	no base level

[a]The i may be omitted. For instance, you can type ib2.group or b2.group.
[b]For example, ib(#2). means to use the second value as the base.

Thus, if you want to use group = 3 as the base, you can type ib3.group. You can type

```
. regress y  i.sex ib3.group sex#ib3.group age sex#c.age
```

or you can type

```
. regress y  i.sex ib3.group sex#group age sex#c.age
```

That is, you only have to set the base once. If you specify the base level more than once, it must be the same base level. You will get an error if you attempt to change base levels in midsentence.

If you type `ib3.group`, the virtual variables become `1.group`, `2.group`, and `b3.group`.

Were you to type `ib(freq).group`, the virtual variables might be `b1.group`, `2.group`, and `3.group`; `1.group`, `b2.group`, and `3.group`; or `1.group`, `2.group`, and `b3.group`, depending on the most frequent group in the data.

11.4.3.3 Setting base levels permanently

You can permanently set the base level by using the `fvset` command; see [R] **fvset**. For example,

```
. fvset base 3  group
```

sets the base for `group` to be 3. The setting is recorded in the data, and if the dataset is resaved, the base level will be remembered in future sessions.

If you want to set the base group back to the default, type

```
. fvset base default  group
```

If you want to set the base levels for a group of variables to be the largest value, you can type

```
. fvset base last  group sex arm
```

See [R] **fvset** for details.

Base levels can be temporarily overridden by using the `ib.` operator regardless of whether they are set explicitly.

11.4.3.4 Selecting levels

Typing `i.group` specifies virtual variables `b1.group`, `2.group`, and `3.group`. Regardless of whether you type `i.group`, you can access those virtual variables. You can, for instance, use them in expressions and `if` statements:

```
. list if 3.group
  (output omitted)
. generate over_age = cond(3.group, age-21, 0)
```

Although throughout this section we have been typing *#*`.group` such as `3.group` as if it is somehow distinctly different from `i.group`, the complete, formal syntax is `i3.group`. You are allowed to omit the `i`. The point is that `i3.group` is just a special case of `i.group`; `i3.group` specifies an indicator for the third level of `group`, and `i.group` specifies the indicators for all the levels of `group`. Anyway, the above commands could be typed as

```
. list if i3.group
  (output omitted)
. generate over_age = cond(i3.group, age-21, 0)
```

Similarly, the virtual variables `b1.group`, `2.group`, and `3.group` more formally would be referred to as `ib1.group`, `i2.group`, and `i3.group`. You are allowed to omit the leading `i` whenever what appears after is a number or a `b` followed by a base specification.

You can select a range of levels—a range of virtual variables—by using the i(*numlist*).*varname*. This can be useful when specifying the model to be fit using estimation commands. You may not omit the i when specifying a numlist.

Examples	Description
i2.cat	single indicator for cat = 2
2.cat	same as i2.cat
i(2 3 4).cat	three indicators, cat = 2, cat = 3, and cat = 4; same as i2.cat i3.cat i4.cat
i(2/4).cat	same as i(2 3 4).cat
2.cat#1.sex	a single indicator that is 1 when cat = 2 and sex = 1 and is 0 otherwise
i2.cat#i1.sex	same as 2.cat#1.sex

11.4.3.5 Applying operators to a group of variables

Factor-variable operators may be applied to groups of variables by using parentheses. You may type, for instance,

```
i.(group sex arm)
```

to mean i.group i.sex i.arm.

In the examples that follow, variables group, sex, arm, and cat are categorical, and variables age, wt, and bp are continuous:

Examples	Expansion
i.(group sex arm)	i.group i.sex i.arm
group#(sex arm cat)	group#sex group#arm group#cat
group##(sex arm cat)	i.group i.sex i.arm i.cat group#sex group#arm group#cat
group#(c.age c.wt c.bp)	group#c.age group#c.wt group#c.bp
group#c.(age wt bp)	same as group#(c.age c.wt c.bp)

Parentheses can shorten what you type and make it more readable. For instance,

```
. regress y  i.sex i.group sex#group age sex#c.age c.age#c.age sex#c.age#c.age
```

is easier to understand when written as

```
. regress y  sex##(group c.age c.age#c.age)
```

11.4.3.6 Using factor variables with time-series operators

Factor-variable operators may be combined with the `L.` and `F.` time-series operators, so you may specify lags and leads of factor variables in time-series applications. You could type `iL.group` or `Li.group`; the order of the operators does not matter. You could type `L.group#L.arm` or `L.group#c.age`.

Examples include

```
. regress y  b1.sex##(i(2/4).group cL.age cL.age#cL.age)
. regress y  2.arm#(sex#i(2/4)b3.group cL.age)
. regress y  2.arm##cat##(sex##i(2/4)b3.group cL.age#c.age) c.bp
>            c.bp#c.bp c.bp#c.bp#c.bp sex##c.bp#c.age
```

11.4.4 Time-series varlists

Time-series varlists are a variation on varlists of existing variables. When a command allows a time-series varlist, you may include time-series operators. For instance, `L.gnp` refers to the lagged value of variable gnp. The time-series operators are

Operator	Meaning
`L.`	lag x_{t-1}
`L2.`	2-period lag x_{t-2}
...	
`F.`	lead x_{t+1}
`F2.`	2-period lead x_{t+2}
...	
`D.`	difference $x_t - x_{t-1}$
`D2.`	difference of difference $x_t - x_{t-1} - (x_{t-1} - x_{t-2}) = x_t - 2x_{t-1} + x_{t-2}$
...	
`S.`	"seasonal" difference $x_t - x_{t-1}$
`S2.`	lag-2 (seasonal) difference $x_t - x_{t-2}$
...	

Time-series operators may be repeated and combined. `L3.gnp` refers to the third lag of variable gnp. So do `LLL.gnp`, `LL2.gnp`, and `L2L.gnp`. `LF.gnp` is the same as gnp. `DS12.gnp` refers to the one-period difference of the 12-period difference. `LDS12.gnp` refers to the same concept, lagged once.

`D1.` = `S1.`, but `D2.` ≠ `S2.`, `D3.` ≠ `S3.`, and so on. `D2.` refers to the difference of the difference. `S2.` refers to the two-period difference. If you wanted the difference of the difference of the 12-period difference of gnp, you would write `D2S12.gnp`.

Operators may be typed in uppercase or lowercase. Most users would type `d2s12.gnp` instead of `D2S12.gnp`.

You may type operators however you wish; Stata internally converts operators to their canonical form. If you typed `ld2ls12d.gnp`, Stata would present the operated variable as `L2D3S12.gnp`.

In addition to using *operator#*, Stata understands *operator*(*numlist*) to mean a set of operated variables. For instance, typing `L(1/3).gnp` in a varlist is the same as typing `L.gnp L2.gnp L3.gnp`. The operators can also be applied to a list of variables by enclosing the variables in parentheses; for example,

```
. use http://www.stata-press.com/data/r12/gxmpl1
. list year L(1/3).(gnp cpi)
```

	year	L.gnp	L2.gnp	L3.gnp	L.cpi	L2.cpi	L3.cpi
1.	1989	.	.	.	.	.	.
2.	1990	5837.9	.	.	124	.	.
3.	1991	6026.3	5837.9	.	130.7	124	.
4.	1992	6367.4	6026.3	5837.9	136.2	130.7	124
5.	1993	6689.3	6367.4	6026.3	140.3	136.2	130.7
6.	1994	7098.4	6689.3	6367.4	144.5	140.3	136.2
7.	1995	7433.4	7098.4	6689.3	148.2	144.5	140.3
8.	1996	7851.9	7433.4	7098.4	152.4	148.2	144.5

The parentheses notation may be used with any operator. Typing `D(1/3).gnp` would return the first through third differences.

The parentheses notation may be used in operator lists with multiple operators, such as `L(0/3)D2S12.gnp`.

Operator lists may include up to one set of parentheses, which may enclose a *numlist*; see [U] **11.1.8 numlist**.

The time-series operators `L.` and `F.` may be combined with factor variables. If we want to lag the indicator variables for the levels of the factor variable `region`, we would type `iL.region`. We could also say that we are specifying the level indicator variables for the lag of the region variables. They are equivalent statements.

The numlists and parentheses notation from both factor varlists and time-series operators may be combined. For example, `iL(1/3).region` specifies the first three lags of the level indicators for `region`. If `region` has four levels, this is equivalent to typing `i1L1.region i2L1.region i3L1.region i4L1.region i1L2.region i2L2.region i3L2.region i4L2.region i1L3.region i2L3.region i3L3.region i4L3.region`. Pushing the notation further, `i(1/2)L(1/3).(region education)` specifies the first three lags of the level 1 and level 2 indicator variables for both the `region` and `education` variables.

❑ Technical note

The `D.` and `S.` time-series operators may not be combined with factor variables because such combinations could have two meanings. `iD.a` could be the level indicators for the difference of the variable `a` from its prior period, or it could be the level indicators differenced between the two periods. These are generally not the same values, nor even the same number of indicators. Moreover, they are rarely interesting.

❑

Before you can use time-series operators in varlists, you must set the time variable by using the `tsset` command:

```
. list l.gnp
time variable not set
r(111);
. tsset time
  (output omitted)
. list l.gnp
  (output omitted)
```

See [TS] **tsset**. The time variable must take on integer values. Also, the data must be sorted on the time variable. `tsset` handles this, but later you might encounter

```
. list l.mpg
not sorted
r(5);
```

Then type `sort time` or type `tsset` to reestablish the order.

The time-series operators respect the time variable. `L2.gnp` refers to $\texttt{gnp}_{t-2}$, regardless of missing observations in the dataset. In the following dataset, the observation for 1992 is missing:

```
. use http://www.stata-press.com/data/r12/gxmpl2

. list year gnp l2.gnp, separator(0)
```

	year	gnp	L2. gnp
1.	1989	5837.9	.
2.	1990	6026.3	.
3.	1991	6367.4	5837.9
4.	1993	7098.4	6367.4
5.	1994	7433.4	.
6.	1995	7851.9	7098.4

← note, filled in correctly

Operated variables may be used in expressions:

```
. generate gnplag2 = l2.gnp
(3 missing values generated)
```

Stata also understands cross-sectional time-series data. If you have cross sections of time series, you indicate this when you `tsset` the data:

```
. tsset country year
```

See [TS] **tsset**. In fact, you can type that, or you can type

```
. xtset country year
```

`xtset` is how you set panel data just as `tsset` is how you set time-series data and here the two commands do the same thing. Some panel datasets are not cross-sectional time series, however, in that the second variable is not time, so `xtset` also allows

```
. xtset country
```

See [XT] **xtset**.

11.5 by varlist: construct

by *varlist*: *command*

The by prefix causes *command* to be repeated for each unique set of values of the variables in the *varlist*. *varlist* may contain numeric, string, or a mixture of numeric and string variables. (*varlist* may not contain time-series operators.)

by is an optional prefix to perform a Stata command separately for each group of observations where the values of the variables in the *varlist* are the same.

During each iteration, the values of the system variables _n and _N are set in relation to the first observation in the by-group; see [U] **13.7 Explicit subscripting**. The in *range* qualifier cannot be used with by *varlist*: because ranges specify absolute rather than relative observation numbers.

❑ Technical note

The inability to combine in and by is not really a constraint because if provides all the functionality of in and a bit more. If you wanted to perform *command* for the first three observations in each of the by-groups, you could type

```
. by varlist: command if _n<=3
```

❑

The results of *command* would be the same as if you had formed separate datasets for each group of observations, saved them, used each separately, and issued *command*.

▷ Example 18

We provide some examples using by in [U] **11.1.2 by varlist:** above. We demonstrate the effect of by on _n, _N, and explicit subscripting in [U] **13.7 Explicit subscripting**.

by requires that the data first be sorted. For instance, if we had data on the average January and July temperatures in degrees Fahrenheit for 420 cities located in the Northeast and West and wanted to obtain the averages, by region, across those cities, we might type

```
. use http://www.stata-press.com/data/r12/citytemp3, clear
(City Temperature Data)
. by region: summarize tempjan tempjuly
not sorted
r(5);
```

Stata refused to honor our request because the data are not sorted by region. We must either sort the data by region first (see [D] **sort**) or specify by's sort option (which has the same effect):

```
. by region, sort: summarize tempjan tempjuly

-> region = NE
    Variable |       Obs        Mean    Std. Dev.       Min        Max
-------------+--------------------------------------------------------
     tempjan |       164    27.88537    3.543096       16.6       31.8
    tempjuly |       164       73.35    2.361203       66.5       76.8

-> region = N Cntrl
    Variable |       Obs        Mean    Std. Dev.       Min        Max
-------------+--------------------------------------------------------
     tempjan |       284    21.69437    5.725392        2.2       32.6
    tempjuly |       284    73.46725    3.103187       64.5       81.4

-> region = South
    Variable |       Obs        Mean    Std. Dev.       Min        Max
-------------+--------------------------------------------------------
     tempjan |       250     46.1456    10.38646       28.9         68
    tempjuly |       250     80.9896     2.97537         71       87.4

-> region = West
    Variable |       Obs        Mean    Std. Dev.       Min        Max
-------------+--------------------------------------------------------
     tempjan |       256    46.22539    11.25412         13       72.6
    tempjuly |       256    72.10859    6.483131       58.1       93.6
```

◁

▷ Example 19

Using the same data as in the example above, we estimate regressions, by `region`, of average January temperature on average July temperature. Both temperatures are specified in degrees Fahrenheit.

```
. by region: regress tempjan tempjuly

-> region = NE
      Source |       SS       df       MS              Number of obs =     164
-------------+------------------------------           F(  1,   162) =  479.82
       Model |  1529.74026     1  1529.74026           Prob > F      =  0.0000
    Residual |  516.484453   162  3.18817564           R-squared     =  0.7476
-------------+------------------------------           Adj R-squared =  0.7460
       Total |  2046.22471   163  12.5535258           Root MSE      =  1.7855

------------------------------------------------------------------------------
     tempjan |      Coef.   Std. Err.      t    P>|t|     [95% Conf. Interval]
-------------+----------------------------------------------------------------
    tempjuly |   1.297424   .0592303    21.90   0.000     1.180461    1.414387
       _cons |  -67.28066   4.346781   -15.48   0.000    -75.86431     -58.697
------------------------------------------------------------------------------
```

```
-> region = N Cntrl

      Source |       SS       df       MS              Number of obs =     284
-------------+------------------------------           F(  1,   282) =  115.89
       Model |  2701.97917     1  2701.97917           Prob > F      =  0.0000
    Residual |  6574.79175   282  23.3148644           R-squared     =  0.2913
-------------+------------------------------           Adj R-squared =  0.2887
       Total |  9276.77092   283  32.7801093           Root MSE      =  4.8285

------------------------------------------------------------------------------
     tempjan |      Coef.   Std. Err.      t    P>|t|     [95% Conf. Interval]
-------------+----------------------------------------------------------------
    tempjuly |   .9957259   .0924944    10.77   0.000     .8136589    1.177793
       _cons |  -51.45888   6.801344    -7.57   0.000    -64.84673   -38.07103
------------------------------------------------------------------------------

-> region = South

      Source |       SS       df       MS              Number of obs =     250
-------------+------------------------------           F(  1,   248) =   95.17
       Model |  7449.51623     1  7449.51623           Prob > F      =  0.0000
    Residual |  19412.2231   248  78.2750933           R-squared     =  0.2773
-------------+------------------------------           Adj R-squared =  0.2744
       Total |  26861.7394   249  107.878471           Root MSE      =  8.8473

------------------------------------------------------------------------------
     tempjan |      Coef.   Std. Err.      t    P>|t|     [95% Conf. Interval]
-------------+----------------------------------------------------------------
    tempjuly |    1.83833   .1884392     9.76   0.000     1.467185    2.209475
       _cons |    -102.74   15.27187    -6.73   0.000    -132.8191   -72.66089
------------------------------------------------------------------------------

-> region = West

      Source |       SS       df       MS              Number of obs =     256
-------------+------------------------------           F(  1,   254) =    2.84
       Model |  357.161728     1  357.161728           Prob > F      =  0.0932
    Residual |  31939.9031   254   125.74765           R-squared     =  0.0111
-------------+------------------------------           Adj R-squared =  0.0072
       Total |  32297.0648   255  126.655156           Root MSE      =  11.214

------------------------------------------------------------------------------
     tempjan |      Coef.   Std. Err.      t    P>|t|     [95% Conf. Interval]
-------------+----------------------------------------------------------------
    tempjuly |   .1825482   .1083166     1.69   0.093    -.0307648    .3958613
       _cons |    33.0621    7.84194     4.22   0.000     17.61859     48.5056
------------------------------------------------------------------------------
```

The regressions show that a 1-degree increase in the average July temperature in the Northeast corresponds to a 1.3-degree increase in the average January temperature. In the West, however, it corresponds to a 0.18-degree increase, which is only marginally significant.

◁

❑ Technical note

by has a second syntax that is especially useful when you want to play it safe:

by *varlist*$_1$ (*varlist*$_2$): *command*

This says that Stata is to verify that the data are sorted by *varlist*$_1$ *varlist*$_2$ and then, assuming that is true, perform *command* by *varlist*$_1$. For instance,

```
. by subject (time): gen finalval = val[_N]
```

By typing this, we want to create new variable `finalval`, which contains, in each observation, the final observed value of `val` for each subject in the data. The final value will be the last value if, within subject, the data are sorted by time. The above command verifies that the data are sorted by `subject` and `time` and then, if they are, performs

```
. by subject: gen finalval = val[_N]
```

If the data are not sorted properly, an error message will instead be issued. Of course, we could have just typed

```
. by subject: gen finalval = val[_N]
```

after verifying for ourselves that the data were sorted properly, as long as we were careful to look.

`by`'s second syntax can be used with `by`'s `sort` option, so we can also type

```
. by subject (time), sort: gen finalval = val[_N]
```

which is equivalent to

```
. sort subject time
. by subject: gen finalval = val[_N]
```

❑

See Mitchell (2010, chap. 7) for numerous examples of processing groups using the `by:` construct.

11.6 Filenaming conventions

Some commands require that you specify a *filename*. Filenames are specified in the way natural for your operating system:

Windows	Unix	Mac
mydata	mydata	mydata
mydata.dta	mydata.dta	mydata.dta
c:mydata.dta	~friend/mydata.dta	~friend/mydata.dta
"my data"	"my data"	"my data"
"my data.dta"	"my data.dta"	"my data.dta"
myproj\mydata	myproj/mydata	myproj/mydata
"my project\my data"	"my project/my data"	"my project/my data"
C:\analysis\data\mydata	~/analysis/data/mydata	~/analysis/data/mydata
"C:\my project\my data"	"~/my project/my data"	"~/my project/my data"
..\data\mydata	../data/mydata	../data/mydata
"..\my project\my data"	"../my project/my data"	"../my project/my data"

In most cases, where *filename* is a file that you are loading, *filename* may also be a URL. For instance, we might specify `use http://www.stata-press.com/data/r12/nlswork`.

Usually (the exceptions being `copy`, `dir`, `ls`, `erase`, `rm`, and `type`), Stata automatically provides a file extension if you do not supply one. For instance, if you type `use mydata`, Stata assumes that you mean `use mydata.dta` because `.dta` is the file extension Stata normally uses for data files.

Stata provides 22 default file extensions that are used by various commands:

`.ado`	automatically loaded do-files
`.dct`	text data dictionary
`.do`	do-file
`.dta`	Stata-format dataset
`.dtasig`	`datasignature` file
`.gph`	`graph`
`.grec`	Graph Editor recording (text format)
`.irf`	impulse–response function datasets
`.log`	log file in text format
`.mata`	Mata source code
`.mlib`	Mata library
`.mmat`	Mata matrix
`.mo`	Mata object file
`.out`	file saved by `outsheet`
`.raw`	text-format data
`.smcl`	log file in SMCL format
`.stbcal`	business calendars
`.ster`	saved estimates
`.sthlp`	help files
`.stptrace`	parameter-trace file; see [MI] **mi ptrace**
`.sum`	checksum files to verify network transfers
`.zip`	zip file

You do not have to name your data files with the `.dta` extension—if you type an explicit file extension, it will override the default. For instance, if your dataset was stored as `myfile.dat`, you could type `use myfile.dat`. If your dataset was stored as simply `myfile` with no file extension, you could type the period at the end of the filename to indicate that you are explicitly specifying the null extension. You would type `use myfile.` to use this dataset.

All operating systems allow blanks in filenames, and so does Stata. However, if the filename includes a blank, you must enclose the filename in double quotes. Typing

```
. save "my data"
```

would create the file `my data.dta`. Typing

```
. save my data
```

would be an error.

❑ Technical note

Stata also uses 12 other file extensions. These files are of interest only to advanced programmers or are for Stata's internal use. They are

`.class`	class file for object-oriented programming; see [P] **class**
`.dlg`	dialog resource file
`.idlg`	dialog resource include file
`.ihlp`	help include file
`.key`	`search`'s keyword database file
`.maint`	maintenance file (for Stata's internal use only)
`.mnu`	menu file (for Stata's internal use only)
`.pkg`	user-site package file
`.plugin`	compiled addition (DLL)
`.scheme`	control file for a graph scheme
`.style`	graph style file
`.toc`	user-site description file

❑

11.6.1 A special note for Mac users

Have you seen the notation `myfolder/myfile` before? This notation is called a path and describes the location of a file or folder (also called a directory).

You do not have to use this notation if you do not like it. You could instead restrict yourself to using files only in the current folder. If that turns out to be too restricting, Stata for Mac provides enough menus and buttons that you can probably get by. You may, however, find the notation convenient. If you do, here is the rest of the definition.

The character / is called a path delimiter and delimits folder names and filenames in a path. If the path starts with no path delimiter, the path is relative to the current folder.

For example, the path `myfolder/myfile` refers to the file `myfile` in the folder `myfolder`, which is contained in the current folder.

The characters `..` refer to the folder containing the current folder. Thus `../myfile` refers to `myfile` in the folder containing the current folder, and `../nextdoor/myfile` refers to `myfile` in the folder `nextdoor` in the folder containing the current folder.

If a path starts with a path delimiter, the path is called an absolute path and describes a fixed location of a file or folder name, regardless of what the current folder is. The leading / in an absolute path refers to the root directory, which is the main hard drive from which the operating system is booted. For example, the path `/myfolder/myfile` refers to the file `myfile` in the folder `myfolder`, which is contained in the main hard drive.

The character ~ refers to the user's home directory. Thus the path `~/myfolder/myfile` refers to `myfile` in the folder `myfolder` in the user's home directory.

11.6.2 A special note for Unix users

Stata understands ~ to mean your home directory. Stata understands this, even if you do not use `csh(1)` as your shell.

11.7 References

Baum, C. F. 2009. *An Introduction to Stata Programming.* College Station, TX: Stata Press.

Cox, N. J. 2009. Stata tip 79: Optional arguments to options. *Stata Journal* 9: 504.

Kolev, G. I. 2006. Stata tip 31: Scalar or variable? The problem of ambiguous names. *Stata Journal* 6: 279–280.

Mitchell, M. N. 2010. *Data Management Using Stata: A Practical Handbook.* College Station, TX: Stata Press.

Ryan, P. 2005. Stata tip 22: Variable name abbreviation. *Stata Journal* 5: 465–466.

12 Data

Contents

12.1 Data and datasets

Data form a rectangular table of numeric and string values in which each row is an observation on all the variables and each column contains the observations on one variable. Variables are designated by *variable names*. Observations are numbered sequentially from 1 to _N. The following example of data contains the first five odd and first five even positive integers, along with a string variable:

```
      odd  even   name
 1.     1     2   Bill
 2.     3     4   Mary
 3.     5     6    Pat
 4.     7     8  Roger
 5.     9    10   Sean
```

The observations are numbered 1 to 5, and the variables are named odd, even, and name. Observations are referred to by number, and variables by name.

A *dataset* is *data* plus labelings, formats, notes, and characteristics.

All aspects of *data* and *datasets* are defined here. Long (2009) offers a long-time Stata user's hard-won advice on how to manage data in Stata to promote accurate, replicable research.

12.2 Numbers

A *number* may contain a sign, an integer part, a decimal point, a fraction part, an e or E, and a signed integer exponent. Numbers may *not* contain commas; for example, the number 1,024 must be typed as 1024 (or 1024. or 1024.0). The following are examples of valid numbers:

```
5
-5
5.2
.5
5.2e+2
5.2e-2
```

❑ Technical note

Stata also allows numbers to be represented in a hexadecimal/binary format, defined as

[+|-]0.0[⟨*zeros*⟩]{X|x}-3ff

or

[+|-]1.⟨*hexdigit*⟩[⟨*hexdigits*⟩]{X|x}{+|-}⟨*hexdigit*⟩[⟨*hexdigits*⟩]

The lead digit is always 0 or 1; it is 0 only when the number being expressed is zero. A maximum of 13 digits to the right of the hexadecimal point are allowed. The power ranges from -3ff to +3ff. The number is expressed in hexadecimal (base 16) digits; the number aX+b means $a \times 2^b$. For instance, 1.0X+3 is 2^3 or 8. 1.8X+3 is 12 because 1.8_{16} is $1 + 8/16 = 1.5$ in decimal and the number is thus $1.5 \times 2^3 = 1.5 \times 8 = 12$.

Stata can also display numbers using this format; see [U] **12.5.1 Numeric formats**. For example,

```
. di 1.81x+2
6.015625

. di %21x 6.015625
+1.8100000000000X+002
```

This hexadecimal format is of special interest to numerical analysts.

❑

12.2.1 Missing values

A number may also take on the special value *missing*, denoted by a period (.). You specify a missing value anywhere that you may specify a number. Missing values differ from ordinary numbers in one respect: any arithmetic operation on a missing value yields a missing value.

In fact, there are 27 missing values in Stata: '.', the one just discussed, as well as .a, .b, ..., and .z, which are known as extended missing values. The missing value '.' is known as the default or system missing value. In any case, some people use extended missing values to indicate why a certain value is unknown—the question was not asked, the person refused to answer, etc. Other people have no use for extended missing values and just use '.'.

Stata's default or system missing value will be returned when you perform an arithmetic operation on missing values or when the arithmetic operation is not defined, such as division by zero, or the logarithm of a nonpositive number.

```
. display 2/0
.

. list

     +----+
     |  a |
     |----|
  1. | .b |
  2. |  . |
  3. | .a |
  4. |  3 |
  5. |  6 |
     +----+

. generate x = a + 1
(3 missing values generated)

. list

     +--------+
     |  a   x |
     |--------|
  1. | .b   . |
  2. |  .   . |
  3. | .a   . |
  4. |  3   4 |
  5. |  6   7 |
     +--------+
```

Numeric missing values are represented by "large positive values". The ordering is

$$\text{all numbers} < \texttt{.} < \texttt{.a} < \texttt{.b} < \cdots < \texttt{.z}$$

Thus the expression

$$\texttt{age} > 60$$

is true if variable age is greater than 60 or is missing. Similarly,

$$\texttt{gender}\,!= 0$$

is true if gender is not zero or is missing.

The way to exclude missing values is to ask whether the value is less than '.', and the way to detect missing values is to ask whether the value is greater than or equal to '.'. For instance,

```
. list if age>60 & age<.
. generate agegt60 = 0 if age<=60
. replace agegt60 = 1 if age>60 & age<.
. generate agegt60 = (age>60) if age<.
```

❑ Technical note

Before Stata 8, Stata only had one representation for missing values, the period (`.`).

To ensure that old programs and do-files continue to work properly, when `version` is set less than 8, all missing values are treated as being the same. Thus `. == .a == .b == .z`, and so '*exp*`==.`' and '*exp*`!=.`' work just as they previously did.

❑

▷ Example 1

We have data on the income of husbands and wives recorded in the variables `hincome` and `wincome`, respectively. Typing the `list` command, we see that your data contain

```
. use http://www.stata-press.com/data/r12/gxmpl3
. list

     +-------------------+
     | hincome   wincome |
     |-------------------|
  1. |   32000         0 |
  2. |   35000     34000 |
  3. |   47000        .b |
  4. |      .z     50000 |
  5. |      .a         . |
     +-------------------+
```

The values of `wincome` in the third and fifth observations are *missing*, as distinct from the value of `wincome` in the first observation, which is known to be zero.

If we use the `generate` command to create a new variable, `income`, that is equal to the sum of `hincome` and `wincome`, three missing values would be produced:

```
. generate income = hincome + wincome
(3 missing values generated)
. list

     +------------------------------+
     | hincome   wincome   income   |
     |------------------------------|
  1. |   32000         0    32000   |
  2. |   35000     34000    69000   |
  3. |   47000        .b        .   |
  4. |      .z     50000        .   |
  5. |      .a         .        .   |
     +------------------------------+
```

`generate` produced a warning message that 3 missing values were created, and when we list the data, we see that 47,000 plus *missing* yields *missing*.

◁

❑ Technical note

Stata stores numeric missing values as the largest 27 numbers allowed by the particular storage type; see [U] **12.2.2 Numeric storage types**. There are two important implications. First, if you `sort` on a variable that has missing values, the missing values will be placed last, and the sort order of any missing values will follow the rule regarding the properties of missing values stated above.

```
. sort wincome

. list wincome

     +---------+
     | wincome |
     |---------|
  1. |       0 |
  2. |   34000 |
  3. |   50000 |
  4. |       . |
  5. |      .b |
     +---------+
```

The second implication concerns relational operators and missing values. Do not forget that a missing value will be larger than any numeric value.

```
. list if wincome > 40000

     +------------------------------+
     | hincome   wincome    income  |
     |------------------------------|
  3. |      .z     50000         .  |
  4. |      .a         .         .  |
  5. |   47000        .b         .  |
     +------------------------------+
```

Observations 4 and 5 are listed because '`.`' and '`.b`' are both missing and thus are greater than 40,000. Relational operators are discussed in detail in [U] **13.2.3 Relational operators**.

❑

▷ Example 2

In producing statistical output, Stata ignores observations with missing values. Continuing with the example above, if we request summary statistics on `hincome` and `wincome` by using the `summarize` command, we obtain

```
. summarize hincome wincome

    Variable |       Obs        Mean    Std. Dev.       Min        Max
-------------+--------------------------------------------------------
     hincome |         3       38000    7937.254      32000      47000
     wincome |         3       28000    25534.29          0      50000
```

Some commands discard the entire observation (known as *casewise deletion*) if one of the variables in the observation is missing. If we use the `correlate` command to obtain the correlation between `hincome` and `wincome`, for instance, we obtain

```
. correlate hincome wincome
(obs=2)

             |  hincome  wincome
-------------+------------------
     hincome |   1.0000
     wincome |   1.0000   1.0000
```

The correlation coefficient is calculated over two observations.

◁

12.2.2 Numeric storage types

Numbers can be stored in one of five variable types: byte, int, long, float (the default), or double. bytes are, naturally, stored in 1 byte. ints are stored in 2 bytes, longs and floats in 4 bytes, and doubles in 8 bytes. The table below shows the minimum and maximum values for each storage type.

Storage type	Minimum	Maximum	Closest to 0 without being 0	Bytes
byte	-127	100	± 1	1
int	$-32{,}767$	32,740	± 1	2
long	$-2{,}147{,}483{,}647$	2,147,483,620	± 1	4
float	$-1.70141173319 \times 10^{38}$	$1.70141173319 \times 10^{38}$	$\pm 10^{-38}$	4
double	$-8.9884656743 \times 10^{307}$	$+8.9884656743 \times 10^{307}$	$\pm 10^{-323}$	8

Do not confuse the term *integer*, which is a characteristic of a number, with int, which is a storage type. For instance, the number 5 is an integer, no matter how it is stored; thus, if you read that an argument must be an integer, that does not mean that it must be stored as an int.

12.3 Dates and times

Stata has nine date, time, and date-and-time numeric encodings known collectively as %t variables or values. They are

%tC	calendar date and time, adjusted for leap seconds
%tc	calendar date and time, ignoring leap seconds
%td	calendar date
%tw	week
%tm	calendar month
%tq	financial quarter
%th	financial half-year
%ty	calendar year
%tb	business calendars

All except %ty and %tb are based on 0 = beginning of January 1960. %tc and %tC record the number of milliseconds since then. %td records the number of days. The others record the numbers of weeks, months, quarters, or half-years. %ty simply records the year, and %tb records a user-defined business calendar format.

For a full discussion of working with dates and times, see [U] **24 Working with dates and times**.

12.4 Strings

A *string* is a sequence of printable characters and is typically enclosed in double quotes. The quotes are not considered a part of the string but delimit the beginning and end of the string. The following are examples of valid strings:

```
"Hello, world"
"String"
"string"
" string"
"string "
""
"x/y+3"
"1.2"
```

All the strings above are distinct; that is, `"String"` is different from `"string"`, which is different from `" string"`, which is different from `"string "`. Also, `"1.2"` is a string and not a number because it is enclosed in quotes.

All strings in Stata are of varying length, which means that Stata internally records the length of the string and never loses track. There is never a circumstance in which a string cannot be delimited by quotes, but there are rare instances where strings do not have to be delimited by quotes, such as during data input. In those cases, nondelimited strings are stripped of their leading and trailing blanks. Delimited strings are always accepted as is.

The special string `""`, often called *null string*, is considered by Stata to be a *missing*. No special meaning is given to the string containing one period, `"."`.

In addition to double quotes for enclosing strings, Stata also allows compound double quotes: `` `" `` and `"'`. You can type "*string*" or you can type `"*string*"', although users seldom type `"*string*"'. Compound double quotes are of special interest to programmers because they nest and provide a way for a quoted string to itself contain double quotes (either simple or compound). See [U] **18.3.5 Double quotes**.

12.4.1 Strings containing identifying data

String variables often contain identifying information, such as the patient's name or the name of the city or state. Such strings are typically listed but are not used directly in statistical analysis, although the data might be sorted on the string or datasets might be merged on the basis of one or more string variables.

12.4.2 Strings containing categorical data

Occasionally, strings contain information that is to be used directly in analysis, such as the patient's sex, which might be coded `"male"` or `"female"`. Stata shows a decided preference for such information to be numerically encoded and stored in numeric variables. Stata's statistical routines treat string variables as if every observation records a numeric missing value.

Stata provides two commands for converting string variables into numeric codes and back again: `encode` and `decode`; see [U] **23.2 Categorical string variables** and [U] **11.4.3 Factor variables**. Also see [D] **destring** for information on mapping string variables to numeric and vice versa.

12.4.3 Strings containing numeric data

If a string variable contains the character representation of a number—for instance, `myvar` contains `"1"`, `"1.2"`, and `"-5.2"`—you can convert it directly into a numeric variable by using the `real()` function or the `destring` command, for example, `generate newvar=real(myvar)`.

Similarly, if you want to convert a numeric variable to its string representation, you can use the `string()` function or the `tostring` command, for example, `generate as_str=string(numvar)`. See [D] **functions**.

12.4.4 String storage types

Strings are stored in string variables with storage types `str1`, `str2`, ..., `str244`. The storage type merely sets the maximum length of the string, not its actual length; thus, `"example"` has length 7 whether it is stored as a `str7`, a `str10`, or even a `str244`. On the other hand, an attempt to assign the string `"example"` to a `str6` would result in `"exampl"`.

The maximum length of a string in Stata is 244. String literals may exceed 244 characters, but only the first 244 characters are significant.

12.5 Formats: Controlling how data are displayed

Formats describe how a number or string is to be presented. For instance, how is the number 325.24 to be presented? As `325.2`, or `325.24`, or `325.240`, or `3.2524e+02`, or `3.25e+02`, or some other way? The *display format* tells Stata exactly how to present such data. You do not have to specify display formats because Stata always makes reasonable assumptions about how to display a variable, but you always have the option.

12.5.1 Numeric formats

A Stata numeric format is formed by

first type	`%`	to indicate the start of the format
then optionally type	`-`	if you want the result left-aligned
then optionally type	`0`	if you want to retain leading zeros (1)
then type	a number w	stating the width of the result
then type	`.`	
then type	a number d	stating the number of digits to follow the decimal point
then type		
either	`e`	for scientific notation, e.g., 1.00e+03
or	`f`	for fixed format, e.g., 1000.0
or	`g`	for general format; Stata chooses based on the number being displayed
then optionally type	`c`	to indicate comma format (not allowed with `e`)

(1) Specifying `0` to mean "include leading zeros" will be honored only with the `f` format.

For example,

```
%9.0g    general format, 9 columns wide
                   sqrt(2) =  1.414214
                     1,000 =      1000
                10,000,000 =  1.00e+07
%9.0gc   general format, 9 columns wide, with commas
                   sqrt(2) =  1.414214
                     1,000 =     1,000
                10,000,000 =  1.00e+07
%9.2f    fixed format, 9 columns wide, 2 decimal places
                   sqrt(2) =      1.41
                     1,000 =   1000.00
                10,000,000 = 10000000.00
%9.2fc   fixed format, 9 columns wide, 2 decimal places, with commas
                   sqrt(2) =      1.41
                     1,000 =  1,000.00
                10,000,000 = 10,000,000.00
%9.2e    exponential format, 9 columns wide
                   sqrt(2) =  1.41e+00
                     1,000 =  1.00e+03
                10,000,000 =  1.00e+07
```

Stata has three numeric format types: `e`, `f`, and `g`. The formats are denoted by a leading percent sign (%) followed by the string *w.d*, where *w* and *d* stand for two integers. The first integer, *w*, specifies the width of the format. The second integer *d* specifies the number of digits that are to follow the decimal point. *d* must be less than *w*. Finally, a character denotes the format type (`e`, `f`, or `g`), and to that may optionally be appended a `c` indicating that commas are to be included in the result (`c` is not allowed with `e`.)

By default, every numeric variable is given a `%`*w*`.0g` format, where *w* is large enough to display the largest number of the variable's type. The `%`*w*`.0g` format is a set of formatting rules that present the values in as readable a fashion as possible without sacrificing precision. The `g` format changes the number of decimal places displayed whenever it improves the readability of the current value.

The default formats for each of the numeric variable types are

```
byte      %8.0g
int       %8.0g
long     %12.0g
float     %9.0g
double   %10.0g
```

You can change the format of a variable by using the `format` *varname* `%`*fmt* command.

In addition to `%`*w*`.0g`, allowed is `%`*w*`.0gc` to display numbers with commas. "One thousand" is displayed as 1000 in `%9.0g` format and as 1,000 in `%9.0gc` format.

In addition to using `%`*w*`.0g` and `%`*w*`.0gc`, you can use `%`*w*`.`*d*`g` and `%`*w*`.`*d*`gc`, $d > 0$. For example, `%9.4g` and `%9.4gc`. The 4 means to display approximately four significant digits. For instance, the number 3.14159265 in `%9.4g` format is displayed as 3.142, 31.4159265 as 31.42, 314.159265 as 314.2, and 3141.59265 as 3142. The format is not exactly a significant-digit format because 31415.9265 is displayed as 31416, not as 3.142e+04.

Under the `f` format, values are always displayed with the same number of decimal places, even if this results in a loss in the displayed precision. Thus the `f` format is similar to the C `f` format. Stata's `f` format is also similar to the Fortran `F` format, but, unlike the Fortran `F` format, it switches to `g` whenever a number is too large to be displayed in the specified `f` format.

In addition to `%`*w*`.`*d*`f`, the format `%`*w*`.`*d*`fc` can display numbers with commas.

The e format is similar to the C e and the Fortran E format. Every value is displayed as a leading digit (with a minus sign, if necessary), followed by a decimal point, the specified number of digits, the letter e, a plus sign or a minus sign, and the power of 10 (modified by the preceding sign) that multiplies the displayed value. When the e format is specified, the width must exceed the number of digits that follow the decimal point by at least seven to accommodate the leading sign and digit, the decimal point, the e, and the signed power of 10.

▷ Example 3

Below we have a 5-observation dataset with three variables: e_fmt, f_fmt, and g_fmt. All three variables have the same values stored in them; only the display format varies. describe shows the display format to the right of the variable type:

```
. use http://www.stata-press.com/data/r12/format, clear

. describe

Contains data from http://www.stata-press.com/data/r12/format.dta
  obs:             5
 vars:             3                          12 Mar 2011 15:18
 size:            60
--------------------------------------------------------------------------------
              storage  display     value
variable name   type   format      label      variable label
--------------------------------------------------------------------------------
e_fmt           float  %9.2e
f_fmt           float  %10.2f
g_fmt           float  %9.0g
--------------------------------------------------------------------------------
Sorted by:
```

The formats for each of these variables were set by typing

```
. format e_fmt %9.2e

. format f_fmt %10.2f
```

It was not necessary to set the format for the g_fmt variable because Stata automatically assigned it the %9.0g format. Nevertheless, we could have typed format g_fmt %9.0g if we wished. Listing the data results in

```
. list

     +--------------------------------------+
     |     e_fmt        f_fmt        g_fmt  |
     |--------------------------------------|
  1. |  2.80e+00         2.80     2.801785  |
  2. |  3.96e+06   3962322.50      3962323  |
  3. |  4.85e+00         4.85     4.852834  |
  4. | -5.60e-06        -0.00    -5.60e-06  |
  5. |  6.26e+00         6.26     6.264982  |
     +--------------------------------------+
```

◁

❑ Technical note

The discussion above is incomplete. There is one other format available that will be of interest to numerical analysts. The %21x format displays base 10 numbers in a hexadecimal (base 16) format. The number is expressed in hexadecimal (base 16) digits; the number aX+b means $a \times 2^b$. For example,

```
. display %21x 1234.75
+1.34b0000000000X+00a
```

Thus the base 10 number 1,234.75 has a base 16 representation of `1.34bX+0a`, meaning

$$\left(1 + 3 \cdot 16^{-1} + 4 \cdot 16^{-2} + 11 \cdot 16^{-3}\right) \times 2^{10}$$

Remember, the hexadecimal–decimal equivalents are

hexadecimal	decimal
0	0
1	1
2	2
3	3
4	4
5	5
6	6
7	7
8	8
9	9
a	10
b	11
c	12
d	13
e	14
f	15

See [U] **12.2 Numbers**.

❑

12.5.2 European numeric formats

The three numeric formats `e`, `f`, and `g` will use '`,`' to indicate the decimal symbol if you specify their width and depth as *w*`,`*d* rather than *w*`.`*d*. For instance, the format `%9,0g` will display what Stata would usually display as 1.5 as `1,5`.

If you use the European specification with `fc` or `gc`, the "comma" will be presented as a period. For instance, `%9,0gc` would display what Stata would usually display as 1,000.5 as `1.000,5`.

If this way of presenting numbers appeals to you, consider using Stata's `set dp comma` command. `set dp comma` tells Stata to interpret nearly all `%`*w*`.`*d*{`g`|`f`|`e`} formats as `%`*w*`,`*d*{`g`|`f`|`e`} formats. Most of Stata is written using a period to represent the decimal symbol, and that means that, even if you set the appropriate `%`*w*`,`*d*{`g`|`f`|`e`} format for your data, it will affect only displays of the data. For instance, if you type `summarize` to obtain summary statistics or `regress` to obtain regression results, the decimal will still be shown as a period.

`set dp comma` changes that and affects all of Stata. With `set dp comma`, it does not matter whether your data are formatted `%`*w*`.`*d*{`g`|`f`|`e`} or `%`*w*`,`*d*{`g`|`f`|`e`}. All results will be displayed using a comma as the decimal character:

```
. use http://www.stata-press.com/data/r12/auto
(1978 Automobile Data)

. set dp comma

. summarize mpg weight foreign

    Variable |       Obs        Mean    Std. Dev.       Min        Max
-------------+--------------------------------------------------------
         mpg |        74     21,2973    5,785503         12         41
      weight |        74    3019,459    777,1936       1760       4840
     foreign |        74    ,2972973    ,4601885          0          1

. regress mpg weight foreign

      Source |       SS       df       MS              Number of obs =      74
-------------+------------------------------           F(  2,    71) =   69,75
       Model |  1619,2877     2  809,643849           Prob > F      =  0,0000
    Residual |  824,171761    71   11,608053           R-squared     =  0,6627
-------------+------------------------------           Adj R-squared =  0,6532
       Total |  2443,45946    73  33,4720474           Root MSE      =  3,4071

------------------------------------------------------------------------------
         mpg |      Coef.   Std. Err.      t    P>|t|     [95% Conf. Interval]
-------------+----------------------------------------------------------------
      weight |  -,0065879   ,0006371   -10,34   0,000    -,0078583   -,0053175
     foreign |  -1,650029   1,075994    -1,53   0,130      -3,7955    ,4954422
       _cons |    41,6797   2,165547    19,25   0,000     37,36172    45,99768
------------------------------------------------------------------------------
```

You can switch the decimal character back to a period by typing `set dp period`.

❑ Technical note

`set dp comma` makes drastic changes inside Stata, and we mention this because some older, user-written programs may not be able to deal with those changes. If you are using an older user-written program, you might `set dp comma` and then find that the program does not work and instead presents some sort of syntax error.

If, using any program, you do get an unanticipated error, try setting `dp` back to `period`. See [D] **format** for more information.

Also understand that `set dp comma` affects how Stata outputs numbers, not how it inputs them. You must still use the period to indicate the decimal point on all input. Even with `set dp comma`, you type

```
. replace x=1.5 if x==2
```

❑

12.5.3 Date and time formats

Date and time formats are really a numeric format because Stata stores dates as the number of milliseconds, days, weeks, months, quarters, half-years, or years from 01jan1960; see [U] **24 Working with dates and times**.

The syntax of the `%t` format is

first type	`%`	to indicate the start of the format
then optionally type	`-`	if you want the result left-aligned
then type	`t`	
then type	*character*	to indicate the units
then optionally type	*other characters*	to indicate how the date/time is to be displayed

The letter you type to specify the units is

`C`	milliseconds from 01jan1960, adjusted for leap seconds
`c`	milliseconds from 01jan1960, ignoring leap seconds
`d`	days from 01jan1960
`w`	weeks from 1960-w1
`m`	calendar months from jan1960
`q`	quarters from 1960-q1
`h`	half years from 1960-h1

There are many codes you can type after that to specify exactly how the date/time is to be displayed, but usually, you do not. Most users use the default `%tc` for date/times and `%td` for dates. See [D] **datetime display formats** for details.

12.5.4 String formats

The syntax for a string format is

first type	`%`	to indicate the start of the format
then optionally type	`-`	if you want the result left-aligned
then type	a number	indicating the width of the result
then type	`s`	

For instance, `%10s` represents a string format of width 10.

For `str`*w*, the default format is `%`*w*`s` or `%9s`, whichever is wider. For example, a `str10` variable receives a `%10s` format. Strings are displayed right-justified in the field, unless the minus sign is coded; `%-10s` would display the string left-aligned.

▷ Example 4

Our automobile data contain a string variable called `make`.

```
. use http://www.stata-press.com/data/r12/auto
(1978 Automobile Data)

. describe make

              storage  display     value
variable name   type   format      label      variable label
-------------------------------------------------------------------

make            str18  %-18s                  Make and Model

. list make in 63/67
```

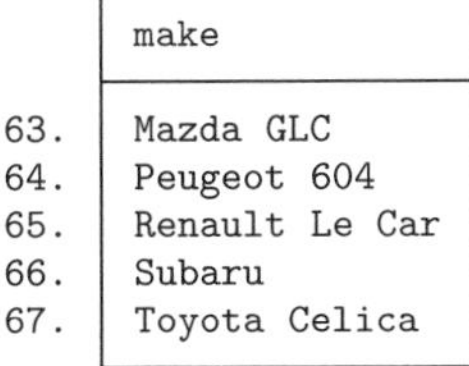

```
      +------------------+
      | make             |
      |------------------|
 63.  | Mazda GLC        |
 64.  | Peugeot 604      |
 65.  | Renault Le Car   |
 66.  | Subaru           |
 67.  | Toyota Celica    |
      +------------------+
```

These values are left-aligned because `make` has a display format of `%-18s`. If we want to right-align the values, we could change the format:

```
. format %18s make
. list make in 63/67

     +----------------+
     |           make |
     |----------------|
 63. |      Mazda GLC |
 64. |    Peugeot 604 |
 65. | Renault Le Car |
 66. |         Subaru |
 67. |  Toyota Celica |
     +----------------+
```

◁

12.6 Dataset, variable, and value labels

Labels are strings used to label elements in Stata, such as labels for datasets, variables, and values.

12.6.1 Dataset labels

Associated with every dataset is an 80-character *dataset label*, which is initially set to blanks. You can use the `label data "`*text*`"` command to define the dataset label.

▷ Example 5

We have just entered 1980 state data on marriage rates, divorce rates, and median ages. The `describe` command will describe the data in memory:

```
. describe
Contains data
  obs:            50
 vars:             4
 size:         1,200
--------------------------------------------------------------------------------
              storage  display     value
variable name   type   format      label      variable label
--------------------------------------------------------------------------------
state           str8   %9s
median_age      float  %9.0g
marriage_rate   long   %12.0g
divorce_rate    long   %12.0g
--------------------------------------------------------------------------------
Sorted by:
     Note:  dataset has changed since last saved
```

`describe` shows that there are 50 observations on four variables named `state`, `median_age`, `marriage_rate`, and `divorce_rate`. `state` is stored as a `str8`; `median_age` is stored as a `float`; and `marriage_rate` and `divorce_rate` are both stored as `long`s. Each variable's display format (see [U] **12.5 Formats: Controlling how data are displayed**) is shown. Finally, the data are not in any particular sort order, and the dataset has changed since it was last saved on disk.

We can label the data by typing `label data "1980 state data"`. We type this and then type `describe` again:

```
. label data "1980 state data"

. describe

Contains data
  obs:            50                      1980 state data
 vars:             4
 size:         1,200
--------------------------------------------------------------------------------
              storage  display     value
variable name   type   format      label      variable label
--------------------------------------------------------------------------------
state           str8   %9s
median_age      float  %9.0g
marriage_rate   long   %12.0g
divorce_rate    long   %12.0g
--------------------------------------------------------------------------------
Sorted by:
     Note:  dataset has changed since last saved
```

◁

The dataset label is displayed by the `describe` and `use` commands.

12.6.2 Variable labels

In addition to the name, every variable has associated with it an 80-character *variable label*. The variable labels are initially set to blanks. You use the `label variable` *varname* "*text*" command to define a new variable label.

▷ Example 6

We have entered data on four variables: `state`, `median_age`, `marriage_rate`, and `divorce_rate`. `describe` portrays the data we entered:

```
. describe

Contains data from states.dta
  obs:            50                      1980 state data
 vars:             4
 size:         1,200
--------------------------------------------------------------------------------
              storage  display     value
variable name   type   format      label      variable label
--------------------------------------------------------------------------------
state           str8   %9s
median_age      float  %9.0g
marriage_rate   long   %12.0g
divorce_rate    long   %12.0g
--------------------------------------------------------------------------------
Sorted by:
     Note:  dataset has changed since last saved
```

We can associate labels with the variables by typing

```
. label variable median_age "Median Age"

. label variable marriage_rate "Marriages per 100,000"

. label variable divorce_rate "Divorces per 100,000"
```

From then on, the result of `describe` will be

```
. describe
Contains data
  obs:            50                          1980 state data
 vars:             4
 size:         1,200

              storage  display     value
variable name   type   format      label      variable label

state           str8   %9s
median_age      float  %9.0g                  Median Age
marriage_rate   long   %12.0g                 Marriages per 100,000
divorce_rate    long   %12.0g                 Divorces per 100,000

Sorted by:
     Note:  dataset has changed since last saved
```

◁

Whenever Stata produces output, it will use the variable labels rather than the variable names to label the results if there is room.

12.6.3 Value labels

Value labels define a correspondence or mapping between numeric data and the words used to describe what those numeric values represent. Mappings are named and defined by the `label define` *lblname* # "*string*" # "*string*"... command. The maximum length for the *lblname* is 32 characters. # must be an integer or an extended missing value (`.a`, `.b`, ..., `.z`). The maximum length of *string* is 32,000 characters. Named mappings are associated with variables by the `label values` *varname* *lblname* command.

▷ Example 7

The definition makes value labels sound more complicated than they are in practice. We create a dataset on individuals in which we record a person's sex, coding 0 for males and 1 for females. If our dataset also contained an employee number and salary, it might resemble the following:

```
. use http://www.stata-press.com/data/r12/gxmpl4
(2007 Employee data)

. describe

Contains data from http://www.stata-press.com/data/r12/gxmpl4.dta
  obs:             7                          2007 Employee data
 vars:             3                          11 Feb 2011 15:31
 size:            84

              storage  display     value
variable name   type   format      label      variable label

empno           float  %9.0g                  Employee number
sex             float  %9.0g                  Sex
salary          float  %8.0fc                 Annual salary, exclusive of
                                                bonus

Sorted by:
```

```
. list

     ┌────────────────────────┐
     │ empno   sex    salary  │
     ├────────────────────────┤
  1. │ 57213     0    34,000  │
  2. │ 47229     1    37,000  │
  3. │ 57323     0    34,000  │
  4. │ 57401     0    34,500  │
  5. │ 57802     1    37,000  │
     ├────────────────────────┤
  6. │ 57805     1    34,000  │
  7. │ 57824     0    32,500  │
     └────────────────────────┘
```

We could create a mapping called `sexlabel` defining 0 as "Male" and 1 as "Female", and then associate that mapping with the variable `sex` by typing

```
. label define sexlabel 0 "Male" 1 "Female"
. label values sex sexlabel
```

From then on, our data would appear as

```
. describe
Contains data from http://www.stata-press.com/data/r12/gxmpl4.dta
  obs:             7                          2007 Employee data
 vars:             3                          11 Feb 2011 15:31
 size:            84
--------------------------------------------------------------------------------
              storage  display     value
variable name   type   format      label      variable label
--------------------------------------------------------------------------------
empno           float  %9.0g                  Employee number
sex             float  %9.0g       sexlabel   Sex
salary          float  %8.0fc                 Annual salary, exclusive of
                                                bonus
--------------------------------------------------------------------------------
Sorted by:
. list

     ┌───────────────────────────┐
     │ empno      sex    salary  │
     ├───────────────────────────┤
  1. │ 57213     Male    34,000  │
  2. │ 47229   Female    37,000  │
  3. │ 57323     Male    34,000  │
  4. │ 57401     Male    34,500  │
  5. │ 57802   Female    37,000  │
     ├───────────────────────────┤
  6. │ 57805   Female    34,000  │
  7. │ 57824     Male    32,500  │
     └───────────────────────────┘
```

Notice not only that the value label is used to produce words when we `list` the data but also that the association of the variable `sex` with the value label `sexlabel` is shown by the `describe` command.

◁

❑ Technical note

Value labels and variables may share the same name. For instance, rather than calling the value label `sexlabel` in the example above, we could just as well have named it `sex`. We would then type `label values sex sex` to associate the value label named `sex` with the variable named `sex`.

❑

▷ Example 8

Stata's `encode` and `decode` commands provide a convenient way to go from string variables to numerically coded variables and back again. Let's pretend that, in the example above, rather than coding 0 for males and 1 for females, we created a string variable recording either `"male"` or `"female"`.

```
. use http://www.stata-press.com/data/r12/gxmpl5
(2007 Employee data)

. describe

Contains data from http://www.stata-press.com/data/r12/gxmpl5
  obs:             7                          2007 Employee data
 vars:             3                          11 Feb 2011 15:37
 size:            98
--------------------------------------------------------------------------------
              storage  display     value
variable name   type   format      label      variable label
--------------------------------------------------------------------------------
empno           float  %9.0g                  Employee number
sex             str6   %9s                    Sex
salary          float  %8.0fc                 Annual salary, exclusive of
                                                bonus
--------------------------------------------------------------------------------
Sorted by:

. list

     +--------------------------+
     | empno      sex   salary  |
     |--------------------------|
  1. | 57213     male   34,000  |
  2. | 47229   female   37,000  |
  3. | 57323     male   34,000  |
  4. | 57401     male   34,500  |
  5. | 57802   female   37,000  |
     |--------------------------|
  6. | 57805   female   34,000  |
  7. | 57824     male   32,500  |
     +--------------------------+
```

We now want to create a numerically encoded variable—we will call it `gender`—from the string variable. We want to do this, say, because we typed `anova salary sex` to perform a one-way ANOVA of salary on sex, and we were told that there were "no observations". We then remembered that all Stata's statistical commands treat string variables as if they contain nothing but missing values. The statistical commands work only with numerically coded data.

```
. encode sex, generate(gender)
. describe
Contains data from http://www.stata-press.com/data/r12/gxmpl5.dta
  obs:             7                          2007 Employee data
 vars:             4                          11 Feb 2011 15:37
 size:           126
-------------------------------------------------------------------------------
              storage  display     value
variable name   type   format      label      variable label
-------------------------------------------------------------------------------
empno           float  %9.0g                  Employee number
sex             str6   %9s                    Sex
salary          float  %8.0fc                 Annual salary, exclusive of
                                                bonus
gender          long   %8.0g       gender     Sex
-------------------------------------------------------------------------------
Sorted by:
     Note:  dataset has changed since last saved
```

`encode` adds a new `long` variable called `gender` to the data and defines a new value label called `gender`. The value label `gender` maps 1 to the string `male` and 2 to `female`, so if we were to `list` the data, we could not tell the difference between the `gender` and `sex` variables. However, they are different. Stata's statistical commands know how to deal with `gender` but do not understand the `sex` variable. See [D] **encode**.

❑

❑ Technical note

Perhaps rather than employee data, our data are on persons undergoing sex-change operations. There would therefore be two sex variables in our data, sex before the operation and sex after the operation. Assume that the variables are named `presex` and `postsex`. We can associate the *same* value label to each variable by typing

```
. label define sexlabel 0 "Male" 1 "Female"
. label values presex sexlabel
. label values postsex sexlabel
```

❑

❑ Technical note

Stata's input commands (`input` and `infile`) can switch from the words in a value label back to the numeric codes. Remember that `encode` and `decode` can translate a string to a numeric mapping and vice versa, so we can map strings to numeric codes either at the time of input or later.

For example,

```
. label define sexlabel 0 "Male" 1 "Female"
. input empno sex:sexlabel salary, label
          empno         sex     salary
  1. 57213 Male 34000
  2. 47229 Female 37000
  3. 57323 0 34000
  4. 57401 Male 34500
  5. 57802 Female 37000
  6. 57805 Female 34000
  7. 57824 Male 32500
  8. end
```

The `label define` command defines the value label `sexlabel`. `input empno sex:sexlabel salary, label` tells Stata to input three variables from the keyboard (`empno`, `sex`, and `salary`), attach the value label `sexlabel` to the `sex` variable, and look up any words that are typed in the value label to try to convert them to numbers. To prove that it works, we `list` the data that we recently entered:

```
. list

     +-------------------------+
     | empno      sex   salary |
     |-------------------------|
  1. | 57213     Male    34000 |
  2. | 47229   Female    37000 |
  3. | 57323     Male    34000 |
  4. | 57401     Male    34500 |
  5. | 57802   Female    37000 |
     |-------------------------|
  6. | 57805   Female    34000 |
  7. | 57824     Male    32500 |
     +-------------------------+
```

Compare the information we typed for observation 3 with the result listed by Stata. We typed 57323 0 34000. Thus the value of `sex` in the third observation is 0. When Stata listed the observation, it indicated the value is `Male` because we told Stata in our `label define` command that zero is equivalent to `Male`.

Let's now add one more observation to our data:

```
. input, label
         empno         sex      salary
  8. 67223 FEmale 33000
'FEmale' cannot be read as a number
  8. 67223 Female 33000
  9. end
```

At first we typed 67223 `FEmale` 33000, and Stata responded with "'FEmale' cannot be read as a number". Remember that Stata always respects case, so `FEmale` is not the same as `Female`. Stata prompted us to type the line again, and we did so, this time correctly.

❑

❑ Technical note

Coupled with the `automatic` option, Stata not only can go from words to numbers but also can create the mapping. Let's input the data again, but this time, rather than typing the data, let's read the data from a file. Assume that we have a text file named `employee.raw` stored on our disk that contains

```
57213 Male 34000
47229 Female 37000
57323 Male 34000
57401 Male 34500
57802 Female 37000
57805 Female 34000
57824 Male 32500
```

The `infile` command can read these data and create the mapping automatically:

```
. label list sexlabel
value label sexlabel not found
r(111);
```

```
. infile empno sex:sexlabel salary using employee, automatic
(7 observations read)
```

Our first command, `label list sexlabel`, is only to prove that we had not previously defined the value label `sexlabel`. Stata `infiled` the data without complaint. We now have

```
. list

     +------------------------+
     | empno      sex   salary |
     |------------------------|
  1. | 57213     Male    34000 |
  2. | 47229   Female    37000 |
  3. | 57323     Male    34000 |
  4. | 57401     Male    34500 |
  5. | 57802   Female    37000 |
     |------------------------|
  6. | 57805   Female    34000 |
  7. | 57824     Male    32500 |
     +------------------------+
```

Of course, `sex` is just another numeric variable; it does not actually take on the values `Male` and `Female`—it takes on numeric codes that have been automatically mapped to `Male` and `Female`. We can find out what that mapping is by using the `label list` command:

```
. label list sexlabel
sexlabel:
             1 Male
             2 Female
```

We discover that Stata attached the codes 1 to `Male` and 2 to `Female`. Anytime we want to see what our data really look like, ignoring the value labels, we can use the `nolabel` option:

```
. list, nolabel

     +---------------------+
     | empno   sex   salary |
     |---------------------|
  1. | 57213     1    34000 |
  2. | 47229     2    37000 |
  3. | 57323     1    34000 |
  4. | 57401     1    34500 |
  5. | 57802     2    37000 |
     |---------------------|
  6. | 57805     2    34000 |
  7. | 57824     1    32500 |
     +---------------------+
```

❑

12.6.4 Labels in other languages

A dataset can contain labels—data, variable, and value—in up to 100 languages. To discover the languages available for the dataset in memory, type `label language`. You will see this

```
. label language

Language for variable and value labels

    In this dataset, value and variable labels have been defined in only one
    language:  default

    To create new language:          . label language <name>, new
    To rename current language:      . label language <name>, rename
```

or something like this:

```
. label language

Language for variable and value labels

    Available languages:
            de
            en
            sp

    Currently set is:                  . label language sp

    To select different language:      . label language <name>

    To create new language:            . label language <name>, new
    To rename current language:        . label language <name>, rename
```

Right now, the example dataset is set with sp (Spanish) labels:

```
. describe

Contains data
  obs:            74                          Automóviles, 1978
 vars:            12                          3 Oct 2010 13:53
 size:         3,478

              storage  display     value
variable name   type   format      label      variable label

make            str18  %-18s                  Marca y modelo
price           int    %8.0gc                 Precio
mpg             int    %8.0g                  Consumo de combustible
rep78           int    %8.0g                  Historia de reparaciones
headroom        float  %6.1f                  Cabeza adelante
trunk           int    %8.0g                  Volumen del maletero
weight          int    %8.0gc                 Peso
length          int    %8.0g                  Longitud
turn            int    %8.0g                  Radio de giro
displacement    int    %8.0g                  Cilindrada
gear_ratio      float  %6.2f                  Relación de cambio
foreign         byte   %8.0g                  Extranjero

Sorted by:  foreign
```

To create labels in more than one language, you set the new language and then define the labels in the standard way; see [D] **label language**.

12.7 Notes attached to data

A dataset may contain notes, which are nothing more than little bits of text that you define and review with the notes command. Typing note, a colon, and the text defines a note:

```
. note:  Send copy to Bob once verified.
```

You can later display whatever notes you have previously defined by typing notes:

```
. notes

_dta:
  1.  Send copy to Bob once verified.
```

Notes are saved with the data, so once you save your dataset, you can replay this note in the future, too.

You can add more notes:

```
. note: Mary wants a copy, too.
. notes
_dta:
  1.  Send copy to Bob once verified.
  2.  Mary wants a copy, too.
```

The notes you have added so far are attached to the data generically, which is why Stata prefixes them with `_dta` when it lists them. You can attach notes to variables:

```
. note state: verify values for Nevada.
. note state: what about the two missing values?
. notes
_dta:
  1.  Send copy to Bob once verified.
  2.  Mary wants a copy, too.
state:
  1.  verify values for Nevada.
  2.  what about the two missing values?
```

When you `describe` your data, you can see whether notes are attached to the dataset or to any of the variables:

```
. describe
Contains data from states.dta
  obs:            50                          1980 state data
 vars:             4
 size:         1,200                          (_dta has notes)
-------------------------------------------------------------------------------
              storage  display     value
variable name   type   format      label      variable label
-------------------------------------------------------------------------------
state           str8   %9s                  *
median_age      float  %9.0g                  Median Age
marriage_rate   long   %12.0g                 Marriages per 100,000
divorce_rate    long   %12.0g                 Divorces per 100,000
                                              * indicated variables have notes
-------------------------------------------------------------------------------
Sorted by:
     Note:  dataset has changed since last saved
```

See [D] **notes** for a complete description of this feature.

12.8 Characteristics

Characteristics are an arcane feature of Stata but are of great use to Stata programmers. In fact, the `notes` command described above was implemented using characteristics.

The dataset itself and each variable within the dataset have associated with them a set of characteristics. Characteristics are named and referred to as *varname*`[`*charname*`]`, where *varname* is the name of a variable or `_dta`. The characteristics contain text and are stored with the data in the Stata-format `.dta` dataset, so they are recalled whenever the data are loaded.

How are characteristics used? The [XT] **xt** commands need to know the name of the panel variable, and some of these commands also need to know the name of the time variable. `xtset` is used to specify the panel variable and optionally the time variable. Users need `xtset` their data only once. Stata then remembers this information, even from a different Stata session. Stata does this with characteristics: _dta[iis] contains the name of the panel variable and _dta[tis] contains the name of the time variable. When an xt command is issued, the command checks these characteristics to obtain the panel and time variables' names. If this information is not found, then the data have not previously been xtset and an error message is issued. This use of characteristics is hidden from the user—no mention is made of how the commands remember the identity of the panel variable and the time variable.

As a Stata user, you need understand only how to set and clear a characteristic for the few commands that explicitly reveal their use of characteristics. You set a variable *varname*'s characteristic *charname* to *x* by typing

```
. char varname[charname] x
```

You set the data's characteristic *charname* to be *x* by typing

```
. char _dta[charname] x
```

You clear a characteristic by typing

```
. char varname[charname]
```

where *varname* is either a variable name or _dta. You can clear a characteristic, even if it has never been set.

The most important feature of characteristics is that Stata remembers them from one session to the next; they are saved with the data.

❑ Technical note

Programmers will want to know more. A technical description is found in [P] **char**, but for an overview, you may refer to *varname*'s *charname* characteristic by embedding its name in single quotes and typing `*varname*[*charname*]'; see [U] **18.3.13 Referring to characteristics**.

You can fetch the names of all characteristics associated with *varname* by typing

```
. local macname : char varname[ ]
```

The maximum length of the contents of a characteristic is 8,681 characters for Small Stata and 67,784 characters for Stata/IC, Stata/SE, and Stata/MP. The association of names with characteristics is by convention. If you, as a programmer, wish to create new characteristics for use in your ado-files, do so, but include at least one capital letter in the characteristic name. The current convention reserves all lowercase names for "official" Stata.

❑

12.9 Data Editor and Variables Manager

We have spent most of this chapter writing about data management performed from Stata's command line. However, Stata provides two powerful features in its interface to help you examine and manage your data: the Data Editor and the Variables Manager.

The Data Editor is a spreadsheet-style data editor that allows you to enter new data, edit existing data, safely browse your data in a read-only mode, and perform almost any data-management task you desire in a reproducible manner using a graphical interface. To open the Data Editor, select **Data** > **Data Editor** > **Data Editor (Edit)** or **Data** > **Data Editor** > **Data Editor (Browse)**. See [GS] **6 Using the Data Editor** (GSM, GSU, or GSW) for a tutorial discussion of the Data Editor. See [D] **edit** for technical details.

The Variables Manager is a tool that lists and allows you to manage all the properties of the variables in your data. Variable properties include the name, label, storage type, format, value label, and notes. The Variables Manager allows you to sort and filter your variables, something that you will find to be very useful if you work with datasets having many variables. The Variables Manager also can be used to create varlists for the Command window. To open the Variables Manager, select **Data > Variables Manager**. See [GS] **7 Using the Variables Manager** (GSM, GSU, or GSW) for a tutorial discussion of the Variables Manager.

Both the Data Editor and Variables Manager submit commands to Stata to perform any changes that you request. This lets you see a log of what changes were made, and it also allows you to work interactively while still building a list of commands you can execute later to reproduce your analysis.

12.10 References

Cox, N. J. 2006. Stata tip 33: Sweet sixteen: Hexadecimal formats and precision problems. *Stata Journal* 6: 282–283.

——. 2010a. Stata tip 84: Summing missings. *Stata Journal* 10: 157–159.

——. 2010b. Stata tip 85: Looping over nonintegers. *Stata Journal* 10: 160–163.

Long, J. S. 2009. *The Workflow of Data Analysis Using Stata.* College Station, TX: Stata Press.

Rising, B. 2010. Stata tip 86: The missing() function. *Stata Journal* 10: 303–304.

13 Functions and expressions

Contents

If you have not read [U] **11 Language syntax**, please do so before reading this entry.

13.1 Overview

Examples of expressions include

```
2+2
miles/gallons
myv+2/oth
(myv+2)/oth
ln(income)
age<25 & income>50000
age<25 | income>50000
age==25
name=="M Brown"
fname + " " + lname
substr(name,1,10)
val[_n-1]
L.gnp
```

Expressions like those above are allowed anywhere *exp* appears in a syntax diagram. One example is [D] **generate**:

generate *newvar* = *exp* [*if*] [*in*]

The first *exp* specifies the contents of the new variable, and the optional second expression restricts the subsample over which it is to be defined. Another is [R] **summarize**:

summarize [*varlist*] [*if*] [*in*]

The optional expression restricts the sample over which summary statistics are calculated.

Algebraic and string expressions are specified in a natural way using the standard rules of hierarchy. You may use parentheses freely to force a different order of evaluation.

▷ Example 1

`myv+2/oth` is interpreted as `myv+(2/oth)`. If you wanted to change the order of the evaluation, you could type `(myv+2)/oth`.

◁

13.2 Operators

Stata has four different classes of operators: arithmetic, string, relational, and logical. Each type is discussed below.

13.2.1 Arithmetic operators

The *arithmetic operators* in Stata are + (addition), - (subtraction), * (multiplication), / (division), ^ (raise to a power), and the prefix - (negation). Any arithmetic operation on a missing value or an impossible arithmetic operation (such as division by zero) yields a missing value.

▷ Example 2

The expression -(x+y^(x-y))/(x*y) denotes the formula

$$-\frac{x+y^{x-y}}{x \cdot y}$$

and evaluates to *missing* if x or y is missing or zero.

◁

13.2.2 String operators

The + sign is also used as a string operator for the *concatenation* of two strings. Stata determines by context whether + means addition or concatenation. If + appears between two strings, Stata concatenates them. If + appears between two numeric values, Stata adds them.

▷ Example 3

The expression `"this"+"that"` results in the string `"thisthat"`, whereas the expression `2+3` results in the number `5`. Stata issues the error message "type mismatch" if the arguments on either side of the `+` sign are not of the same type. Thus the expression `2+"this"` is an error, as is `2+"3"`.

The expressions on either side of the `+` can be arbitrarily complex:

```
substr(string(20+2),1,1) + upper(substr("rf",1+1,1))
```

The result of the above expression is the string `"2F"`. See [D] **functions** below for a description of the `substr()`, `string()`, and `upper()` functions.

◁

13.2.3 Relational operators

The *relational operators* are `>` (greater than), `<` (less than), `>=` (greater than or equal), `<=` (less than or equal), `==` (equal), and `!=` (not equal). Observe that the relational operator for equality is a pair of equal signs. This convention distinguishes relational equality from the *=exp* assignment phrase.

❑ Technical note

You may use `~` anywhere `!` would be appropriate to represent the logical operator "not". Thus the not-equal operator may also be written as `~=`.

❑

Relational expressions are either *true* or *false*. Relational operators may be used on either numeric or string subexpressions; thus, the expression `3>2` is *true*, as is `"zebra">"cat"`. In the latter case, the relation merely indicates that `"zebra"` comes after the word `"cat"` in the dictionary. All uppercase letters precede all lowercase letters in Stata's book, so `"cat">"Zebra"` is also *true*.

Missing values may appear in relational expressions. If `x` were a numeric variable, the expression `x>=.` is *true* if `x` is missing and *false* otherwise. A missing value is greater than any nonmissing value; see [U] **12.2.1 Missing values**.

▷ Example 4

You have data on `age` and `income` and wish to list the subset of the data for persons aged 25 years or less. You could type

```
. list if age<=25
```

If you wanted to list the subset of data of persons aged exactly 25, you would type

```
. list if age==25
```

Note the double equal sign. It would be an error to type `list if age=25`.

◁

Although it is convenient to think of relational expressions as evaluating to *true* or *false*, they actually evaluate to numbers. A result of *true* is defined as 1 and *false* is defined as 0.

▷ Example 5

The definition of *true* and *false* makes it easy to create indicator, or dummy, variables. For instance,

```
generate incgt10k=income>10000
```

creates a variable that takes on the value 0 when `income` is less than or equal to $10,000, and 1 when `income` is greater than $10,000. Because missing values are greater than all nonmissing values, the new variable `incgt10k` will also take on the value 1 when `income` is *missing*. It would be safer to type

```
generate incgt10k=income>10000 if income<.
```

Now, observations in which `income` is *missing* will also contain *missing* in `incgt10k`. See [U] **25 Working with categorical data and factor variables** for more examples.

◁

❑ Technical note

Although you will rarely wish to do so, because arithmetic and relational operators both evaluate to numbers, there is no reason you cannot mix the two types of operators in one expression. For instance, `(2==2)+1` evaluates to 2, because `2==2` evaluates to 1, and $1 + 1$ is 2.

Relational operators are evaluated after all arithmetic operations. Thus the expression `(3>2)+1` is equal to 2, whereas `3>2+1` is equal to 0. Evaluating relational operators last guarantees the *logical* (as opposed to the *numeric*) interpretation. It should make sense that `3>2+1` is *false*.

❑

13.2.4 Logical operators

The *logical operators* are `&` (and), `|` (or), and `!` (not). The logical operators interpret any nonzero value (including *missing*) as *true* and zero as *false*.

▷ Example 6

If you have data on `age` and `income` and wish to `list` data for persons making more than $50,000 along with persons under the age of 25 making more than $30,000, you could type

```
list if income>50000 | income>30000 & age<25
```

The `&` takes precedence over the `|`. If you were unsure, however, you could have typed

```
list if income>50000 | (income>30000 & age<25)
```

In either case, the statement will also `list` all observations for which `income` is *missing*, because *missing* is greater than 50,000.

◁

❑ Technical note

Like relational operators, logical operators return 1 for *true* and 0 for *false*. For example, the expression `5 & .` evaluates to 1. Logical operations, except for `!`, are performed after all arithmetic and relational operations; the expression `3>2 & 5>4` is interpreted as `(3>2) & (5>4)` and evaluates to 1.

❑

13.2.5 Order of evaluation, all operators

The order of evaluation (from first to last) of all operators is ! (or ~), ^, - (negation), /, *, - (subtraction), +, != (or ~=), >, <, <=, >=, ==, &, and |.

13.3 Functions

Stata provides mathematical functions, probability and density functions, matrix functions, string functions, functions for dealing with dates and time series, and a set of special functions for programmers. You can find all these documented in [D] **functions**. Stata's matrix programming language, Mata, provides more functions and those are documented in the *Mata Reference Manual*, or online (type `help mata functions`).

Functions are merely a set of rules; you supply the function with arguments, and the function evaluates the arguments according to the rules that define the function. Because functions are essentially subroutines that evaluate arguments and cause no action on their own, functions must be used in conjunction with a Stata command. Functions are indicated by the function name, an open parenthesis, an expression or expressions separated by commas, and a close parenthesis.

For example,

```
. display sqrt(4)
2
```

or

```
. display sqrt(2+2)
2
```

demonstrates the simplest use of a function. Here we have used the mathematical function, `sqrt()`, which takes one number (or expression) as its argument and returns its square root. The function was used with the Stata command `display`. If we had simply typed

```
. sqrt(4)
```

Stata would have returned the error message

```
unrecognized command:  sqrt
r(199);
```

Functions can operate on variables, as well. For example, suppose that you wanted to generate a random variable that has observations drawn from a lognormal distribution. You could type

```
. set obs 5
obs was 0, now 5
. generate y = runiform()
. replace y = invnormal(y)
(5 real changes made)
. replace y = exp(y)
(5 real changes made)
. list

         y
  1.   .686471
  2.  2.380994
  3.  .2814537
  4.  1.215575
  5.  .2920268
```

You could have saved yourself some typing by typing just

```
. generate y = exp(rnormal())
```

Functions accept expressions as arguments.

All functions are defined over a specified domain and return values within a specified range. Whenever an argument is outside a function's domain, the function will return a missing value or issue an error message, whichever is most appropriate. For example, if you supplied the `log()` function with an argument of zero, the `log(0)` would return a missing value because zero is outside the natural logarithm function's domain. If you supplied the `log()` function with a string argument, Stata would issue a "type mismatch" error because `log()` is a numerical function and is undefined for strings. If you supply an argument that evaluates to a value that is outside the function's range, the function will return a missing value. Whenever a function accepts a string as an argument, the string must be enclosed in double quotes, unless you provide the name of a variable that has a string storage type.

13.4 System variables (_variables)

Expressions may also contain *_variables* (pronounced "underscore variables"), which are built-in system variables that are created and updated by Stata. They are called *_variables* because their names all begin with the underscore character, '_'.

The *_variables* are

[*eqno*]`_b`[*varname*] (synonym: [*eqno*]`_coef`[*varname*]) contains the value (to machine precision) of the coefficient on *varname* from the most recently fit model (such as ANOVA, regression, Cox, logit, probit, and multinomial logit). See [U] **13.5 Accessing coefficients and standard errors** below for a complete description.

`_cons` is always equal to the number `1` when used directly and refers to the intercept term when used indirectly, as in `_b[_cons]`.

`_n` contains the number of the current observation.

`_N` contains the total number of observations in the dataset or the number of observations in the current `by()` group.

`_rc` contains the value of the return code from the most recent `capture` command.

[*eqno*]`_se`[*varname*] contains the value (to machine precision) of the standard error of the coefficient on *varname* from the most recently fit model (such as ANOVA, regression, Cox, logit, probit, and multinomial logit). See [U] **13.5 Accessing coefficients and standard errors** below for a complete description.

13.5 Accessing coefficients and standard errors

After fitting a model, you can access the coefficients and standard errors and use them in subsequent expressions. Also see [R] **predict** (and [U] **20 Estimation and postestimation commands**) for an easier way to obtain predictions, residuals, and the like.

13.5.1 Single-equation models

First, let's consider estimation methods that yield one estimated equation with a one-to-one correspondence between coefficients and variables such as logit, ologit, oprobit, probit, regress, and tobit. _b[*varname*] (synonym _coef[*varname*]) contains the coefficient on *varname* and _se[*varname*] contains its standard error, and both are recorded to machine precision. Thus _b[age] refers to the calculated coefficient on the age variable after typing, say, regress response age sex, and _se[age] refers to the standard error on the coefficient. _b[_cons] refers to the constant and _se[_cons] to its standard error. Thus you might type

```
. regress response age sex
. generate asif = _b[_cons] + _b[age]*age
```

13.5.2 Multiple-equation models

The syntax for referring to coefficients and standard errors in multiple-equation models is the same as in the simple-model case, except that _b[] and _se[] are preceded by an equation number in square brackets. There are, however, many alternatives in how you may type requests. The way that you are supposed to type requests is

[*eqno*]_b[*varname*]
[*eqno*]_se[*varname*]

but you may substitute _coef[] for _b[]. In fact, you may omit the _b[] altogether, and most Stata users do:

[*eqno*][*varname*]

You may also omit the second pair of square brackets:

[*eqno*]*varname*

You may retain the _b[] or _se[] and insert a colon between *eqno* and *varname*:

_b[*eqno*:*varname*]

There are two ways to specify the equation number *eqno*: either as an absolute equation number or as an "indirect" equation number. In the absolute form, the number is preceded by a '#' sign. Thus [#1]displ refers to the coefficient on displ in the first equation (and [#1]_se[displ] refers to its standard error). You can even use this form for simple models, such as regress, if you prefer. regress estimates one equation, so [#1]displ refers to the coefficient on displ, just as _b[displ] does. Similarly, [#1]_se[displ] and _se[displ] are equivalent. The logic works both ways—in the multiple-equation context, _b[displ] refers to the coefficient on displ in the first equation and _se[displ] refers to its standard error. _b[*varname*] (_se[*varname*]) is just another way of saying [#1]*varname* ([#1]_se[*varname*]).

Equations may also be referred to indirectly. [res]displ refers to the coefficient on displ in the equation named res. Equations are often named after the corresponding dependent variable name if there is such a concept in the fitted model, so [res]displ might refer to the coefficient on displ in the equation for variable res.

For multinomial logit (mlogit), multinomial probit (mprobit), and similar commands, equations are named after the levels of the single dependent categorical variable. In these models, there is one dependent variable, and there is an equation corresponding to each of the outcomes (values taken on) recorded in that variable, except for the one that is taken to be the base outcome. [res]displ would be interpreted as the coefficient on displ in the equation corresponding to the outcome res. If outcome res is the base outcome, Stata treats [res]displ as zero (and Stata does the same for [res]_se[displ]).

Continuing with the multinomial outcome case: the outcome variable must be numeric. The syntax `[res]displ` would be understood only if there were a value label associated with the numeric outcome variable and `res` were one of the labels. If your data are not labeled, then you can use the usual multiple-equation syntax `[##]`*varname* and `[##]_se[`*varname*`]` to refer to the coefficient and standard error for variable *varname* in the #th equation.

For `mlogit`, if your data are not labeled, you can also use the syntax `[#]`*varname* and `[#]_se[`*varname*`]` (without the '`#`') to refer to the coefficient and standard error for *varname* in the equation for outcome #.

13.5.3 Factor variables and time-series operators

We refer to time-series–operated variables exactly as we refer to normal variables. We type the name of the variable, which for time-series–operated variables includes the operators; see [U] **11.4.4 Time-series varlists**. You might type

```
. regress open L.close LD.volume
. display _b[L.close]
. display _b[LD.volume]
```

We cannot refer to factor variables such as `i.group` in expressions. Assuming that `i.group` has three levels, `i.group` represents three virtual indicator variables—`1b.group`, `2.group`, and `3.group`. We can refer to the indicator variables in expressions by typing, for example, `_b[i2.group]` or just `_b[2.group]`. That is to say, we include the operators and the levels of the factor variables when typing the indicator-variable name. Consider a regression using factor variables:

```
. use http://www.stata-press.com/data/r12/fvex, clear
(Artificial factor variables' data)
. regress y i.sex i.group sex#group age sex#c.age
      Source |       SS       df       MS              Number of obs =    3000
-------------+------------------------------           F(  7,  2992) =   80.84
       Model |  221310.507     7  31615.7868           Prob > F      =  0.0000
    Residual |   1170122.5  2992  391.083723           R-squared     =  0.1591
-------------+------------------------------           Adj R-squared =  0.1571
       Total |  1391433.01  2999  463.965657           Root MSE      =  19.776

------------------------------------------------------------------------------
           y |      Coef.   Std. Err.      t    P>|t|     [95% Conf. Interval]
-------------+----------------------------------------------------------------
       1.sex |   32.29378   3.782064     8.54   0.000     24.87807    39.70949
             |
       group |
          2  |   9.477077   1.624075     5.84   0.000     6.292659    12.66149
          3  |   18.31292   1.776337    10.31   0.000     14.82995    21.79588
             |
   sex#group |
        1 2  |  -6.621804   2.021384    -3.28   0.001    -10.58525   -2.658361
        1 3  |  -10.48293      3.209    -3.27   0.001      -16.775   -4.190858
             |
         age |   -.212332   .0538345    -3.94   0.000    -.3178884   -.1067756
             |
   sex#c.age |
          1  |   -.226838   .0745707    -3.04   0.002    -.3730531   -.0806229
             |
       _cons |   60.48167   2.842955    21.27   0.000     54.90732    66.05601
------------------------------------------------------------------------------
```

If we want to use the coefficient for level 2 of `group` in an expression, we type `_b[2.group]`; for level 3, we type `_b[3.group]`. To refer to the coefficient of an interaction of two levels of two factor variables, we specify the interaction operator and the level of each variable. For example, to use the coefficient for `sex` = 1 and `group` = 2, we type `_b[1.sex#2.group]`. When one of the variables in an interaction is continuous, we can either make that explicit, `_b[1.sex#c.age]`, or we can leave off the `c.`, `_b[1.sex#age]`.

Referring to interactions is more challenging than referring to normal variables. It is also more challenging to refer to coefficients from estimators that use multiple equations. If you find it difficult to know what to type for a coefficient, replay your estimation results using the `coeflegend` option.

```
. regress, coeflegend
      Source |       SS       df       MS              Number of obs =    3000
-------------+------------------------------           F(  7,  2992) =   80.84
       Model |  221310.507     7  31615.7868           Prob > F      =  0.0000
    Residual |   1170122.5  2992  391.083723           R-squared     =  0.1591
-------------+------------------------------           Adj R-squared =  0.1571
       Total |  1391433.01  2999  463.965657           Root MSE      =  19.776

------------------------------------------------------------------------------
           y |      Coef.  Legend
-------------+----------------------------------------------------------------
       1.sex |   32.29378  _b[1.sex]
             |
       group |
          2  |   9.477077  _b[2.group]
          3  |   18.31292  _b[3.group]
             |
   sex#group |
        1 2  |  -6.621804  _b[1.sex#2.group]
        1 3  |  -10.48293  _b[1.sex#3.group]
             |
         age |   -.212332  _b[age]
             |
   sex#c.age |
          1  |   -.226838  _b[1.sex#c.age]
             |
       _cons |   60.48167  _b[_cons]
------------------------------------------------------------------------------
```

The `Legend` column shows you exactly what to type to refer to any coefficient in the estimation.

If your estimation results have both equations and factor variables, nothing changes from what we said in [U] **13.5.2 Multiple-equation models** above. What you type for *varname* is just a little more complicated.

13.6 Accessing results from Stata commands

Most Stata commands—not just estimation commands—save results so that you can access them in subsequent expressions. You do that by referring to `e(`*name*`)`, `r(`*name*`)`, `s(`*name*`)`, or `c(`*name*`)`.

```
. summarize age
. generate agedev = age-r(mean)
. regress mpg weight
. display "The number of observations used is " e(N)
```

Most commands are categorized as r-class, meaning that they save results in `r()`. The returned results—such as `r(mean)`—are available immediately following the command, and if you are going to refer to them, you need to refer to them soon because the next command will probably replace what is in `r()`.

e-class commands are Stata's estimation commands—commands that fit models. Results in `e()` remain available until the next model is fit.

s-class commands are parsing commands—commands used by programmers to interpret commands you type. Few commands save anything in `s()`.

There are no c-class commands. `c()` contains values that are always available, such as `c(current_date)` (today's date), `c(pwd)` (the current directory), `C(N)` (the number of observations), and so on. There are many `c()` values and they are documented in [P] **creturn**.

Every command of Stata is designated r-class, e-class, or s-class, or, if the command saves nothing, n-class. r stands for return as in returned results, e stands for estimation as in estimation results, s stands for string, and, admittedly, this last acronym is weak, n stands for null.

You can find out what is stored where by looking in the *Saved results* section for the particular command in the *Reference* manual. If you know the class of a command—and it is easy enough to guess—you can also see what is stored by typing `return list`, `ereturn list`, or `sreturn list`:

See [R] **saved results** and [U] **18.8 Accessing results calculated by other programs**.

13.7 Explicit subscripting

Individual observations on variables can be referred to by subscripting the variables. Explicit subscripts are specified by following a variable name with square brackets that contain an expression. The result of the subscript expression is truncated to an integer, and the value of the variable for the indicated observation is returned. If the value of the subscript expression is less than `1` or greater than `_N`, a missing value is returned.

13.7.1 Generating lags and leads

When you type something like

```
. generate y = x
```

Stata interprets it as if you typed

```
. generate y = x[_n]
```

which means that the first observation of `y` is to be assigned the value from the first observation of `x`, the second observation of `y` is to be assigned the value from the second observation on `x`, and so on. If you instead typed

```
. generate y = x[1]
```

you would set each observation of `y` equal to the first observation on `x`. If you typed

```
. generate y = x[2]
```

you would set each observation of `y` equal to the second observation on `x`. If you typed

```
. generate y = x[0]
```

Stata would merely copy a missing value into every observation of `y` because observation 0 does not exist. The same would happen if you typed

```
. generate y = x[100]
```

and you had fewer than 100 observations in your data.

When you type the square brackets, you are specifying explicit subscripts. Explicit subscripting combined with the _variable _n can be used to create lagged values on a variable. The lagged value of a variable x can be obtained by typing

```
. generate xlag = x[_n-1]
```

If you are really interested in lags and leads, you probably have time-series data and would be better served by using the time-series operators, such as L.x. Time-series operators can be used with varlists and expressions and they are safer because they account for gaps in the data; see [U] **11.4.4 Time-series varlists** and [U] **13.9 Time-series operators**. Even so, it is important that you understand how the above works.

The built-in underscore variable _n is understood by Stata to mean the observation number of the current observation. That is why

```
. generate y = x[_n]
```

results in observation 1 of x being copied to observation 1 of y and similarly for the rest of the observations. Consider

```
. generate xlag = x[_n-1]
```

_n-1 evaluates to the observation number of the previous observation. For the first observation, _n-1 = 0 and therefore xlag[1] is set to missing. For the second observation, _n-1 = 1 and xlag[2] is set to the value of x[1], and so on.

Similarly, the lead of x can be created by

```
. generate xlead = x[_n+1]
```

Here the last observation on the new variable xlead will be *missing* because _n+1 will be greater than _N (_N is the total number of observations in the dataset).

13.7.2 Subscripting within groups

When a command is preceded by the by *varlist*: prefix, subscript expressions and the underscore variables _n and _N are evaluated relative to the subset of the data currently being processed. For example, consider the following (admittedly not very interesting) data:

```
. use http://www.stata-press.com/data/r12/gxmpl6

. list

     +----------------+
     | bvar   oldvar  |
     |----------------|
  1. |    1      1.1  |
  2. |    1      2.1  |
  3. |    1      3.1  |
  4. |    2      4.1  |
  5. |    2      5.1  |
     +----------------+
```

To see how _n, _N, and explicit subscripting work, let's create three new variables demonstrating each and then list their values:

```
. generate small_n = _n

. generate big_n = _N

. generate newvar = oldvar[1]
```

```
. list
```

	bvar	oldvar	small_n	big_n	newvar
1.	1	1.1	1	5	1.1
2.	1	2.1	2	5	1.1
3.	1	3.1	3	5	1.1
4.	2	4.1	4	5	1.1
5.	2	5.1	5	5	1.1

small_n (which is equal to _n) goes from 1 to 5, and big_n (which is equal to _N) is 5. This should not be surprising; there are 5 observations in the data, and _n is supposed to count observations, whereas _N is the total number. newvar, which we defined as oldvar[1], is 1.1. Indeed, we see that the first observation on oldvar is 1.1.

Now, let's repeat those same three steps, only this time preceding each step with the prefix by bvar:. First, we will drop the old values of small_n, big_n, and newvar so that we start fresh:

```
. drop small_n big_n newvar
. by bvar, sort: generate small_n=_n
. by bvar: generate big_n =_N
. by bvar: generate newvar=oldvar[1]
. list
```

	bvar	oldvar	small_n	big_n	newvar
1.	1	1.1	1	3	1.1
2.	1	2.1	2	3	1.1
3.	1	3.1	3	3	1.1
4.	2	4.1	1	2	4.1
5.	2	5.1	2	2	4.1

The results are different. Remember that we claimed that _n and _N are evaluated relative to the subset of data in the by-group. Thus small_n (_n) goes from 1 to 3 for bvar = 1 and from 1 to 2 for bvar = 2. big_n (_N) is 3 for the first group and 2 for the second. Finally, newvar (oldvar[1]) is 1.1 and 4.1.

▷ Example 7

You now know enough to do some amazing things.

Suppose that you have data on individual states and you have another variable in your data called region that divides the states into the four census regions. You have a variable x in your data, and you want to make a new variable called avgx to include in your regressions. This new variable is to take on the average value of x for the region in which the state is located. Thus, for California, you will have the observation on x and the observation on the average value in the region, avgx. Here is how:

```
. by region, sort: generate avgx=sum(x)/_n
. by region: replace avgx=avgx[_N]
```

First, by region, we generate avgx equal to the running sum of x divided by the number of observations so far. The , sort ensures that the data are in region order. We have, in effect, created the running average of x within region. It is the last observation of this running average, the overall average within the region, that interests us. So, by region, we replace every avgx observation in a region with the last observation within the region, avgx[_N].

Here is what we will see when we type these commands:

```
. use http://www.stata-press.com/data/r12/gxmpl7, clear
. by region, sort: generate avgx=sum(x)/_n
. by region: replace avgx=avgx[_N]
(46 real changes made)
```

In our example, there are no missing observations on x. If there had been, we would have obtained the wrong answer. When we created the running average, we typed

```
. by region, sort: generate avgx=sum(x)/_n
```

The problem is not with the sum() function. When sum() encounters a missing, it adds zero to the sum. The problem is with _n. Let's assume that the second observation in the first region has recorded a missing for x. When Stata processes the third observation in that region, it will calculate the sum of two elements (remember that one is missing) and then divide the sum by 3 when it should be divided by 2. There is an easy solution:

```
. by region: generate avgx=sum(x)/sum(x<.)
```

Rather than divide by _n, we divide by the total number of nonmissing observations seen on x so far, namely, the sum(x<.).

If our goal were simply to obtain the mean, we could have more easily accomplished it by typing egen avgx=mean(x), by(region); see [D] **egen**. egen, however, is written in Stata, and the above is how egen's mean() function works. The general principles are worth understanding.

◁

▷ Example 8

You have some patient data recording vital signs at various times during an experiment. The variables include patient, an ID number or name of the patient; time, a variable recording the date or time or epoch of the vital-sign reading; and vital, a vital sign. You probably have more than one vital sign, but one is enough to illustrate the concept. Each observation in your data represents a patient-time combination.

Let's assume that you have 1,000 patients and, for every observation on the same patient, you want to create a new variable called orig that records the patient's initial value of this vital sign.

```
. use http://www.stata-press.com/data/r12/gxmpl8, clear
. sort patient time
. by patient: generate orig=vital[1]
```

Observe that vital[1] refers not to the first reading on the first patient but to the first reading on the current patient, because we are performing the generate command by patient.

◁

▷ Example 9

Let's do one more example with these patient data. Suppose that we want to create a new dataset from our patient data that record not only the patient's identification, the time of the reading of the first vital sign, and the first vital sign reading itself, but also the time of the reading of the last vital sign and its value. We want 1 observation per patient. Here's how:

```
. sort patient time
. by patient: generate lasttime=time[_N]
. by patient: generate lastvital=vital[_N]
. by patient: drop if _n!=1
```

◁

See Mitchell (2010, chap. 7) for numerous examples of subscripting and subscripting within groups

13.8 Indicator values for levels of factor variables

Stata's factor-variable features let us access virtual indicator variables for categorical variables and their interactions; see [U] **11.4.3 Factor variables** and [U] **25 Working with categorical data and factor variables**. We can use those virtual indicator variables in expressions just as though the virtual variables existed in our data. If you have not read about factor-variable varlists in [U] **11.4.3 Factor variables**, do so now.

If `group` is a categorical variable taking on the value 1, 2, or 3, consider the expression

```
. generate group1 = 1.group
```

We have taken the virtual indicator variable that is 1 when `group` = 1 and 0 when `group` ≠ 1 and made it into a real variable—`group1`. That is strictly true only if `group` is never missing. If group can be missing, we need to add that `1.group` is missing when `group` is missing.

These virtual variables extend to interactions. If we also have a variable, `sex`, that is 0 for males and 1 for females, then

```
. generate sex0grp2 = 0.sex#2.group
```

creates the variable `sex0grp2`, which is `1` when `sex` = 0 and `group` = 2, `.` (missing) when `sex` or `group` is missing, and `0` otherwise.

Virtual indicator variables can be used in any expression, including `if` expressions.

❑ Technical note

We have been using the shorthand notation for virtual indicators that drops the `i` prefix. We have written `2.group` rather than `i2.group`. There are three cases where we cannot drop the `i` prefix—when our variable name is `e`, `d`, or `x`. These three letters can be used to construct numbers such as `1e-3`, which can also be typed `1.e-3`. If we have a variable named `e`, are we to interpret `1.e-3` as the number 0.001 or as the virtual indicator variable `1.e` with the number 3 subtracted? Because of longstanding precedent, it is interpreted as the number 0.001. If we want `1.e` interpreted as a virtual indicator, we must include the `i` prefix—`i1.e`.

❑

13.9 Time-series operators

Time-series operators allow you to refer to the lag of gnp by typing `L.gnp`, the second lag by typing `L2.gnp`, etc. There are also operators for lead (`F`), difference `D`, and seasonal difference `S`.

Time-series operators can be used with varlists and with expressions. See [U] **11.4.4 Time-series varlists** if you have not read it already. This section has to do with using time-series operators in expressions such as with `generate`. You do not have to create new variables; you can use the time-series operated variables directly.

13.9.1 Generating lags, leads, and differences

In a time-series context, referring to L2.gnp is better than referring to gnp[_n-2] because there might be missing observations. Pretend that observation 4 contains data for $t = 25$ and observation 5 data for $t = 27$. L2.gnp will still produce correct answers; L2.gnp for observation 5 will be the value from observation 4 because the time-series operators look at t to find the relevant observation. The more mechanical gnp[_n-2] just goes 2 observations back, which, here, would not produce the desired result.

This same idea holds for differences. In our example, D.gnp will produce a missing value in observation 5 ($t = 27$) because there is no data recorded for $t = 26$, and therefore there is no first difference for $t = 27$.

Time-series operators can be used with varlists or with expressions, so you can type

```
. regress val L.gnp r
```

or

```
. generate gnplagged = L.gnp
. regress val gnplagged
```

Before you can type either one, however, you must use the tsset command to tell Stata the identity of the time variable; see [TS] **tsset**. Once you have tsset the data, anyplace you see an *exp* in a syntax diagram, you may type time series–operated variables, so you can type

```
. summarize r if F.gnp < gnp
```

or

```
. generate grew = 1 if gnp > L.gnp & L.gnp < .
. replace grew = 0 if grew >= . & L.gnp < .
```

or

```
. generate grew = (gnp > L.gnp) if L.gnp < .
```

13.9.2 Time-series operators and factor variables

As with varlists, factor variables may be combined with the L. (lag) and F. (lead) time-series operators in expressions. We can generate a variable containing the lag of the level 2 indicator of group (group = 2) by typing

```
. generate lag2group = 2L.group
```

The operators can be combined anywhere expressions are allowed. We can select observations for which the lag of the second level of group is 1 by typing if i2L.group.

They can be combined in interactions. We can generate the lag of the interaction of sex = 1 with group = 3 by typing

```
. generate lag1sexX3grp = 1L.sex#2L.group
```

See [U] **11.4.3 Factor variables** and [U] **11.4.4 Time-series varlists** for more on factor variables and time-series operators.

13.9.3 Operators within groups

Stata also understands panel or cross-sectional time-series data. For instance, if you type

```
. tsset country time
```

you are declaring that you have time-series data. The time variable is `time`, and you have time-series data for separate countries.

Once you have `tsset` both cross-sectional and time identifiers, you proceed just as you would if you had a simple time series.

```
. generate grew = (gnp > L.gnp) if L.gnp < .
```

would produce correct results. The `L.` operator will not confuse the observation at the end of one panel with the beginning of the next.

13.10 Label values

If you have not read [U] **12.6 Dataset, variable, and value labels**, please do so. You may use labels in an expression in place of the numeric values with which they are associated. To use a label in this way, type the label in double quotes followed by a colon and the name of the value label.

▷ Example 10

If the value label `yesno` associates the label `yes` with 1 and `no` with 0, then `"yes":yesno` (said aloud as the value of `yes` under `yesno`) is evaluated as 1. If the double-quoted label is not defined in the indicated value label, or if the value label itself is not found, a missing value is returned. Thus the expression `"maybe":yesno` is evaluated as *missing*.

```
. use http://www.stata-press.com/data/r12/gxmpl9, clear
. list

     +--------------------+
     |      name   answer |
     |--------------------|
  1. |   Mikulin       no |
  2. |    Gaines       no |
  3. |     Hilbe      yes |
  4. |    DeLeon       no |
  5. |      Cain       no |
     |--------------------|
  6. |      Wann      yes |
  7. | Schroeder       no |
  8. |       Cox       no |
  9. |    Bishop       no |
 10. |    Hardin      yes |
     |--------------------|
 11. | Lancaster      yes |
 12. |     Poole       no |
     +--------------------+

. list if answer=="yes":yesno

     +--------------------+
     |      name   answer |
     |--------------------|
  3. |     Hilbe      yes |
  6. |      Wann      yes |
 10. |    Hardin      yes |
 11. | Lancaster      yes |
     +--------------------+
```

In the above example, the variable `answer` is not a string variable; it is a numeric variable that has the associated value label `yesno`. Because `yesno` associates `yes` with 1 and `no` with 0, we could have typed `list if answer==1` instead of what we did type. We could not have typed `list if answer=="yes"` because `answer` is not a string variable. If we had, we would have received the error message "type mismatch".

◁

13.11 Precision and problems therein

Examine the following short Stata session:

```
. drop _all
. input x y
              x          y
  1. 1 1.1
  2. 2 1.2
  3. 3 1.3
  4. end
. count if x==1
    1
. count if y==1.1
    0
. list

     +---------+
     | x     y |
     |---------|
  1. | 1   1.1 |
  2. | 2   1.2 |
  3. | 3   1.3 |
     +---------+
```

We created a dataset containing two variables, `x` and `y`. The first observation has `x` equal to 1 and `y` equal to 1.1. When we asked Stata to `count` the number of times that the variable `x` took on the value 1, we were told that it occurred once. Yet when we asked Stata to `count` the number of times `y` took on the value 1.1, we were told zero—meaning that it never occurred. What has gone wrong? When we `list` the data, we see that the first observation has `y` equal to 1.1.

Despite appearances, Stata has not made a mistake. Stata stores numbers internally in binary form, and the number 1.1 has no exact binary representation—that is, there is no finite string of binary digits that is equal to 1.1.

❑ Technical note

The number 1.1 in binary form is 1.0001100110011 . . . , where the period represents the binary point. The problem binary computers have with storing numbers like 1/10 is much like the problem we base-10 users have in precisely writing 1/11, which is 0.0909090909 . . . .

For detailed information about precision on binary computers and how Stata stores binary floating-point numbers, see Gould (2011).

❑

The number that appears as 1.1 in the listing above is actually 1.1000000238419, which is off by roughly 2 parts in 10^8. Unless we tell Stata otherwise, it stores all numbers as `floats`, which are also known as *single-precision* or *4-byte reals*. On the other hand, Stata performs all internal calculations in `double`, which is also known as *double-precision* or *8-byte reals*. This is what leads to the difficulty.

In the above example, we compared the number 1.1, stored as a `float`, with the number 1.1 stored as a `double`. The double-precision representation of 1.1 is more accurate than the single-precision representation, but it is also different. Those two numbers are not equal.

There are several ways around this problem. The problem with 1.1 apparently not equaling 1.1 would never arise if the storage precision and the precision of the internal calculations were the same. Thus you could store all your data as `doubles`. This takes more computer memory, however, and it is unlikely that your data are really that accurate and the extra digits would meaningfully affect any calculated result, even if the data were that accurate.

❑ Technical note

This is unlikely to affect any calculated result because Stata performs all internal calculations in double precision. This is all rather ironic, because the problem would also not arise if we had designed Stata to use single precision for its internal calculations. Stata would be less accurate, but the problem would have been completely disguised from the user, making this entry unnecessary. ❑

Another solution is to use the `float()` function. `float(`*x*`)` rounds *x* to its `float` representation. If we had typed `count if y==float(1.1)` in the above example, we would have been informed that there is one such value.

13.12 References

Cox, N. J. 2006. Stata tip 33: Sweet sixteen: Hexadecimal formats and precision problems. *Stata Journal* 6: 282–283.

——. 2011. Stata tip 96: Cube roots. *Stata Journal* 11: 149–154.

Gould, W. W. 2006. Mata Matters: Precision. *Stata Journal* 6: 550–560.

——. 2011. How to read the %21x format, part 2. The Stata Blog: Not Elsewhere Classified. http://blog.stata.com/2011/02/02/how-to-read-the-percent-21x-format-part-2/

Linhart, J. M. 2008. Mata Matters: Overflow, underflow and the IEEE floating-point format. *Stata Journal* 8: 255–268.

Mitchell, M. N. 2010. *Data Management Using Stata: A Practical Handbook*. College Station, TX: Stata Press.

Weiss, M. 2009. Stata tip 80: Constructing a group variable with specified group sizes. *Stata Journal* 9: 640–642.

14 Matrix expressions

Contents

14.1 Overview

Stata has two matrix programming languages, one that might be called Stata's older matrix language and another that is called Mata. Stata's Mata is the new one, and there is an uneasy relationship between the two.

Below we discuss Stata's older language and leave the newer one to another manual—the *Mata Reference Manual* ([M])—or you can learn about the newer one by typing `help mata`.

We admit that the newer language is better in almost every way than the older language, but the older one still has a use because it is the one that Stata truly and deeply understands. Even when Mata wants to talk to Stata, matrixwise, it is the older language that Mata must use, so you must learn to use the older language as well as the new.

This is not nearly as difficult, or messy, as you might imagine because Stata's older language is remarkably easy to use, and really, there is not much to learn. Just remember that for heavy-duty programming, it will be worth your time to learn Mata, too.

14.1.1 Definition of a matrix

Stata's definition of a matrix includes a few details that go beyond the mathematics. To Stata, a matrix is a named entity containing an $r \times c$ ($0 < r \leq$ `matsize`, $0 < c \leq$ `matsize`) rectangular array of double-precision numbers (including missing values) that is bordered by a row and a column of names.

```
. matrix list A

A[3,2]
    c1  c2
r1   1   2
r2   3   4
r3   5   6
```

Here we have a 3×2 matrix named A containing elements 1, 2, 3, 4, 5, and 6. Row 1, column 2 (written $A_{1,2}$ in math and A[1,2] in Stata) contains 2. The columns are named c1 and c2 and the rows, r1, r2, and r3. These are the default names Stata comes up with when it cannot do better. The names do not play a role in the mathematics, but they are of great help when it comes to labeling the output.

The names are operated on just as the numbers are. For instance,

```
. matrix B=A'*A

. matrix list B

symmetric B[2,2]
    c1  c2
c1  35
c2  44  56
```

We defined $\mathbf{B} = \mathbf{A}'\mathbf{A}$. The row and column names of $\mathbf{B}$ are the same. Multiplication is defined for any $a \times b$ and $b \times c$ matrices, the result being $a \times c$. Thus the row and column names of the result are the row names of the first matrix and the column names of the second matrix. We formed $\mathbf{A}'\mathbf{A}$, using the transpose of $\mathbf{A}$ for the first matrix—which also interchanged the names—and so obtained the names shown.

14.1.2 matsize

Matrices are limited to being no larger than matsize $\times$ matsize. The default value of matsize is 400 for Stata/MP, Stata/SE, and Stata/IC, but you can reset this with the set matsize command; see [R] **matsize**.

The maximum value of matsize is 800 for Stata/IC, so matrices are not suitable for holding many data. This restriction does not prove a limitation because terms that appear in statistical formulas are of the form $(\mathbf{X}'\mathbf{W}\mathbf{Z})$ and Stata provides a command, matrix accum, for efficiently forming such matrices; see [U] **14.6 Creating matrices by accumulating data** below. The maximum value of matsize is 11,000 for Stata/MP and Stata/SE, so performing matrix operations directly on many data is more feasible. The matsize limit does not apply to Mata matrices; see the *Mata Reference Manual*.

14.2 Row and column names

Matrix rows and columns always have names. Stata is smart about setting these names when the matrix is created, and the matrix commands and operators manipulate these names throughout calculations, so the names typically are set correctly at the conclusion of matrix calculations.

For instance, consider the matrix calculation $\mathbf{b} = (\mathbf{X}'\mathbf{X})^{-1}\mathbf{X}'\mathbf{y}$ performed on real data:

```
. use http://www.stata-press.com/data/r12/auto
(1978 Automobile Data)

. matrix accum XprimeX = weight foreign
(obs=74)

. matrix vecaccum yprimeX = mpg weight foreign

. matrix b = invsym(XprimeX)*yprimeX'

. matrix list b

b[3,1]
               mpg
 weight  -.00658789
foreign  -1.6500291
  _cons   41.679702
```

These names were produced without our ever having given a special command to place the names on the result. When we formed matrix XprimeX, Stata produced the result

```
. matrix list XprimeX

symmetric XprimeX[3,3]
             weight    foreign      _cons
 weight  7.188e+08
foreign      50950         22
  _cons     223440         22         74
```

matrix accum forms $\mathbf{X}'\mathbf{X}$ matrices from data and sets the row and column names to the variable names used. The names are correct in the sense that, for instance, the (1,1) element is the sum across the observations of squares of weight and the (2,1) element is the sum of the product of weight and foreign.

Similarly, matrix vecaccum forms $\mathbf{y}'\mathbf{X}$ matrices, and it sets the row and column names to the variable names used, so matrix vecaccum yprimeX = mpg weight foreign resulted in

```
. matrix list yprimeX

yprimeX[1,3]
      weight  foreign    _cons
mpg  4493720      545     1576
```

The final step, matrix b = invsym(XprimeX)*yprimeX', manipulated the names, and, if you think carefully, you can derive the rules for yourself. invsym() (inversion) is much like transposition, so row and column names must be swapped. Here, however, the matrix was symmetric, so that amounted to leaving the names as they were. Multiplication amounts to taking the column names of the first matrix and the row names of the second. The final result is

```
. matrix list b

b[3,1]
               mpg
 weight  -.00658789
foreign  -1.6500291
  _cons   41.679702
```

and the interpretation is $\texttt{mpg} = -0.00659\,\texttt{weight} - 1.65\,\texttt{foreign} + 41.68 + e$.

Researchers realized long ago that using matrix notation simplifies the description of complex calculations. What they may not have realized is that, corresponding to each mathematical definition of a matrix operator, there is a definition of the operator's effect on the names that can be used to carry the names forward through long and complex matrix calculations.

14.2.1 The purpose of row and column names

Mostly, matrices in Stata are used in programming estimators, and Stata uses row and column names to produce pretty output. Say that we wrote code—interactively or in a program—that produced the following coefficient vector b and covariance matrix V:

```
. matrix list b

b[1,3]
         weight  displacement         _cons
y1   -.00656711     .00528078     40.084522

. matrix list V

symmetric V[3,3]
                   weight  displacement          _cons
      weight    1.360e-06
displacement    -.0000103     .00009741
       _cons   -.00207455     .01188356      4.0808455
```

We could now produce standard estimation output by coding two more lines:

```
. ereturn post b V
. ereturn display
```

	Coef.	Std. Err.	z	P>|z|	[95% Conf. Interval]	
weight	-.0065671	.0011662	-5.63	0.000	-.0088529	-.0042813
displacement	.0052808	.0098696	0.54	0.593	-.0140632	.0246248
_cons	40.08452	2.02011	19.84	0.000	36.12518	44.04387

Stata's `ereturn` command knew to produce this output because of the row and column names on the coefficient vector and variance matrix. Moreover, we usually do nothing special in our code that produces `b` and `V` to set the row and column names because, given how matrix names work, they work themselves out.

Also, sometimes row and column names help us detect programming errors. Assume that we wrote code to produce matrices `b` and `V` but made a mistake. Sometimes our mistake will result in the wrong row and column names. Rather than the `b` vector we previously showed you, we might produce

```
. matrix list b
b[1,3]
         weight           c2        _cons
y1  -.00656711        42.23    40.084522
```

If we `posted` our estimation results now, Stata would refuse because it can tell by the names that there is a problem:

```
. ereturn post b V
name conflict
r(507);
```

Understand, however, that Stata follows the standard rules of matrix algebra; the names are just along for the ride. Matrices are summed by position, meaning that a directive to form $\mathbf{C} = \mathbf{A} + \mathbf{B}$ results in $C_{11} = A_{11} + B_{11}$, regardless of the names, and it is not an error to sum matrices with different names:

```
. matrix list a
symmetric a[3,3]
              c1          c2          c3
   mpg     14419
weight   1221120   1.219e+08
 _cons       545       50950          22
. matrix list b
symmetric b[3,3]
                   c1          c2          c3
displacement  3211055
         mpg   227102       22249
       _cons    12153        1041          52
. matrix c = a + b
. matrix list c
symmetric c[3,3]
                      c1           c2           c3
displacement     3225474
         mpg     1448222    1.219e+08
       _cons       12698        51991           74
```

Matrix row and column names are used to label output; they do not affect how matrix algebra is performed.

14.2.2 Two-part names

Row and column names have two parts separated by a colon: *equation_name*:*opvarname*.

In the examples shown so far, the *equation_name* has been blank and the *opvarnames* have been simple variable names without factor-variable or time-series operators. A blank *equation_name* is typical. Run any single-equation model (such as `regress`, `probit`, or `logistic`), and if you fetch the resulting matrices, you will find that they have row and column names that use only *opvarnames*.

Those who work with time-series data will find matrices with row and column names of the form *opvarname*. For time-series variables, *opvarname* is the variable name prefixed by a time-series operator such as `L.`, `D.`, or `L2D.`; see [U] **11.4.4 Time-series varlists**. For example,

```
. matrix list example1

symmetric example1[3,3]
                           L.
               rate      rate      _cons
   rate   3.0952534
 L.rate    .0096504  .00007742
  _cons  -2.8413483 -.01821928  4.8578916
```

We obtained this matrix by running a linear regression on `rate` and `L.rate` and then fetching the covariance matrix. Think of the row and column name `L.rate` no differently from how you think of `rate` or, in the previous examples, `r1`, `r2`, `c1`, `c2`, `weight`, and `foreign`.

Those who work with factor variables will also find row and column names of the *opvarname* form. For factor variables, *opvarname* is any factor-variable construct that references a single virtual indicator variable. For example, `3.group` refers to the virtual variable that is `1` when `group` = 3 and is `0` otherwise, `1.sex#3.group` refers to the virtual variable that is `1` when `sex` = 1 and `group` = 3 and is `0` otherwise, and `1.sex#c.age` refers to the virtual variable that takes on the values of `age` when `sex` = 1 and is `0` otherwise. For example,

```
. matrix list example2

symmetric example2[5,5]
                      0b.          1.     0b.sex#      1.sex#
                      sex         sex       c.age       c.age       _cons
      0b.sex            0
       1.sex            0   7.7785864
     0b.sex#
       c.age            0   .08350827   .00231307
 1.sex#c.age            0  -.09705697  -1.977e-16   .00223195
       _cons            0  -3.2868185  -.08350827   7.688e-15   3.2868185
```

`1.sex#c.age` is a row name and column name just like `rate` or `L.rate` in the prior example. For details on factor variables and valid factor-variable constructs see [U] **11.4.3 Factor variables**, [U] **25 Working with categorical data and factor variables**, [U] **13.8 Indicator values for levels of factor variables**, and [U] **20.11 Accessing estimated coefficients**.

Factor-variable operators may be combined with the time-series operators `L.` and `F.`, leading to *opvarnames* such as `1L.sex` (the first lag of the level 1 indicator of `sex`) and `3L2.group` (the second lag of the level 3 indicator of `group`).

Equation names are used to label partitioned matrices and, in estimation, occur in the context of multiple equations. Here is a matrix with *equation_names* and simple (unoperated) *opvarnames*.

```
. matrix list example3

symmetric example2[5,5]
                     mpg:        mpg:        mpg:        mpg:        mpg:
                  foreign       displ       _cons     foreign       _cons
   mpg:foreign  1.6483972
     mpg:displ    .004747   .00003876
     mpg:_cons -1.4266352  -.00905773   2.4341021
weight:foreign -51.208454  -4.665e-19   15.224135   24997.727
  weight:_cons  15.224135   2.077e-17  -15.224135  -7431.7565   7431.7565
```

Here is an example with *equation_names* and operated variable names:

```
. matrix list example4

symmetric example3[5,5]
                     val:        val:        val:     weight:     weight:
                                   L.
                     rate        rate       _cons     foreign       _cons
      val:rate  2.2947268
    val:L.rate   .00385216   .0000309
     val:_cons -1.4533912   -.0072726   2.2583357
weight:foreign -163.86684   7.796e-17   49.384526   25351.696
  weight:_cons  49.384526  -1.566e-16  -49.384526   -7640.237    7640.237
```

`val:L.rate` is a column name, just as, in the previous section, `c2` and `foreign` were column names.

Say that this last matrix is the variance matrix produced by a program we wrote and that our program also produced a coefficient vector, `b`:

```
. matrix list b

b[1,5]
          val:        val:        val:     weight:     weight:
                        L.
          rate        rate       _cons     foreign       _cons
y1   4.5366753  -.00316923    20.68421  -1008.7968   3324.7059
```

Here is the result of posting and displaying the results:

```
. ereturn post b example4

. ereturn display
```

	Coef.	Std. Err.	z	P>|z|	[95% Conf.	Interval]
val						
rate						
—	4.536675	1.514836	2.995	0.003	1.567652	7.505698
L1	-.0031692	.0055591	-0.570	0.569	-.0140648	.0077264
_cons	20.68421	1.502776	13.764	0.000	17.73882	23.6296
weight						
foreign	-1008.797	159.2222	-6.336	0.000	-1320.866	-696.7271
_cons	3324.706	87.40845	38.036	0.000	3153.388	3496.023

We have been using `matrix list` to see the row and column names on our matrices because `matrix list` works on all matrices. There is a better way to see the names when we are working with estimation results because estimation results have the same names on the rows and columns of the variance matrix, and those same names are also the column names for the coefficient vector. That better way is the `coeflegend` display option available on almost every estimation command. For example,

```
. sureg (y = sex##group) (distance = d.age il2.sex)
  (output omitted)

. sureg, coeflegend

Seemingly unrelated regression

Equation             Obs  Parms        RMSE    "R-sq"       chi2        P

y                   2998      5    20.03657    0.1343     464.08   0.0000
distance            2998      2    181.3797    0.0005       0.92   0.6314


                   Coef.  Legend

y
       1.sex    21.59726  _b[y:1.sex]

       group
           2    11.42832  _b[y:2.group]
           3     21.6461  _b[y:3.group]

   sex#group
         1 2   -4.892653  _b[y:1.sex#2.group]
         1 3   -6.220653  _b[y:1.sex#3.group]

       _cons     50.5957  _b[y:_cons]

distance
         age
         D1.    .2230927  _b[distance:D.age]

      L2.sex
           1    1.300898  _b[distance:1L2.sex]
       _cons    57.96172  _b[distance:_cons]
```

We could have used `matrix list e(V)` or `matrix list e(b)` to see the names, but the limited space available to `matrix list` to write the names would have made the names more difficult to read. With `coeflegend`, the names are neatly arrayed in their own `Legend` column. One difference between `matrix list` and the `coeflegend` option is that `coeflegend` brackets the names with `_b[]`. That is because `coeflegend`'s primary use is to show us how to type coefficients in expressions and postestimation commands; see [U] **13.5 Accessing coefficients and standard errors** and [U] **20.11 Accessing estimated coefficients**. There the `_b[]` is required.

14.2.3 Setting row and column names

You reset row and column names by using the `matrix rownames` and `matrix colnames` commands.

Before resetting the names, use `matrix list` to verify that the names are not set correctly; often, they already are. When you enter a matrix by hand, however, the row names are unimaginatively set to `r1`, `r2`, ..., and the column names to `c1`, `c2`, ...

```
. matrix a = (1,2,3\4,5,6)

. matrix list a

a[2,3]
    c1  c2  c3
r1   1   2   3
r2   4   5   6
```

Regardless of the current row and column names, `matrix rownames` and `matrix colnames` reset them:

```
. matrix colnames a = foreign alpha _cons
. matrix rownames a = one two
. matrix list a
a[2,3]
     foreign    alpha    _cons
one        1        2        3
two        4        5        6
```

You may set the *operator* as part of the *opvarname*,

```
. matrix colnames a = foreign l.rate _cons
. matrix list a
a[2,3]
                    L.
     foreign     rate    _cons
one        1        2        3
two        4        5        6
```

The names you specify may be any virtual factor-variable indicators, and those names may include the base (`b.`) and omitted (`o.`) operators,

```
. matrix colnames b = 0b.sex 2o.arm 1.sex#c.age 1.sex#3.group#2.arm
. matrix list b
b[2,4]
                                         1.sex#
             0b.        2o.    1.sex#  3.group#
             sex        arm     c.age     2.arm
one            1          2         3         3
two            5          6         7         8
```

See [U] **11.4.3 Factor variables** for more about factor-variable operators.

You may set equation names:

```
. matrix colnames a = this:foreign this:l.rate that:_cons
. matrix list a
a[2,3]
        this:     this:     that:
                    L.
     foreign     rate    _cons
one        1        2        3
two        4        5        6
```

See [P] **matrix rownames** for more information.

14.2.4 Obtaining row and column names

`matrix list` displays the matrix with its row and column names. In a programming context, you can fetch the row and column names into a macro using

`local` ... : `rowfullnames` *matname*
`local` ... : `colfullnames` *matname*
`local` ... : `rownames` *matname*
`local` ... : `colnames` *matname*
`local` ... : `roweq` *matname*
`local` ... : `coleq` *matname*

`rowfullnames` and `colfullnames` return the full names (*equation_name*:*opvarnames*) listed one after the other.

`rownames` and `colnames` omit the equations and return *opvarnames*, listed one after the other.

`roweq` and `coleq` return the equation names, listed one after the other.

See [P] **macro** and [P] **matrix define** for more information.

14.3 Vectors and scalars

Stata does not have vectors as such—they are considered special cases of matrices and are handled by the `matrix` command.

Stata does have scalars, although they are not strictly necessary because they, too, could be handled as special cases. See [P] **scalar** for a description of scalars.

14.4 Inputting matrices by hand

You input matrices using

`matrix input` *matname* `= (...)`

or

`matrix` *matname* `= (...)`

In either case, you enter the matrices by row. You separate one element from the next by using commas (`,`) and one row from the next by using backslashes (\). If you omit the word `input`, you are using the expression parser to input the matrix:

```
. matrix a = (1,2\3,4)

. matrix list a

a[2,2]
    c1  c2
r1   1   2
r2   3   4
```

This has the advantage that you can use expressions for any of the elements:

```
. matrix b = (1, 2+3/2 \ cos(_pi), _pi)

. matrix list b

b[2,2]
            c1          c2
r1           1         3.5
r2          -1   3.1415927
```

The disadvantage is that the matrix must be small, say, no more than 50 elements (regardless of the value of `matsize`).

`matrix input` has no such restriction, but you may not use subexpressions for the elements:

```
. matrix input c  = (1,2\3,4)

. matrix input d = (1, 2+3/2 \ cos(_pi), _pi)
invalid syntax
r(198);
```

Either way, after inputting the matrix, you will probably want to set the row and column names; see [U] **14.2.3 Setting row and column names** above.

For small matrices, you may prefer entering them in a dialog box. Launch the dialog box from the menu **Data > Matrices, ado language > Input matrix by hand**, or by typing `db matrix_input`. The dialog box is particularly convenient for small symmetric matrices.

14.5 Accessing matrices created by Stata commands

Some Stata commands—including all estimation commands—leave behind matrices that you can subsequently use. After executing an estimation command, type `ereturn list` to see what is available:

```
. use http://www.stata-press.com/data/r12/auto
(1978 Automobile Data)
. probit foreign mpg weight
  (output omitted)
. ereturn list
scalars:
                  e(rank) =  3
                     e(N) =  74
                    e(ic) =  5
                     e(k) =  3
                  e(k_eq) =  1
                  e(k_dv) =  1
             e(converged) =  1
                    e(rc) =  0
                    e(ll) =  -26.84418900579868
            e(k_eq_model) =  1
                  e(ll_0) =  -45.03320955699139
                  e(df_m) =  2
                  e(chi2) =  36.37804110238542
                     e(p) =  1.26069126402e-08
                 e(N_cdf) =  0
                 e(N_cds) =  0
                  e(r2_p) =  .4039023807124773
macros:
               e(cmdline) : "probit foreign mpg weight"
                   e(cmd) : "probit"
             e(estat_cmd) : "probit_estat"
               e(predict) : "probit_p"
                 e(title) : "Probit regression"
              e(chi2type) : "LR"
                   e(opt) : "moptimize"
                   e(vce) : "oim"
                  e(user) : "mopt__probit_d2()"
             e(ml_method) : "d2"
             e(technique) : "nr"
                 e(which) : "max"
                e(depvar) : "foreign"
            e(properties) : "b V"
```

```
matrices:
                  e(b) :  1 x 3
                  e(V) :  3 x 3
                e(mns) :  1 x 3
              e(rules) :  1 x 4
               e(ilog) :  1 x 20
           e(gradient) :  1 x 3
functions:
             e(sample)
```

Most estimation commands leave behind `e(b)` (the coefficient vector) and `e(V)` (the variance–covariance matrix of the estimator):

```
. matrix list e(b)
e(b)[1,3]
       foreign:    foreign:    foreign:
           mpg      weight       _cons
y1  -.10395033  -.00233554    8.275464
```

You can refer to `e(b)` and `e(V)` in any matrix expression:

```
. matrix myb = e(b)
. matrix list myb
myb[1,3]
       foreign:    foreign:    foreign:
           mpg      weight       _cons
y1  -.10395033  -.00233554    8.275464
. matrix c = e(b)*invsym(e(V))*e(b)'
. matrix list c
symmetric c[1,1]
           y1
y1  22.440542
```

14.6 Creating matrices by accumulating data

In programming estimators, matrices of the form $\mathbf{X}'\mathbf{X}$, $\mathbf{X}'\mathbf{Z}$, $\mathbf{X}'\mathbf{W}\mathbf{X}$, and $\mathbf{X}'\mathbf{W}\mathbf{Z}$ often occur, where $\mathbf{X}$ and $\mathbf{Z}$ are data matrices. `matrix accum`, `matrix glsaccum`, `matrix vecaccum`, and `matrix opaccum` produce such matrices; see [P] **matrix accum**.

We recommend that you not load the data into a matrix and use the expression parser directly to form such matrices, although see [P] **matrix mkmat** if that is your interest. If that is your interest, be sure to read the technical note at the end of [P] **matrix mkmat**. There is much to recommend learning how to use the `matrix accum` commands.

14.7 Matrix operators

You can create new matrices or replace existing matrices by typing

`matrix` *matname* = *matrix_expression*

For instance,

```
. matrix A = invsym(R*V*R')
. matrix IAR = I(rowsof(A)) - A*R
. matrix beta = b*IAR' + r*A'
. matrix C = -C'
. matrix D = (A, B \ B', A)
. matrix E = (A+B)*C'
. matrix S = (S+S')/2
```

The following operators are provided:

Operator	Symbol
Unary operators	
negation	-
transposition	'
Binary operators	
(lowest precedence)	
row join	\
column join	,
addition	+
subtraction	-
multiplication	*
division by scalar	/
Kronecker product	#
(highest precedence)	

Parentheses may be used to change the order of evaluation.

Note in particular that `,` and `\` are operators; `(1,2)` creates a 1×2 matrix (vector), and `(A,B)` creates a `rowsof(A)` $\times$ `colsof(A)+colsof(B)` matrix, where `rowsof(A)` $=$ `rowsof(B)`. `(1\2)` creates a 2×1 matrix (vector), and `(A\B)` creates a `rowsof(A)+rowsof(B)` $\times$ `colsof(A)` matrix, where `colsof(A)` $=$ `colsof(B)`. Thus expressions of the form

```
matrix R = (A,B)*Vinv*(A,B)'
```

are allowed.

14.8 Matrix functions

In addition to the functions listed below, see [P] **matrix svd** for singular value decomposition, [P] **matrix symeigen** for eigenvalues and eigenvectors of symmetric matrices, and see [P] **matrix eigenvalues** for eigenvalues of nonsymmetric matrices. For a full description of the matrix functions, see [D] **functions**.

Matrix functions returning matrices:

`cholesky(`M`)`	`I(`n`)`	`nullmat(`*matname*`)`
`corr(`M`)`	`inv(`M`)`	`sweep(`M`,`i`)`
`diag(`v`)`	`invsym(`M`)`	`vec(`M`)`
`get(`*systemname*`)`	`J(`r`,`c`,`z`)`	`vecdiag(`M`)`
`hadamard(`M`,`N`)`	`matuniform(`r`,`c`)`	

Matrix functions returning scalars:

`colnumb(`M`,`s`)`	`el(`M`,`i`,`j`)`	`rownumb(`M`,`s`)`
`colsof(`M`)`	`issymmetric(`M`)`	`rowsof(`M`)`
`det(`M`)`	`matmissing(`M`)`	`trace(`M`)`
`diag0cnt(`M`)`	`mreldif(`X`,`Y`)`	

14.9 Subscripting

1. In matrix and scalar expressions, you may refer to *matname*[r,c], where r and c are scalar expressions, to obtain one element of *matname* as a scalar.

   Examples:
   ```
   matrix A = A / A[1,1]
   generate newvar = oldvar / A[2,2]
   ```

2. In matrix expressions, you may refer to *matname*[s_r,s_c], where s_r and s_c are string expressions, to obtain a submatrix with one element. The element returned is based on searching the row and column names.

   Examples:
   ```
   matrix B = V["price","price"]
   generate sdif = dif / sqrt(V["price","price"])
   ```

3. In matrix expressions, you may mix these two syntaxes and refer to *matname*[r,s_c] or to *matname*[s_r,c].

   Example:
   ```
   matrix b = b * R[1,"price"]
   ```

4. In matrix expressions, you may use *matname*[r_1..r_2,c_1..c_2] to refer to submatrices; r_1, r_2, c_1, and c_2 may be scalar expressions. If r_2 evaluates to missing, it is taken as referring to the last row of *matname*; if c_2 evaluates to missing, it is taken as referring to the last column of *matname*. Thus *matname*[r_1...,c_1...] is allowed.

   Examples:
   ```
   matrix S = Z[1..4, 1..4]
   matrix R = Z[5..., 5...]
   ```

5. In matrix expressions, you may refer to *matname*[s_{r1}..s_{r2},s_{c1}..s_{c2}] to refer to submatrices where s_{r1}, s_{r2}, s_{c1}, and s_{c2}, are string expressions. The matrix returned is based on looking up the row and column names.

   If the string evaluates to an equation name only, all the rows or columns for the equation are returned.

   Examples:
   ```
   matrix S = Z["price".."weight", "price".."weight"]
   matrix L = D["mpg:price".."mpg:weight", "mpg:price".."mpg:weight"]
   matrix T1 = C["mpg:", "mpg:"]
   matrix T2 = C["mpg:", "price:"]
   ```

6. In matrix expressions, any of the above syntaxes may be combined.

   Examples:
   ```
   matrix T1 = C["mpg:", "price:weight".."price:displ"]
   matrix T2 = C["mpg:", "price:weight"...]
   matrix T3 = C["mpg:price", 2..5]
   matrix T4 = C["mpg:price", 2]
   ```

7. When defining an element of a matrix, use

   `matrix` *matname*`[`*i*`,`*j*`]` = *expression*

   where i and j are scalar expressions. The matrix *matname* must already exist.

   Example:
   ```
   matrix A = J(2,2,0)
   matrix A[1,2] = sqrt(2)
   ```

8. To replace a submatrix within a matrix, use the same syntax. If the expression on the right evaluates to a scalar or 1×1 matrix, the element is replaced. If it evaluates to a matrix, the submatrix with top-left element at (i, j) is replaced. The matrix *matname* must already exist.

   Example:
   ```
   matrix A = J(4,4,0)
   matrix A[2,2] = C'*C
   ```

14.10 Using matrices in scalar expressions

Scalar expressions are documented as *exp* in the Stata manuals:

`generate` *newvar* = *exp* `if` *exp* ...
`replace` *newvar* = *exp* `if` *exp* ...
`regress` ... `if` *exp* ...
`if` *exp* `{`... `}`
`while` *exp* `{`... `}`

Most importantly, scalar expressions occur in `generate` and `replace`, in the `if` *exp* modifier allowed on the end of many commands, and in the `if` and `while` commands for program control.

You will rarely need to refer to a matrix in any of these situations except when using the `if` and `while` commands.

In any case, you may refer to matrices in any of these situations, but the expression cannot require evaluation of matrix expressions returning matrices. Thus you could refer to `trace(A)` but not to `trace(A+B)`.

It can be difficult to predict when an evaluation of an expression requires evaluating a matrix; even experienced users can be surprised. If you get the error message "matrix operators that return matrices not allowed in this context", r(509), you have encountered such a situation.

The solution is to split the line in two. For instance, you would change

```
if trace(A+B)==0 {
        ...
}
```

to

```
matrix AplusB = A+B
if trace(AplusB)==0 {
        ...
}
```

or even to

```
matrix Trace = trace(A+B)
if Trace[1,1]==0 {
        ...
}
```

15 Saving and printing output—log files

Contents

15.1 Overview

Stata can record your session into a file called a log file but does not start a log automatically; you must tell Stata to record your session. By default, the resulting log file contains what you type and what Stata produces in response, recorded in a format called Stata Markup and Control Language (SMCL); see [P] **smcl**. The file can be printed or converted to plain text for incorporation into documents you create with your word processor.

To start a log:	. `log using` *filename*
Your session is now being recorded in file *filename*`.smcl`.	
To temporarily stop logging:	
Temporarily stop:	. `log off`
Resume:	. `log on`
To stop logging and close the file:	. `log close`
You can now print *filename*`.smcl` or type: to create *filename*`.log` that you can load into your word processor.	. `translate` *filename*`.smcl` *filename*`.log`
You can also create a PDF of *filename*`.smcl` on Windows or Mac:	. `translate` *filename*`.smcl` *filename*`.pdf`
Alternative ways to start logging:	
append to an existing log:	. `log using` *filename*`, append`
replace an existing log:	. `log using` *filename*`, replace`
Using the GUI:	
To start a log:	click on the **Log** button
To temporarily stop logging:	click on the **Log** button, and choose **Suspend**
To resume:	click on the **Log** button, and choose **Resume**
To stop logging and close the file:	click on the **Log** button, and choose **Close**
To print previous or current log:	select **File > View**, choose file, right-click on the Viewer, and select **Print**

Also, `cmdlog` will produce logs containing solely what you typed—logs that, although not containing your results, are sufficient to recreate the session.

To start a command-only log:	. `cmdlog using` *filename*
To stop logging and close the file:	. `cmdlog close`
To recreate your session:	. `do` *filename*`.txt`

15.1.1 Starting and closing logs

With great foresight, you begin working in Stata and type `log using session` (or click on the **Log** button) before starting your work:

```
. log using session
────────────────────────────────────────────────────────────────────────────────
      name:  <unnamed>
       log:  C:\example\session.smcl
  log type:  smcl
 opened on:  17 Mar 2011, 12:35:08

. use http://www.stata-press.com/data/r12/census
(1980 Census data by state)

. tabulate reg [freq=pop]

     Census |
     region |      Freq.     Percent        Cum.
------------+-----------------------------------
         NE | 49,135,283       21.75       21.75
    N Cntrl | 58,865,670       26.06       47.81
      South | 74,734,029       33.08       80.89
       West | 43,172,490       19.11      100.00
------------+-----------------------------------
      Total |225,907,472      100.00

. summarize medage

    Variable |       Obs        Mean    Std. Dev.       Min        Max
-------------+--------------------------------------------------------
      medage |        50       29.54    1.693445       24.2       34.7

. log close
      name:  <unnamed>
       log:  C:\example\session.smcl
  log type:  smcl
 closed on:  17 Mar 2011, 12:35:38
────────────────────────────────────────────────────────────────────────────────
```

There is now a file named `session.smcl` on your disk. If you were to look at it in a text editor or word processor, you would see something like this:

```
{smcl}
{com}{sf}{ul off}{txt}{.-}
      name:  {res}<unnamed>
       {txt}log:  {res}C:\example\session.smcl
  {txt}log type:  {res}smcl
 {txt}opened on:  {res}17 Mar 2011, 12:35:08

{com}. use http://www.stata-press.com/data/r12/census
{txt}(1980 Census data by state)

{com}. tabulate reg [freq=pop]

     {txt}Census {c |}
     region {c |}      Freq.     Percent        Cum.
{hline 12}{c +}{hline 35}
         NE {c |}{res} 49,135,283       21.75       21.75
{txt}    N Cntrl {c |}{res} 58,865,670       26.06       47.81
  (output omitted)
```

What you are seeing is SMCL, which Stata understands. Here is the result of typing the file using Stata's `type` command:

```
. type session.smcl
─────────────────────────────────────────────────────────────────────────────
      name:  <unnamed>
       log:  C:\example\session.smcl
  log type:  smcl
 opened on:  17 Mar 2011, 12:35:08
. use http://www.stata-press.com/data/r12/census
(1980 Census data by state)
. tabulate reg [freq=pop]
     Census │
     region │      Freq.     Percent        Cum.
────────────┼───────────────────────────────────
         NE │ 49,135,283       21.75       21.75
    N Cntrl │ 58,865,670       26.06       47.81
      South │ 74,734,029       33.08       80.89
       West │ 43,172,490       19.11      100.00
────────────┼───────────────────────────────────
      Total │225,907,472      100.00
. summarize medage
    Variable │       Obs        Mean    Std. Dev.       Min        Max
─────────────┼───────────────────────────────────────────────────────
      medage │        50       29.54    1.693445       24.2       34.7
. log close
      name:  <unnamed>
       log:  C:\example\session.smcl
  log type:  smcl
 closed on:  17 Mar 2011, 12:35:38
─────────────────────────────────────────────────────────────────────────────

. _
```

What you will see is a perfect copy of what you previously saw. If you use Stata to print the file, you will get a perfect printed copy, too.

SMCL files can be translated to plain text, which is a format more useful for inclusion into a word processing document. If you type `translate` *filename*`.smcl` *filename*`.log`, Stata will translate *filename*`.smcl` to text and store the result in *filename*`.log`:

```
. translate session.smcl session.log
```

The resulting file `session.log` looks like this:

```
-------------------------------------------------------------------------------
      name:  <unnamed>
       log:  C:\example\session.smcl
  log type:  smcl
 opened on:  17 Mar 2011, 12:35:08
. use http://www.stata-press.com/data/r12/census
(1980 Census data by state)
. tabulate reg [freq=pop]
     Census |
     region |      Freq.     Percent        Cum.
------------+-----------------------------------
         NE | 49,135,283       21.75       21.75
    N Cntrl | 58,865,670       26.06       47.81
      South | 74,734,029       33.08       80.89
  (output omitted)
```

When you use `translate` to create *filename*`.log` from *filename*`.smcl`, *filename*`.log` must not already exist:

```
. translate session.smcl session.log
file session.log already exists
r(602);
```

If the file does already exist and you wish to overwrite the existing copy, you can specify the `replace` option:

```
. translate session.smcl session.log, replace
```

See [R] **translate** for more information.

On Windows and Mac, you can also convert your SMCL file to a PDF to share it more easily with others:

```
. translate session.smcl session.pdf
```

See [R] **translate** for more information.

If you prefer, you can skip the SMCL and create text logs directly, either by specifying that you want the log in `text` format,

```
. log using session, text
```

or by specifying that the file to be created be a `.log` file:

```
. log using session.log
```

15.1.2 Appending to an existing log

Stata never lets you accidentally write over an existing log file. If you have an existing log file and you want to continue logging, you have three choices:

- create a new log file
- append the new log onto the existing log file by typing `log using` *logname*`, append`
- replace the existing log file by typing `log using` *logname*`, replace`

For example, if you have an existing log file named `session.smcl`, you might type

```
. log using session, append
```

to append the new log to the end of the existing log file, `session.smcl`.

15.1.3 Temporarily suspending and resuming logging

Once you have started logging your session, you can turn logging on and off. When you turn logging off, Stata temporarily stops recording your session but leaves the log file open. When you turn logging back on, Stata continues to record your session, appending the additional record to the end of the file.

Say that the first time something interesting happens, you type `log using results` (or click on **Log** and open `results.smcl`). You then retype the command that produced the interesting result (or double-click on the command in the Review window, or use the *PgUp* key to retrieve the command; see [U] **10 Keyboard use**). You now have a copy of the interesting result saved in the log file.

You are now reasonably sure that nothing interesting will occur, at least for a while. Rather than type `log close`, however, you type `log off`, or you click on **Log** and choose **Suspend**. From now on, nothing goes into the file. The next time something interesting happens, you type `log on` (or click on **Log** and choose **Resume**) and reissue the (interesting) command. After that, you type `log off`. You keep working like this—toggling the log on and off.

15.2 Placing comments in logs

Stata treats lines starting with a "*" as comments and ignores them. Thus, if you are working interactively and wish to make a comment, you can type "*" followed by your comment:

```
. * check that all the spells are completed
. _
```

Stata ignores your comment, but if you have a log going the comment now appears in the file.

❑ Technical note

`log` can be combined with `#review` (see [U] **10 Keyboard use**) to bail you out when you have not adequately planned ahead. Say that you have been working in front of your computer, and you now realize that you have done what you wanted to do. Unfortunately, you are not sure exactly what it is you have done. Did you make a mistake? Could you reproduce the result? Unfortunately, you have not been logging your output. Typing `#review` will allow you to look over what commands you have issued, and, combined with `log`, will allow you to make a record. You can also see the commands that you have issued in the Review window. You can save those commands to a file by right-clicking on the Review window and selecting **Save Review Contents...**.

Type `log using` *filename*. Type `#review 100`. Stata will list the last 100 commands you gave, or however many it has stored. Because `log` is making a record, that list will also be stored in the file. Finally, type `log close`.

❑

15.3 Logging only what you type

Log files record everything that happens during a session, both what you type and what Stata produces in response.

Stata can also produce command log files—files that contain only what you type. These files are perfect for later going back and creating a Stata do-file.

`cmdlog` creates command log files, and its basic syntax is

`cmdlog using` *filename* [`, append replace`]	creates *filename*`.txt`
`cmdlog off`	temporarily suspends command logging
`cmdlog on`	resumes command logging
`cmdlog close`	closes the command log file

See [R] **log** for all the details.

Command logs are plain text files. If you typed

```
. cmdlog using session
(cmdlog C:\example\session.txt opened)
. use http://www.stata-press.com/data/r12/census
(Census Data)
. tabulate reg [freq=pop]
  (output omitted)
. summarize medage
  (output omitted)
. cmdlog close
(cmdlog C:\example\session.txt closed)
```

file `mycmds.txt` would contain

```
use http://www.stata-press.com/data/r12/census
tabulate reg [freq=pop]
summarize medage
```

You can create both kinds of logs—full session logs and command logs—simultaneously, if you wish. A command log file can later be used as a do-file; see [R] **do**.

15.4 The log-button alternative

The capabilities of the `log` command (but not the `cmdlog` command) are available from Stata's GUI interface; just click on the **Log** button or select **Log** from the **File** menu.

You can use the Viewer to view logs, even logs that are in the process of being created. Just select **File > View**. If you are currently logging, the filename to view will already be filled in with the current log file, and all you need to do is click on **OK**. Periodically, you can click on the **Refresh** button to bring the Viewer up to date.

You can also use the Viewer to view previous logs.

You can access the Viewer by selecting **File > View**, or you can use the `view` command:

```
. view myoldlog.smcl
```

15.5 Printing logs

You print logs from the Viewer. Select **File > View**, or type `view` *logfilename* from the command line to load the log into the Viewer, and then right-click on the Viewer and select **Print**.

You can also print logs by other means; see [R] **translate**.

15.6 Creating multiple log files simultaneously

Programmers or advanced users may wish to create more than one log file simultaneously. For example, you may wish to create a log file of your whole session but also create a separate log file for part of your session.

You can create multiple logs by using `log`'s `name()` option; see [R] **log**.

16 Do-files

Contents

16.1 Description

Rather than typing commands at the keyboard, you can create a text file containing commands and instruct Stata to execute the commands stored in that file. Such files are called *do-files* because the command that causes them to be executed is `do`.

A do-file is a standard text file that is executed by Stata when you type `do` *filename*. You can use any text editor or the built-in Do-file Editor to create do-files; see [GSW] **13 Using the Do-file Editor—automating Stata**. Using do-files rather than typing commands with the keyboard or using dialog boxes offers several advantages. By writing the steps you take to manage and analyze your data in the form of a do-file, you can reproduce your work later. Also, writing a do-file makes the inevitable debugging process much easier. If you decide to change one part of your analysis, changing the relevant commands in your do-file is much easier than having to start back at square one, as is often necessary when working interactively. In this chapter, we describe the mechanics of do-files. Long (2009) cogently argues that do-files should be used in all research projects and offers an abundance of time-tested advice in how to manage data and statistical analysis.

▷ Example 1

You can use do-files to create a batchlike environment in which you place all the commands you want to perform in a file and then instruct Stata to `do` that file. Assume that you use your text editor or word processor to create a file called `myjob.do` that contains these three lines:

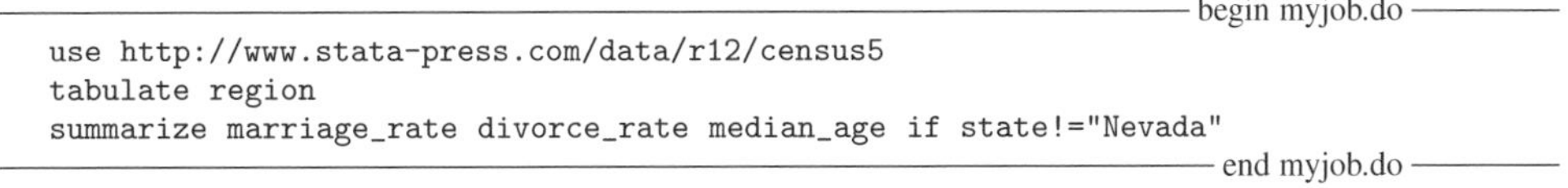

```
——————————————————————————————— begin myjob.do ————
use http://www.stata-press.com/data/r12/census5
tabulate region
summarize marriage_rate divorce_rate median_age if state!="Nevada"
——————————————————————————————— end myjob.do ————
```

You then enter Stata and instruct Stata to do the file:

```
. do myjob

. use http://www.stata-press.com/data/r12/census5
(1980 Census data by state)

. tabulate region

     Census |
     region |      Freq.     Percent        Cum.
------------+-----------------------------------
         NE |          9       18.00       18.00
    N Cntrl |         12       24.00       42.00
      South |         16       32.00       74.00
       West |         13       26.00      100.00
------------+-----------------------------------
      Total |         50      100.00

. summarize marriage_rate divorce_rate median_age if state !="Nevada"

    Variable |       Obs        Mean    Std. Dev.       Min        Max
-------------+--------------------------------------------------------
marriage_r~e |        49    .0106791    .0021746   .0074654   .0172704
divorce_rate |        49    .0054268    .0015104   .0029436    .008752
  median_age |        49    29.52653    1.708286       24.2       34.7
```

You typed only `do myjob` to produce this output. Because you did not specify the file extension, Stata assumed you meant `do myjob.do`; see [U] **11.6 Filenaming conventions**.

◁

16.1.1 Version

We recommend that the first line in your do-file declare the Stata release you used when you wrote the do-file; `myjob.do` would read better as

begin myjob.do

```
version 12
use http://www.stata-press.com/data/r12/census
tabulate region
summarize marriage_rate divorce_rate median_age if state!="Nevada"
```

end myjob.do

We admit that we do not always follow our own advice, as you will see many examples in this manual that do not include the `version 12` line.

If you intend to keep the do-file, however, you should include this line because it ensures that your do-file will continue to work with future versions of Stata. Stata is under continual development, and sometimes things change in surprising ways.

For instance, in Stata 3.0, a new syntax for specifying the weights was introduced. If you had an old do-file written for Stata 2.1 that analyzed weighted data and did not have `version 2.1` at the top, you would find that today's Stata would flag some of the file's lines as syntax errors. If you had the `version 2.1` line, it would work just as it used to.

Skipping ahead to Stata 10, we introduced `xtset` and declared that, to use the `xt` commands, you must `xtset` your data first. Previously, you specified options on the end of each `xt` command that identified the group and, optionally, the time variables. Despite this change, if you include `version 9` or earlier at the top of your do-file, the `xt` commands will continue to work the old way.

For an overview of versioning and an up-to-date list of the issues that versioning does not address automatically, see `help version`.

When running an old do-file that includes a `version` statement, you need not worry about setting the version back after it has completed. Stata automatically restores the previous value of `version` when the do-file completes.

16.1.2 Comments and blank lines in do-files

You may freely include blank lines in your do-file. In the previous example, the do-file could just as well have read

begin myjob.do

```
version 12

use http://www.stata-press.com/data/r12/census
tabulate region
summarize marriage_rate divorce_rate median_age if state!="Nevada"
```

end myjob.do

There are four ways to include comments in a do-file.

1. Begin the line with a '*'; Stata ignores such lines. * cannot be used within Mata.
2. Place the comment in /* */ delimiters.
3. Place the comment after two forward slashes, that is, //. Everything after the // to the end of the current line is considered a comment (unless the // is part of `http://...`).
4. Place the comment after three forward slashes, that is, ///. Everything after the /// to the end of the current line is considered a comment. However, when you use ///, the next line joins with the current line. /// lets you split long lines across multiple lines in the do-file.

❑ Technical note

The /* */, //, and /// comment indicators can be used in do-files and ado-files only; you may not use them interactively. You can, however, use the '*' comment indicator interactively.

❑

`myjob.do` then might read

begin myjob.do

```
* a sample analysis job
version 12

use http://www.stata-press.com/data/r12/census

/* obtain the summary statistics: */
tabulate region
summarize marriage_rate divorce_rate median_age if state!="Nevada"
```

end myjob.do

or equivalently,

begin myjob.do

```
// a sample analysis job
version 12

use http://www.stata-press.com/data/r12/census

// obtain the summary statistics:
tabulate region
summarize marriage_rate divorce_rate median_age if state!="Nevada"
```

end myjob.do

The style of comment indicator you use is up to you. One advantage of the /* */ method is that it can be put at the end of lines:

begin myjob.do

```
* a sample analysis job
version 12
use http://www.stata-press.com/data/r12/census
tabulate region                     /* obtain summary statistics */
summarize marriage_rate divorce_rate median_age if state!="Nevada"
```

end myjob.do

In fact, `/* */` can be put anywhere, even in the middle of a line:

begin myjob.do

```
* a sample analysis job
version 12
use /* confirm this is latest */ http://www.stata-press.com/data/r12/census
tabulate region                     /* obtain summary statistics */
summarize marriage_rate divorce_rate median_age if state!="Nevada"
```

end myjob.do

You can achieve the same results with the `//` and `///` methods:

begin myjob.do

```
// a sample analysis job
version 12
use http://www.stata-press.com/data/r12/census
tabulate region                     // obtain summary statistics
summarize marriage_rate divorce_rate median_age if state!="Nevada"
```

end myjob.do

or

begin myjob.do

```
// a sample analysis job
version 12
use /// confirm this is latest
http://www.stata-press.com/data/r12/census
tabulate region                     // obtain summary statistics
summarize marriage_rate divorce_rate median_age if state!="Nevada"
```

end myjob.do

16.1.3 Long lines in do-files

When you use Stata interactively, you press *Enter* to end a line and tell Stata to execute it. If you need to type a line that is wider than the screen, you simply do it, letting it wrap or scroll.

You can follow the same procedure in do-files—if your editor or word processor will let you—but you can do better. You can change the end-of-line delimiter to ';' by using **#delimit**, you can comment out the line break by using `/* */` comment delimiters, or you can use the `///` line-join indicator.

▷ Example 2

In the following fragment of a do-file, we temporarily change the end-of-line delimiter:

fragment of example.do

```
use mydata
#delimit ;
summarize weight price displ headroom rep78 length turn gear_ratio
          if substr(company,1,4)=="Ford" |
             substr(company,1,2)=="GM", detail ;
gen byte ford = substr(company,1,4)=="Ford" ;
#delimit cr
gen byte gm = substr(company,1,2)=="GM"
```

fragment of example.do

Once we change the line delimiter to semicolon, all lines, even short ones, must end in semicolons. Stata treats carriage returns as no different from blanks. We can change the delimiter back to carriage return by typing `#delimit cr`.

The `#delimit` command is allowed only in do-files—it is not allowed interactively. You need not remember to set the delimiter back to carriage return at the end of a do-file because Stata will reset it automatically.

◁

▷ Example 3

The other way around long lines is to comment out the carriage return by using /* */ comment brackets or to use the /// line-join indicator. Thus our code fragment could also read

fragment of example.do

```
use mydata
summarize weight price displ headroom rep78 length turn gear_ratio /*
      */ if substr(company,1,4)=="Ford" |    /*
      */    substr(company,1,2)=="GM", detail
gen byte ford = substr(company,1,4)=="Ford"
gen byte gm = substr(company,1,2)=="GM"
```

fragment of example.do

or

fragment of example.do

```
use mydata
summarize weight price displ headroom rep78 length turn gear_ratio ///
        if substr(company,1,4)=="Ford" |    ///
           substr(company,1,2)=="GM", detail
gen byte ford = substr(company,1,4)=="Ford"
gen byte gm = substr(company,1,2)=="GM"
```

fragment of example.do

◁

16.1.4 Error handling in do-files

A do-file stops executing when the end of the file is reached, an `exit` is executed, or an error (nonzero *return code*) occurs. If an error occurs, the remaining commands in the do-file are not executed.

If you press *Break* while executing a do-file, Stata responds as though an error has occurred, stopping the do-file. This happens because the return code is nonzero; see [U] **8 Error messages and return codes** for an explanation of return codes.

▷ Example 4

Here is what happens when we execute a do-file and then press *Break*:

```
. do myjob2

. version 12

. use census
(Census data)

. tabulate region

     Census |
     region |      Freq.     Percent        Cum.
—Break—
r(1);

end of do-file
—Break—
r(1);

. _
```

When we pressed *Break*, Stata responded by typing `—Break—` and showed a return code of 1. Stata seemingly repeated itself, typing first "end of do-file", and then `—Break—` and the return code of 1 again. Do not worry about the repeated messages. The first message indicates that Stata was stopping the `tabulate` because you pressed *Break*, and the second message indicates that Stata is stopping the do-file for the same reason.

◁

▷ Example 5

Let's try our example again, but this time, let's introduce an error. We change the file `myjob2.do` to read

begin myjob2.do

```
version 12
use censas
tabulate region
summarize marriage_rate divorce_rate median_age if state!="Nevada"
```

end myjob2.do

To introduce a subtle typographical error, we typed `use censas` when we meant `use census`. We assume that there is no file called `censas.dta`, so now we have an error. Here is what happens when you instruct Stata to `do` the file:

```
. do myjob2
. version 12
. use censas
file censas.dta not found
r(601);
end of do-file
r(601);
. _
```

When Stata was told to use `censas`, it responded with "file censas.dta not found" and a return code of 601. Stata then typed "end of do-file" and repeated the return code of 601. The repeated message occurred for the same reason it did when we pressed *Break* in the previous example. The `use` resulted in a return code of 601, so the do-file itself resulted in the same return code. The important thing to understand is that Stata stopped executing the file because there was an error.

◁

❑ Technical note

We can tell Stata to continue executing the file even if there are errors by typing `do` *filename*`, nostop`. Here is the result:

```
. do myjob2, nostop
. version 12
. use censas
file censas.dta not found
r(601);
. tabulate region
no variables defined
r(111);
summarize marriage_rate divorce_rate median_age if state!="Nevada"
no variables defined
r(111);
end of do-file
. _
```

None of the commands worked because the do-file's first command failed. That is why Stata ordinarily stops. However, if our file had contained anything that could work, it would have worked. In general, we do not recommend coding in this manner, as unintended consequences can result when errors do not stop execution.

❑

16.1.5 Logging the output of do-files

You log the output of do-files just as you would an interactive session; see [U] **15 Saving and printing output—log files**.

Many users include the commands to start and stop the logging in the do-file itself:

begin myjob3.do

```
version 12
log using myjob3, replace
* a sample analysis job
use census
tabulate region                  // obtain summary statistics
summarize marriage_rate divorce_rate median_age if state!="Nevada"
log close
```

end myjob3.do

We chose to open with `log using myjob3, replace`, the important part being the `replace` option. Had we omitted the option, we could not easily rerun our do-file. If `myjob3.smcl` had already existed and `log` was not told that it is okay to replace the file, the do-file would have stopped and instead reported that "file myjob3.smcl already exists". We could get around that, of course, by erasing the log file before running the do-file.

16.1.6 Preventing –more– conditions

Assume that you are running a do-file and logging the output so that you can look at it later. Then Stata's feature of pausing every time the screen is full is just an irritation: it means that you have to sit and watch the do-file run so you can clear the —`more`—.

The way around this is to include the line `set more off` in your do-file. Setting more to `off`, as explained in [U] **7 –more– conditions**, prevents Stata from ever issuing a —`more`—.

16.2 Calling other do-files

Do-files may call other do-files. Say that you wrote `makedata.do`, which infiles your data, generates a few variables, and saves `step1.dta`. Say that you wrote `anlstep1.do`, which performed a little analysis on `step1.dta`. You could then create a third do-file,

begin master.do

```
version 12
do makedata
do anlstep1
```

end master.do

and so in effect combine the two do-files.

Do-files may call other do-files, which, in turn, call other do-files, and so on. Stata allows do-files to be nested 64 deep.

Be not confused: `master.do` above could call 1,000 do-files one after the other, and still the level of nesting would be only two.

16.3 Creating and running do-files

16.3.1 Creating and running do-files for Windows

1. You can execute do-files by typing `do` followed by the filename, as we did above.
2. You can execute do-files by selecting **File > Do...**.
3. You can use the Do-file Editor to compose, save, and execute do-files; see [GSW] **13 Using the Do-file Editor—automating Stata**. To use the Do-file Editor, click on the **Do-file Editor** button, or type `doedit` in the Command window.
4. You can double-click on the icon for the do-file to launch Stata and open the do-file in the Do-file Editor.
5. You can run the do-file in batch mode. See [GSW] **C.5 Stata batch mode** for details, but the short explanation is that you open a Window command window and type

   ```
   C:\data> "C:\Program Files\Stata12\Stata" /s do myjob
   ```

   or

   ```
   C:\data> "C:\Program Files\Stata12\Stata" /b do myjob
   ```

   to run in batch mode, assuming that you have installed Stata in the folder `C:\Program Files\Stata12`. `/b` and `/s` determine the kind of log produced, but put that aside for a second. When you start Stata in these ways, Stata will run in the background. When the do-file completes, the Stata icon on the taskbar will flash. You can then click on it to close Stata. If you want to stop the do-file before it completes, click on the Stata icon on the taskbar, and Stata will ask you if you want to cancel the job. If you want Stata to exit when the do-file is complete rather than flashing on the taskbar, also specify `/e` on the command line.

To log the output, you can start the log before executing the do-file or you can include the `log using` and `log close` in your do-file.

When you run Stata in these ways, Stata takes the following actions:

a. Stata automatically opens a log. If you specified `/s`, Stata will open a SMCL log; if you specified `/b`, Stata will open a plain text log. If your do-file is named *xyz*`.do`, the log will be called *xyz*`.smcl` (`/s`) or *xyz*`.log` (`/b`) in the same directory.
b. If your do-file explicitly opens another log, Stata will save two copies of the output.
c. Stata ignores —`more`— conditions and anything else that would cause the do-file to stop were it running interactively.

16.3.2 Creating and running do-files for Mac

1. You can execute do-files by typing `do` followed by the filename, as we did above.
2. You can execute do-files by selecting **File > Do...**.
3. You can use the Do-file Editor to compose, save, and execute do-files; see [GSM] **13 Using the Do-file Editor—automating Stata**. Click on the **Do-file Editor** button, or type `doedit` in the Command window.
4. You can double-click on the icon for the do-file to open the do-file in the Do-file Editor.
5. Double-clicking on the icon for a do-file named `Stata.do` will launch Stata if it is not already running and set the current working directory to the location of the do-file.

6. You can run the do-file in batch mode. See [GSM] **C.3 Stata batch mode** for details, but the short explanation is that you open a Terminal window and type

```
% /Applications/Stata/Stata.app/Contents/MacOS/Stata -s do myjob
```

or

```
% /Applications/Stata/Stata.app/Contents/MacOS/Stata -b do myjob
```

to run in batch mode, assuming that you have installed Stata/IC in the folder `/Applications/Stata`. `-b` and `-s` determine the kind of log produced, but put that aside for a second. When you start Stata in these ways, Stata will run in the background. When the do-file completes, the Stata icon on the Dock will bounce until you put Stata into the foreground. You can then exit Stata. If you want to stop the do-file before it completes, right-click on the Stata icon on the Dock, and select **Quit**.

To log the output, you can start the log before executing the do-file or you can include the `log using` and `log close` in your do-file.

When you run Stata in these ways, Stata takes the following actions:

a. Stata automatically opens a log. If you specified `-s`, Stata will open a SMCL log; if you specified `-b`, Stata will open a plain text log. If your do-file is named *xyz*`.do`, the log will be called *xyz*`.smcl` (`-s`) or *xyz*`.log` (`-b`) in the same directory.

b. If your do-file explicitly opens another log, Stata will save two copies of the output.

c. Stata ignores `—more—` conditions and anything else that would cause the do-file to stop were it running interactively.

16.3.3 Creating and running do-files for Unix

1. You can execute do-files by typing `do` followed by the filename, as we did above.

2. You can execute do-files by selecting **File > Do...**.

3. You can use the Do-file Editor to compose, save, and execute do-files; see [GSW] **13 Using the Do-file Editor—automating Stata**. Click on the **Do-file Editor** button, or type `doedit` in the Command window.

4. At the Unix prompt, you can type

   `$ xstata do` *filename*

   or

   `$ stata do` *filename*

   to launch Stata and run the do-file. When the do-file completes, Stata will prompt you for the next command just as if you had started Stata the normal way. If you want Stata to exit instead, include `exit, STATA clear` as the last line of your do-file.

To log the output, you can start the log before executing the do-file or you can include the `log using` and `log close` in your do-file.

5. At the Unix prompt, you can type

   $ stata -s do *filename* &

   or

   $ stata -b do *filename* &

   to run the do-file in the background. The above two examples both involve the use of stata, not xstata. Type stata, even if you usually use the GUI version of Stata, xstata. The examples differ only in that one specifies the -s option and the other, the -b option, which determines the kind of log that will be produced. In the above examples, Stata takes the following actions:

   a. Stata automatically opens a log. If you specified -s, Stata will open a SMCL log; if you specified -b, Stata will open a plain text log. If your do-file is named *xyz*.do, the log will be called *xyz*.smcl (-s) or *xyz*.log (-b) in the current directory (the directory from which you issued the stata command).

   b. If your do-file explicitly opens another log, Stata will save two copies of the output.

   c. Stata ignores —more— conditions and anything else that would cause the do-file to stop were it running interactively.

To reiterate: one way to run a do-file in the background and obtain a text log is by typing

```
$ stata -b do myfile &
```

Another way uses standard redirection:

```
$ stata < myfile.do > myfile.log &
```

The first way is slightly more efficient. Either way, Stata knows it is in the background and ignores —more— conditions and anything else that would cause the do-file to stop if it were running interactively. However, if your do-file contains either the #delimit command or the comment characters (/* at the end of one line and */ at the beginning of the next), the second method will not work. We recommend that you use the first method: stata -b do myfile &.

The choice between stata -b do myfile & and stata -s do myfile & is more personal. We prefer obtaining SMCL logs (-s) because they look better when printed, and, in any case, they can always be converted to text format with translate; see [R] **translate**.

16.4 Programming with do-files

This is an advanced topic, and we are going to refer to concepts not yet explained; see [U] **18 Programming Stata** for more information.

16.4.1 Argument passing

Do-files accept arguments, just as Stata programs do; this is described in [U] **18 Programming Stata** and [U] **18.4 Program arguments**. In fact, the logic Stata follows when invoking a do-file is the same as when invoking a program: the local macros are saved, and new ones are defined. Arguments are stored in the local macros '1', '2', and so on. When the do-file completes, the previous definitions are restored, just as with programs.

Thus, if you wanted your do-file to

1. use a dataset of your choosing,
2. tabulate a variable named region, and
3. summarize variables marriage_rate and divorce_rate,

you could write the do-file

begin myxmpl.do

```
use `1'
tabulate region
summarize marriage_rate divorce_rate
```

end myxmpl.do

and you could run this do-file by typing, for instance,

```
. do myxmpl census
  (output omitted)
```

The first command—`use `1'`—would be interpreted as `use census` because `census` was the first argument you typed after `do myxmpl`.

An even better version of the do-file would read

begin myxmpl.do

```
args dsname
use `dsname'
tabulate region
summarize marriage_rate divorce_rate
```

end myxmpl.do

The `args` command merely assigns a better name to the argument passed. `args dsname` does not verify that what we type following `do myxmpl` is a filename—we would have to use the `syntax` command if we wanted to do that—but substituting `` `dsname' `` for `` `1' `` does make the code more readable.

If our program were to receive two arguments, we could refer to them as `` `1' `` and `` `2' ``, or we could put an '`args dsname other`' at the top of our do-file and then refer to `` `dsname' `` and `` `other' ``.

To learn more about argument passing, see [U] **18.4 Program arguments**. Baum (2009) provides many examples and tips related to do-files.

16.4.2 Suppressing output

There is an alternative to typing `do` *filename*; it is `run` *filename*. `run` works in the same way as `do`, except that neither the instructions in the file nor any of the output caused by those instructions is shown on the screen or in the log file.

For instance, with the above `myxmpl.do`, typing `run myxmpl census` results in

```
. run myxmpl census

. _
```

All the instructions were executed, but none of the output was shown.

This is not useful here, but if the do-file contained only the definitions of Stata programs—see [U] **18 Programming Stata**—and you merely wanted to load the programs without seeing the code, `run` would be useful.

16.5 References

Baum, C. F. 2009. *An Introduction to Stata Programming*. College Station, TX: Stata Press.

Long, J. S. 2009. *The Workflow of Data Analysis Using Stata*. College Station, TX: Stata Press.

17 Ado-files

Contents

17.1 Description

Stata is programmable, and even if you never write a Stata program, Stata's programmability is still important. Many of Stata's features are implemented as Stata programs, and new features are implemented every day, both by StataCorp and by others.

1. You can obtain additions from the *Stata Journal*. You subscribe to the printed journal, but the software additions are available free over the Internet.
2. You can obtain additions from the Stata listserver, Statalist, where an active group of users advise each other on how to use Stata, and often, in the process, trade programs. Visit the Stata website, http://www.stata.com, for instructions on how to subscribe; subscribing to the listserver is free.
3. The Boston College Statistical Software Components Archive (SSC) is a distributed database making available a large and constantly growing number of Stata programs. You can browse and search the archive, and you can find links to the archive from http://www.stata.com. Importantly, Stata knows how to access the archive and other places, as well. You can search for additions by using Stata's `search, net` command; see [R] **search**. You can immediately install materials you find with `search, net` by using the hyperlinks that will be displayed by `search` in the Results window or by using the `net` command. A specialized command, `ssc`, has several options available to help you find and install the user-written commands that are available from this site; see [R] **ssc**.
4. You can write your own additions to Stata.

This chapter is written for people who want to use ado-files. All users should read it. If you later decide you want to write ado-files, see [U] **18.11 Ado-files**.

17.2 What is an ado-file?

An ado-file defines a Stata command, but not all Stata commands are defined by ado-files.

When you type summarize to obtain summary statistics, you are using a command built into Stata.

When you type ci to obtain confidence intervals, you are running an ado-file. The results of using a built-in command or an ado-file are indistinguishable.

An ado-file is a text file that contains a Stata program. When you type a command that Stata does not know, it looks in certain places for an ado-file of that name. If Stata finds it, Stata loads and executes it, so it appears to you as if the ado-command is just another command built into Stata.

We just told you that Stata's ci command is implemented as an ado-file. That means that, somewhere, there is a file named ci.ado.

Ado-files usually come with help files. When you type help ci (or select **Help > Stata Command...**, and type ci), Stata looks for ci.sthlp, just as it looks for ci.ado when you use the ci command. A help file is also a text file that tells Stata's help system what to display.

17.3 How can I tell if a command is built in or an ado-file?

You can use the which command to determine whether a file is built in or implemented as an ado-file. For instance, logistic is an ado-file, and here is what happens when you type which logistic:

```
. which logistic
C:\Program Files\Stata12\ado\base\l\logistic.ado
*! version 3.5.0  02mar2011
```

summarize is a built-in command:

```
. which summarize
built-in command:  summarize
```

17.4 How can I look at an ado-file?

When you type which followed by an ado-command, Stata reports where the file is stored:

```
. which logistic
C:\Program Files\Stata12\ado\base\l\logistic.ado
*! version 3.5.0  02mar2011
```

Ado-files are just text files containing the Stata program, so you can type them or view them in Stata's Viewer (or even look at them in your editor or word processor):

```
. type "C:\Program Files\Stata12\ado\base\l\logistic.ado"
*! version 3.5.0  02mar2011
program define logistic, eclass prop(or svyb svyj svyr swml mi) byable(onecall)
        version 6.0, missing
  (output omitted)
end
```

or

```
. viewsource logistic.ado
  (output omitted)
```

The type command displays the contents of a file. The viewsource command searches for a file along the ado directories and displays the file in the Viewer. You can also look at the corresponding help file in raw form if you wish. If there is a help file, it is stored in the same place as the ado-file:

```
. type "C:\Program Files\Stata12\ado\base\l\logistic.sthlp", asis
{smcl}
{* *! version 1.3.4  03may2011}{...}
{viewerdialog logistic "dialog logistic"}{...}
{viewerdialog "svy: logistic" "dialog logistic, message(-svy-)
    name(svy_logistic)"}{...}
{vieweralsosee "[R] logistic" "mansection R logistic"}{...}
  (output omitted)
```

or

```
. viewsource logistic.sthlp
  (output omitted)
```

17.5 Where does Stata look for ado-files?

Stata looks for ado-files in seven places, which can be categorized in three ways:

I. The official ado directories:
   1. (UPDATES), the official updates directory containing updated ado-files from StataCorp
   2. (BASE), the official base directory containing the ado-files shipped with your version of Stata

II. Your personal ado-directories:
   3. (SITE), the directory for ado-files your site might have installed
   4. (PLUS), the directory for ado-files you personally might have installed
   5. (PERSONAL), the directory for ado-files you might have written
   6. (OLDPLACE), the directory where Stata users used to save their personally written ado-files

III. The current directory:
   7. (.), the ado-files you have written just this instant or for just this project

The location of these directories varies from computer to computer, but Stata's sysdir command will tell you where they are on your computer:

```
. sysdir
   STATA:  C:\Program Files\Stata12\
 UPDATES:  C:\Program Files\Stata12\ado\updates\
    BASE:  C:\Program Files\Stata12\ado\base\
    SITE:  C:\Program Files\Stata12\ado\site\
    PLUS:  C:\ado\plus\
PERSONAL:  C:\ado\personal\
OLDPLACE:  C:\ado\
```

17.5.1 Where are the official ado-directories?

These are the directories listed as BASE and UPDATES by sysdir:

```
. sysdir
   STATA:  C:\Program Files\Stata12\
 UPDATES:  C:\Program Files\Stata12\ado\updates\
    BASE:  C:\Program Files\Stata12\ado\base\
    SITE:  C:\Program Files\Stata12\ado\site\
    PLUS:  C:\ado\plus\
PERSONAL:  C:\ado\personal\
OLDPLACE:  C:\ado\
```

1. BASE contains the ado-files we originally shipped to you.
2. UPDATES contains any updates you might have installed since then. You can install these updates by using the update command or by selecting **Help > Check for Updates**; see [U] **17.8 How do I install official updates?**.

17.5.2 Where is my personal ado-directory?

These are the directories listed as PERSONAL, PLUS, SITE, and OLDPLACE by sysdir:

```
. sysdir
   STATA:  C:\Program Files\Stata12\
 UPDATES:  C:\Program Files\Stata12\ado\updates\
    BASE:  C:\Program Files\Stata12\ado\base\
    SITE:  C:\Program Files\Stata12\ado\site\
    PLUS:  C:\ado\plus\
PERSONAL:  C:\ado\personal\
OLDPLACE:  C:\ado\
```

1. PERSONAL is for ado-files you have written. Store your private ado-files here; see [U] **17.7 How do I add my own ado-files?**.
2. PLUS is for ado-files you personally installed but did not write. Such ado-files are usually obtained from the SJ or the SSC, but they are sometimes found in other places, too. You find and install such files by using Stata's net command, or you can select **Help > SJ and User-written Programs**; see [U] **17.6 How do I install an addition?**.
3. SITE is really the opposite of a personal ado directory—it is a public directory corresponding to PLUS. If you are on a networked computer, the site administrator can install ado-files here, and all Stata users will then be able to use them just as if they all found and installed them in their PLUS directory for themselves. Site administrators find and install the ado-files just as you would, using Stata's net command, but they specify an option when they install something that tells Stata to write the files into SITE rather than PLUS; see [R] **net**.
4. OLDPLACE is for old-time Stata users. Prior to Stata 6, all "personal" ado-files, whether personally written or just personally installed, were written in the same directory—OLDPLACE. So that the old-time Stata users do not have to go back and rearrange what they have already done, Stata still looks in OLDPLACE.

17.6 How do I install an addition?

Additions come in four types:

1. User-written additions, which you might find in the SJ, etc.
2. Updates to user-written additions
   See [U] **17.9 How do I install updates to user-written additions?**.
3. Ado-files you have written
   See [U] **17.7 How do I add my own ado-files?** If you have an ado-file obtained from the Stata listserver or a friend, treat it as belonging to this case.
4. Official updates provided by StataCorp
   See [U] **17.8 How do I install official updates?**.

User-written additions you might find in the *Stata Journal* (SJ), etc., are obtained over the Internet. To access them on the Internet,

1. select **Help > SJ and User-written Programs**, and click on one of the links

or

2. type `net from http://www.stata.com`.

What to do next will be obvious, but, in case it is not, see [GSW] **19 Updating and extending Stata—Internet functionality**, [GSM] **19 Updating and extending Stata—Internet functionality**, or [GSU] **19 Updating and extending Stata—Internet functionality**. Also see [U] **28 Using the Internet to keep up to date**, [R] **net**, and [R] **adoupdate**.

17.7 How do I add my own ado-files?

You write a Stata program (see [U] **18 Programming Stata**), store it in a file ending in `.ado`, perhaps write a help file, and copy everything to the directory `sysdir` lists as `PERSONAL`:

```
. sysdir
   STATA:  C:\Program Files\Stata12\
 UPDATES:  C:\Program Files\Stata12\ado\updates\
    BASE:  C:\Program Files\Stata12\ado\base\
    SITE:  C:\Program Files\Stata12\ado\site\
    PLUS:  C:\ado\plus\
PERSONAL:  C:\ado\personal\
OLDPLACE:  C:\ado\
```

Here we would copy the files to `C:\ado\personal`.

While you are writing your ado-file, it is sometimes convenient to store the pieces in the current directory. Do that if you wish; you can move them to your personal ado-directory when the program is debugged.

17.8 How do I install official updates?

Updates are available over the Internet:

1. select **Help > Check for Updates**, and then click on **http://www.stata.com**

or

2. type `update query`.

What to do next should be obvious, but in case it is not, see [GSW] **19 Updating and extending Stata—Internet functionality**, [GSM] **19 Updating and extending Stata—Internet functionality**, or [GSU] **19 Updating and extending Stata—Internet functionality**. Also see [U] **28 Using the Internet to keep up to date** and [R] **net**.

The official updates include bug fixes and new features but do not change the syntax of an existing command or change the way Stata works.

Once you have installed the updates, you can enter Stata and type `help whatsnew` (or select **Help > What's New?**) to learn about what has changed.

17.9 How do I install updates to user-written additions?

If you have previously installed user-written additions, you can check for updates to them by typing `adoupate`. If updates are available, you can install them by typing `adoupdate, update`. See [R] **adoupdate**.

17.10 Reference

Cox, N. J. 2006. Stata tip 30: May the source be with you. *Stata Journal* 6: 149–150.

18 Programming Stata

Contents

Stata programming is an advanced topic. Some Stata users live productive lives without ever programming Stata. After all, you do not need to know how to program Stata to import data, create new variables, and fit models. On the other hand, programming Stata is not difficult—at least if the problem is not difficult—and Stata's programmability is one of its best features. The real power of Stata is not revealed until you program it.

If you are uncertain whether to read this chapter, we recommend that you start reading and then bail out when it gets too arcane for you. You will learn things about Stata that you may find useful even if you never write a Stata program.

If you want even more, we offer courses over the Internet on Stata programming; see [U] **3.7.2 NetCourses**. Baum (2009) provides a wealth of practical knowledge related to Stata programming.

18.1 Description

When you type a command that Stata does not recognize, Stata first looks in its memory for a program of that name. If Stata finds it, Stata executes the program.

There is no Stata command named `hello`,

```
. hello
unrecognized command
r(199);
```

but there could be if you defined a program named `hello`, and after that, the following might happen when you typed `hello`:

```
. hello
hi there
. _
```

This would happen if, beforehand, you had typed

```
. program hello
  1. display "hi there"
  2. end
. _
```

That is how programming works in Stata. A program is defined by

```
program progname
        Stata commands
end
```

and it is executed by typing *progname* at Stata's dot prompt.

18.2 Relationship between a program and a do-file

Stata treats programs the same way it treats do-files. Below we will discuss passing arguments, consuming results from Stata commands, and other topics, but everything we say applies equally to do-files and programs.

Programs and do-files differ in the following ways:

1. You invoke a do-file by typing `do` *filename*. You invoke a program by simply typing the program's name.

2. Programs must be defined (loaded) before they are used, whereas all that is required to run a do-file is that the file exist. There are ways to make programs load automatically, however, so this difference is of little importance.

3. When you type do *filename*, Stata displays the commands it is executing and the results. When you type *progname*, Stata shows only the results, not the display of the underlying commands. This is an important difference in outlook: in a do-file, how it does something is as important as what it does. In a program, the how is no longer important. You might think of a program as a new feature of Stata.

Let's now mention some of the similarities:

1. Arguments are passed to programs and do-files in the same way.

2. Programs and do-files both contain Stata commands. Any Stata command you put in a do-file can be put in a program.

3. Programs may call other programs. Do-files may call other do-files. Programs may call do-files (this rarely happens), and do-files may call programs (this often happens). Stata allows programs (and do-files) to be nested up to 64 deep.

Now here is the interesting thing: programs are typically defined in do-files (or in a variant of do-files called ado-files; we will get to that later).

You can define a program interactively, and that is useful for pedagogical purposes, but in real applications, you will compose your program in a text editor and store its definition in a do-file.

You have already seen your first program:

```
program hello
        display "hi there"
end
```

You could type those commands interactively, but if the body of the program were more complicated, that would be inconvenient. So instead, suppose that you typed the commands into a do-file:

——————————————————— begin hello.do ———

```
program hello
        display "hi there"
end
```

——————————————————— end hello.do ———

Now returning to Stata, you type

```
. do hello

. program hello
  1.          display "hi there"
  2. end

.
end of do-file
```

Do you see that typing `do hello` did nothing but load the program? Typing `do hello` is the same as typing out the program's definition because that is all the do-file contains. The do-file was executed, but the statements in the do-file only defined the program `hello`; they did not execute it. Now that the program is loaded, we can execute it interactively:

```
. hello
hi there
```

So, that is one way you could use do-files and programs together. If you wanted to create new commands for interactive use, you could

1. Write the command as a `program ... end` in a do-file.
2. `do` the do-file before you use the new command.
3. Use the new command during the rest of the session.

There are more convenient ways to do this that would automatically load the do-file, but put that aside. The above method would work.

Another way we could use do-files and programs together is to put the definition of the program and its execution together into a do-file:

begin hello.do
```
program hello
        display "hi there"
end
hello
```
end hello.do

Here is what would happen if we executed this do-file:

```
. do hello

. program hello
  1.        display "hi there"
  2. end

. hello
hi there

.
end of do-file
```

Do-files and programs are often used in such combinations. Why? Say that program `hello` is long and complicated and you have a problem where you need to do it twice. That would be a good reason to write a program. Moreover, you may wish to carry forth this procedure as a step of your analysis and, being cautious, do not want to perform this analysis interactively. You never intended program `hello` to be used interactively—it was just something you needed in the midst of a do-file—so you defined the program and used it there.

Anyway, there are many variations on this theme, but few people actually sit in front of Stata and interactively type `program` and then compose a program. They instead do that in front of their text editor. They compose the program in a do-file and then execute the do-file.

There is one other (minor) thing to know: once a program is defined, Stata does not allow you to redefine it:

```
. program hello
hello already defined
r(110);
```

Thus, in our most recent do-file that defines and executes `hello`, we could not rerun it in the same Stata session:

```
. do hello

. program hello
hello already defined
r(110);

end of do-file
r(110);
```

That problem is solved by typing `program drop hello` before redefining it. We could do that interactively, or we could modify our do-file:

begin hello.do
```
program drop hello
program hello
        display "hi there"
end
hello
```
end hello.do

There is a problem with this solution. We can now rerun our do-file, but the first time we tried to run it in a Stata session, it would fail:

```
. do hello

. program drop hello
hello not found
r(111);

end of do-file
r(111);
```

The way around this conundrum is to modify the do-file:

begin hello.do
```
capture program drop hello
program hello
        display "hi there"
end
hello
```
end hello.do

`capture` in front of a command makes Stata indifferent to whether the command works; see [P] **capture**. In real do-files containing programs, you will often see `capture program drop` before the program's definition.

To learn about the `program` command itself, see [P] **program**. It manipulates programs. `program` can define programs, drop programs, and show you a directory of programs that you have defined.

A program can contain any Stata command, but certain Stata commands are of special interest to program writers; see the *Programming* heading in the subject table of contents in the *Quick Reference and Index*.

18.3 Macros

Before we can begin programming, we must discuss macros, which are the variables of Stata programs.

A *macro* is a string of characters, called the *macroname*, that stands for another string of characters, called the *macro contents*.

Macros can be local or global. We will start with local macros because they are the most commonly used, but nothing really distinguishes one from the other at this stage.

18.3.1 Local macros

Local macro names can be up to 31 (not 32) characters long.

One sets the contents of a local macro with the `local` command. In fact, we can do this interactively. We will begin by experimenting with macros in this way to learn about them. If we type

```
. local shortcut "myvar thisvar thatvar"
```

then `‘shortcut’` is a synonym for “`myvar thisvar thatvar`”. Note the single quotes around `shortcut`. We said that sentence exactly the way we meant to because

if you type	`‘shortcut’`,
i.e.,	left-single-quote `shortcut` right-single-quote,
Stata hears	`myvar thisvar thatvar`.

To access the contents of the macro, we use a left single quote (located at the upper left on most keyboards), the macro name, and a right single quote (located under the " on the right side of most keyboards).

The single quotes bracketing the macroname `shortcut` are called the macro-substitution characters. `shortcut` means `shortcut`. `‘shortcut’` means `myvar thisvar thatvar`.

So, if you typed

```
. list ‘shortcut’
```

the effect would be exactly as if you typed

```
. list myvar thisvar thatvar
```

Macros can be used anywhere in Stata. For instance, if we also defined

```
. local cmd "list"
```

we could type

```
. ‘cmd’ ‘shortcut’
```

to mean `list myvar thisvar thatvar`.

For another example, consider the definitions

```
. local prefix "my"
. local suffix "var"
```

Then

```
. ‘cmd’ ‘prefix’‘suffix’
```

would mean `list myvar`.

One other important note is on the way we use left and right single quotes within Stata, which you will especially deal with when working with macros (see [U] **18.3 Macros**). Single quotes (and double quotes, for that matter) may look different on your keyboard, your monitor, and our printed documentation, making it difficult to determine which key to press on your keyboard to replicate what we have shown you.

For the left single quote, we use the grave accent, which occupies a key by itself on most computer keyboards. On U.S. keyboards, the grave accent is located at the top left, next to the numeral 1. On some non-U.S. keyboards, the grave accent is produced by a dead key. For example, pressing the grave accent dead key followed by the letter a would produce à; to get the grave accent by itself, you would press the grave accent dead key followed by a space. This accent mark appears in our printed documentation as ‘.

For the right single quote, we use the standard single quote, or apostrophe. On U.S. keyboards, the single quote is located on the same key as the double quote, on the right side of the keyboard next to the *Enter* key.

18.3.2 Global macros

Let's put aside why Stata has two kinds of macros—local and global—and focus right now on how global macros work.

Global macros can have names that are up to 32 (not 31) characters long. You set the contents of a global macro by using the `global` rather than the `local` command:

```
. global shortcut "alpha beta"
```

You obtain the contents of a global macro by prefixing its name with a dollar sign: `$shortcut` is equivalent to "`alpha beta`".

In the previous section, we defined a local macro named `shortcut`, which is a different macro. `‘shortcut’` is still "`myvar thisvar thatvar`".

Local and global macros may have the same names, but even if they do, they are unrelated and are still distinguishable.

Global macros are just like local macros except that you set their contents with `global` rather than `local`, and you substitute their contents by prefixing them with a `$` rather than enclosing them in `‘’`.

18.3.3 The difference between local and global macros

The difference between local and global macros is that local macros are private and global macros are public.

Say that you have written a program

```
program myprog
        code using local macro alpha
end
```

The local macro `alpha` in `myprog` is private in that no other program can modify or even look at `alpha`'s contents. To make this point absolutely clear, assume that your program looks like this:

```
program myprog
        code using local macro alpha
        mysub
        more code using local macro alpha
end
program mysub
        code using local macro alpha
end
```

`myprog` calls `mysub`, and both programs use a local macro named `alpha`. Even so, the local macros in each program are different. `mysub`'s `alpha` macro may contain one thing, but that has nothing to do with what `myprog`'s `alpha` macro contains. Even when `mysub` begins execution, its `alpha` macro is different from `myprog`'s. It is not that `mysub`'s inherits `myprog`'s `alpha` macro contents but is then free to change it. It is that `myprog`'s `alpha` and `mysub`'s `alpha` are entirely different things.

When you write a program using local macros, you need not worry that some other program has been written using local macros with the same names. Local macros are just that: local to your program.

Global macros, on the other hand, are available to all programs. If both myprog and mysub use the global macro beta, they are using the same macro. Whatever the contents of $beta are when mysub is invoked, those are the contents when mysub begins execution, and, whatever the contents of $beta are when mysub completes, those are the contents when myprog regains control.

18.3.4 Macros and expressions

From now on, we are going to use local and global macros according to whichever is convenient; whatever is said about one applies to the other.

Consider the definitions

```
. local one 2+2
. local two = 2+2
```

(which we could just as well have illustrated using the global command). In any case, note the equal sign in the second macro definition and the lack of the equal sign in the first. Formally, the first should be

```
. local one "2+2"
```

but Stata does not mind if we omit the double quotes in the local (global) statement.

local one 2+2 (with or without double quotes) copies the string 2+2 into the macro named one.

local two = 2+2 evaluates the expression 2+2, producing 4, and stores 4 in the macro named two.

That is, you type

local *macname contents*

if you want to copy *contents* to *macname*, and you type

local *macname* = *expression*

if you want to evaluate *expression* and store the result in *macname*.

In the second form, *expression* can be numeric or string. 2+2 is a numeric expression. As an example of a string expression,

```
. local res = substr("this",1,2) + "at"
```

stores that in res.

Because the expression can be either numeric or string, what is the difference between the following statements?

```
. local a "example"
. local b = "example"
```

Both statements store example in their respective macros. The first does so by a simple copy operation, whereas the second evaluates the expression "example", which is a string expression because of the double quotes that, here, evaluates to itself.

There is, however, a difference. Stata's expression parser is limited to handling strings of 244 characters. Strings longer than that are truncated.

The copy operation of the first syntax is not limited—it can copy up to the maximum length of a macro, which is currently 165,200 characters for Stata/IC and 8,681 for Small Stata. For Stata/MP and Stata/SE, the limit is $33 *$ c(max_k_theory) $+ 200$ characters, which for the default setting of 5,000 is 165,200 characters.

To a programmer, the length limit for string expressions may seem restrictive, but it is not, because of another feature discussed in [U] **18.3.6 Extended macro functions**.

There are some other issues of using macros and expressions that look a little strange to programmers coming from other languages, at least the first time they see them. Say that the macro `‘i’` contains 5. How would you increment `i` so that it contains $5 + 1 = 6$? The answer is

```
local i = ‘i’ + 1
```

Do you see why the single quotes are on the right but not the left? Remember, `‘i’` refers to the contents of the local macro named `i`, which, we just said, is 5. Thus, after expansion, the line reads

```
local i = 5 + 1
```

which is the desired result.

There is a another way to increment local macros that will be more familiar to some programmers, especially C programmers:

```
local ++i
```

As C programmers would expect, `local ++i` is more efficient (executes more quickly) than `local i = i+1`, but in terms of outcome, it is equivalent. You can decrement a local macro by using

```
local --i
```

`local --i` is equivalent to `local i = i-1` but executes more quickly. Finally,

```
local i++
```

will *not* increment the local macro `i` but instead redefines the local macro `i` to contain `++`. There is, however, a context in which `i++` (and `i--`) do work as expected; see [U] **18.3.7 Macro increment and decrement functions**.

18.3.5 Double quotes

Consider another local macro, `‘answ’`, which might contain `yes` or `no`. In a program that was supposed to do something different on the basis of `answ`'s content, you might code

```
if "‘answ’" == "yes" {
        ...
}
else {
        ...
}
```

Note the odd-looking `"‘answ’"`, and now think about the line after substitution. The line reads either

```
if "yes" == "yes" {
```

or

```
if "no" == "yes" {
```

either of which is the desired result. Had we omitted the double quotes, the line would have read

```
if no == "yes" {
```

(assuming `answ' contains no), and that is not at all the desired result. As the line reads now, no would not be a string but would be interpreted as a variable in the data.

The key to all this is to think of the line after substitution.

Double quotes are used to enclose strings: "yes", "no", "my dir\my file", "`answ'" (meaning that the contents of local macro answ, treated as a string), and so on. Double quotes are used with macros,

```
local a "example"
if "`answ'" == "yes" {
        ...
}
```

and double quotes are used by many Stata commands:

```
. regress lnwage age ed if sex=="female"
. gen outa = outcome if drug=="A"
. use "person file"
```

Do not omit the double quotes just because you are using a "quoted" macro:

```
. regress lnwage age ed if sex=="`x'"
. gen outa = outcome if drug=="`firstdrug'"
. use "`filename'"
```

Stata has two sets of double-quote characters, of which "" is one. The other is `""'. They both work the same way:

```
. regress lnwage age ed if sex==`"female"'
. gen outa = outcome if drug==`"A"'
. use `"person file"'
```

No rational user would use `""' (called compound double quotes) instead of "" (called simple double quotes), but smart programmers do use them:

```
local a `"example"'
if `"`answ'"' == `"yes"' {
        ...
}
```

Why is `"example"' better than "example", `"`answ'"' better than "`answ'", and `"yes"' better than "yes"? The answer is that only `"`answ'"' is better than "`answ'"; `"example"' and `"yes"' are no better—and no worse—than "example" and "yes".

`"`answ'"' is better than "`answ'" because the macro answ might itself contain (simple or compound) double quotes. The really great thing about compound double quotes is that they nest. Say that `answ' contained the string "I "think" so". Then,

Stata would find	if "`answ'"=="yes"
confusing because it would expand to	if "I "think" so"=="yes"
Stata would not find	if `"`answ'"'==`"yes"'
confusing because it would expand to	if `"I "think" so"'==`"yes"'

Open and close double quote in the simple form look the same; open quote is " and so is close quote. Open and close double quote in the compound form are distinguishable; open quote is `" and close quote is "', and so Stata can pair the close with the corresponding open double quote. `"I "think" so"' is easy for Stata to understand, whereas "I "think" so" is a hopeless mishmash. (If you disagree, consider what "A"B"C" might mean. Is it the quoted string A"B"C, or is it quoted string A, followed by B, followed by quoted string C?)

Because Stata can distinguish open from close quotes, even nested compound double quotes are understandable: `` `"I `"think"' so"' ``. (What does `"A"B"C"` mean? Either it means `` `"A`"B"'C"' `` or it means `` `"A"'B`"C"' ``.)

Yes, compound double quotes make you think that your vision is stuttering, especially when combined with the macro substitution `` `' `` characters. That is why we rarely use them, even when writing programs. You do not have to use exclusively one or the other style of quotes. It is perfectly acceptable to code

```
local a "example"
if `"`answ'"' == "yes" {
        ...
}
```

using compound double quotes where it might be necessary (`` `"`answ'"' ``) and using simple double quotes in other places (such as `"yes"`). It is also acceptable to use simple double quotes around macros (for example, `` "`answ'" ``) if you are certain that the macros themselves do not contain double quotes or (more likely) if you do not care what happens if they do.

Sometimes careful programmers should use compound double quotes. Later you will learn that Stata's `syntax` command interprets standard Stata syntax and so makes it easy to write programs that understand things like

```
. myprog mpg weight if strpos(make,"VW")!=0
```

`syntax` works—we are getting ahead of ourselves—by placing the `if` *exp* typed by the user in the local macro `if`. Thus `` `if' `` will contain "`if strpos(make,"VW")!=0`" here. Now say that you are at a point in your program where you want to know whether the user specified an `if` *exp*. It would be natural to code

```
if `"`if'"' != "" {
        // the if exp was specified
        ...
}
else {
        // it was not
        ...
}
```

We used compound double quotes around the macro `` `if' ``. The local macro `` `if' `` might contain double quotes, so we placed compound double quotes around it.

18.3.6 Extended macro functions

In addition to allowing =*exp*, `local` and `global` provide *extended functions*. The use of an extended function is denoted by a colon (`:`) following the macro name, as in

```
local       lbl : variable label myvar
local filenames : dir "." files "*.dta"
local        xi : word `i' of `list'
```

Some macro extended functions access a piece of information. In the first example, the variable label associated with variable `myvar` will be stored in macro `lbl`. Other macro extended functions perform operations to gather the information. In the second example, macro `filenames` will contain the names of all the `.dta` datasets in the current directory. Still other macro extended functions perform an operation on their arguments and return the result. In the third example, `xi` will contain the `` `i' ``th word (element) of `` `list' ``. See [P] **macro** for a list of the macro extended functions.

Another useful source of information is `c()`, documented in [P] **creturn**:

```
local  today "‘c(current_date)’"
local curdir "‘c(pwd)’"
local   newn = c(N)+1
```

`c()` refers to a prerecorded list of values, which may be used directly in expressions or which may be quoted and the result substituted anywhere. `c(current_date)` returns today's date in the form "*dd MON yyyy*". Thus the first example stores in macro `today` that date. `c(pwd)` returns the current directory, such as `C:\data\proj`. Thus the second example stores in macro `curdir` the current directory. `c(N)` returns the number of observations of the data in memory. Thus the third example stores in macro `newn` that number, plus one.

Note the use of quotes with `c()`. We could just as well have coded the first two examples as

```
local  today = c(current_date)
local curdir = c(pwd)
```

`c()` is a Stata function in the same sense that `sqrt()` is a Stata function. Thus we can use `c()` directly in expressions. It is a special property of macro expansion, however, that you may use the `c()` function inside macro-expansion quotes. The same is not true of `sqrt()`.

In any case, whenever you need a piece of information, whether it be about the dataset or about the environment, look in [P] **macro** and [P] **creturn**. It is likely to be in one place or the other, and sometimes, it is in both. You can obtain the current directory by using

```
local curdir = c(pwd)
```

or by using

```
local curdir : pwd
```

When information is in both, it does not matter which source you use.

18.3.7 Macro increment and decrement functions

We mentioned incrementing macros in [U] **18.3.4 Macros and expressions**. The construct

```
command that makes reference to ‘i’
local ++i
```

occurs so commonly in Stata programs that it is convenient (and faster when executed) to collapse both lines of code into one and to increment (or decrement) `i` at the same time that it is referred to. Stata allows this:

```
while (‘++i’ < 1000) {
        ...
}
while (‘i++’ < 1000) {
        ...
}
while (‘--i’ > 0) {
        ...
}
while (‘i--’ > 0) {
        ...
}
```

Above we have chosen to illustrate this by using Stata's `while` command, but `++` and `--` can be used anyplace in any context, just so long as it is enclosed in macro-substitution quotes.

When the `++` or `--` appears before the name, the macro is first incremented or decremented, and then the result is substituted.

When the `++` or `--` appears after the name, the current value of the macro is substituted and then the macro is incremented or decremented.

❑ Technical note

Do not use the inline `++` or `--` operators when a part of the line might not be executed. Consider

```
if (`i'==0) local j = `k++'
```

versus

```
if (`i'==0) {
        local j = `k++'
}
```

The first will not do what you expect because macros are expanded before the line is interpreted. Thus the first will result in `k` always being incremented, whereas the second increments `k` only when `` `i'==0``.

❑

18.3.8 Macro expressions

Typing

command that makes reference to `` `=``*exp*`'`

is equivalent to

`local` *macroname* `=` *exp*
command that makes reference to `` ` ``*macroname*`'`

although the former runs faster and is easier to type. When you use `` `=``*exp*`'` within some larger command, *exp* is evaluated by Stata's expression evaluator, and the results are inserted as a literal string into the larger command. Then the command is executed. For example,

```
summarize u4
summarize u`=2+2'
summarize u`=4*(cos(0)==1)'
```

all do the same thing. *exp* can be any valid Stata expression and thus may include references to variables, matrices, scalars, or even other macros. In the last case, just remember to enclose the submacros in quotes:

```
replace `var' = `group'[`=`j'+1']
```

Also, typing

command that makes reference to `` `: ``*extended macro function*`'`

is equivalent to

`local` *macroname* `:` *extended macro function*
command that makes reference to `` ` ``*macroname*`'`

Thus one might code

```
format y `:format x'
```

to assign to variable `y` the same format as the variable `x`.

❑ Technical note

There is another macro expansion operator, `.` (called dot), which is used in conjunction with Stata's class system; see [P] **class** for more information.

There is also a macro expansion function, `macval()`, which is for use when expanding a macro—`` `macval( ``*name*`)'`—which confines the macro expansion to the first level of *name*, thereby suppressing the expansion of any embedded references to macros within *name*. Only two or three Stata users have or will ever need this, but, if you suspect you are one of them, see [P] **macro** and then see [P] **file** for an example.

❑

18.3.9 Advanced local macro manipulation

This section is really an aside to help test your understanding of macro substitution. The tricky examples illustrated below sometimes occur in real programs.

1. Say that you have macros `x1`, `x2`, `x3`, and so on. Obviously, `` `x1' `` refers to the contents of `x1`, `` `x2' `` to the contents of `x2`, etc. What does `` `x`i'' `` refer to? Suppose that `` `i' `` contains `6`.

   The rule is to expand the inside first:

`` `x`i'' ``	expands to	`` `x6' ``
`` `x6' ``	expands to	the contents of local macro `x6`

   So, there you have a vector of macros.

2. We have already shown adjoining expansions: `` `alpha'`beta' `` expands to `myvar` if `` `alpha' `` contains `my` and `` `beta' `` contains `var`. What does `` `alpha'`gamma'`beta' `` expand to when `gamma` is undefined?

   Stata does not mind if you refer to a nonexistent macro. A nonexistent macro is treated as a macro with no contents. If local macro `gamma` does not exist, then

   `` `gamma' `` expands to nothing

   It is not an error. Thus `` `alpha'`gamma'`beta' `` expands to `myvar`.

3. You clear a local macro by setting its contents to nothing:

   `local` *macname*
   or `local` *macname* `""`
   or `local` *macname* `= ""`

18.3.10 Advanced global macro manipulation

Global macros are rarely used, and when they are used, it is typically for communication between programs. You should never use a global macro where a local macro would suffice.

1. Constructions like `$x$i` are expanded sequentially. If `$x` contained `this` and `$i 6`, then `$x$i` expands to `this6`. If `$x` was undefined, then `$x$i` is just `6` because undefined global macros, like undefined local macros, are treated as containing nothing.
2. You can nest macro expansion by including braces, so if `$i` contains 6, `${x$i}` expands to `${x6}`, which expands to the contents of `$x6` (which would be nothing if `$x6` is undefined).
3. You can mix global and local macros. Assume that local macro `j` contains 7. Then, `${x`j'}` expands to the contents of `$x7`.
4. You also use braces to force the contents of global macros to run up against the succeeding text. For instance, assume that the macro `drive` contains "`d:`". If `drive` were a local macro, you could type

   ```
   `drive'myfile.dta
   ```

   to obtain `b:myfile.dta`. Because `drive` is a global macro, however, you must type

   ```
   ${drive}myfile.dta
   ```

   You could not type

   ```
   $drive myfile.dta
   ```

   because that would expand to `b: myfile.dta`. You could not type

   ```
   $drivemyfile.dta
   ```

   because that would expand to `.dta`.
5. Because Stata uses `$` to mark global-macro expansion, printing a real `$` is sometimes tricky. To display the string $22.15 with the `display` command, you can type `display "\$22.15"`, although you can get away with `display "$22.15"` because Stata is rather smart. Stata would not be smart about `display "$this"` if you really wanted to display `$this` and not the contents of the macro `this`. You would have to type `display "\$this"`. Another alternative would be to use the SMCL code for a dollar sign when you wanted to display it: `display "{c S|}this"`; see [P] **smcl**.
6. Real dollar signs can also be placed into the contents of macros, thus postponing substitution. First, let's understand what happens when we do not postpone substitution; consider the following definitions:

   ```
   global baseset "myvar thatvar"
   global bigset "$baseset thisvar"
   ```

   `$bigset` is equivalent to "`myvar thatvar thisvar`". Now say that we redefine the macro `baseset`:

   ```
   global baseset "myvar thatvar othvar"
   ```

   The definition of `bigset` has not changed—it is still equivalent to "`myvar thatvar thisvar`". It has not changed because `bigset` used the definition of `baseset` that was current at the time it was defined. `bigset` no longer knows that its contents are supposed to have any relation to `baseset`.

   Instead, let's assume that we had defined `bigset` as

   ```
   global bigset "\$baseset thisvar"
   ```

at the outset. Then $bigset is equivalent to "$baseset thisvar", which in turn is equivalent to "myvar thatvar othvar thisvar". Because bigset explicitly depends upon baseset, anytime we change the definition of baseset, we will automatically change the definition of bigset as well.

18.3.11 Constructing Windows filenames by using macros

Stata uses the \ character to tell its parser not to expand macros.

Windows uses the \ character as the directory path separator.

Mostly, there is no problem using a \ in a filename. However, if you are writing a program that contains a Windows path in macro path and a filename in fname, do not assemble the final result as

```
`path'\`fname'
```

because Stata will interpret the \ as an instruction to not expand `fname'. Instead, assemble the final result as

```
`path'/`fname'
```

Stata understands / as a directory separator on all platforms.

18.3.12 Accessing system values

Stata programs often need access to system parameters and settings, such as the value of π, the current date and time, or the current working directory.

System values are accessed via Stata's c-class values. The syntax works much the same as if you were referring to a local macro. For example, a reference to the c-class value for π, `c(pi)', will expand to a literal string containing 3.141592653589793 and could be used to do

```
. display sqrt(2*`c(pi)')
2.5066283
```

You could also access the current time

```
. display "`c(current_time)'"
11:34:57
```

C-class values are designed to provide one all-encompassing way to access system parameters and settings, including system directories, system limits, string limits, memory settings, properties of the data currently in memory, output settings, efficiency settings, network settings, and debugging settings.

See [P] **creturn** for a detailed list of what is available. Typing

```
. creturn list
```

will give you the list of current settings.

18.3.13 Referring to characteristics

Characteristics—see [U] **12.8 Characteristics**—are like macros associated with variables. They have names of the form *varname*[*charname*]—such as mpg[comment]—and you quote their names just as you do macro names to obtain their contents:

To substitute the value of *varname*[*charname*], type	`*varname*[*charname*]'
For example,	`mpg[comment]'

You set the contents using the `char` command:

`char` *varname*`[`*charname*`]` [["]*text*["]]

This is similar to the `local` and `global` commands, except that there is no =*exp* variation. You clear a characteristic by setting its contents to nothing just as you would with a macro:

Type	`char` *varname*`[`*charname*`]`
or	`char` *varname*`[`*charname*`] ""`

What is unique about characteristics is that they are saved with the data, meaning that their contents survive from one session to the next, and they are associated with variables in the data, so if you ever drop a variable, the associated characteristics disappear, too. (Also, `_dta[`*charname*`]` is associated with the data but not with any variable in particular.)

All the standard rules apply: characteristics may be referred to by quotation in any context, and the characteristic's contents are substituted for the quoted characteristic name. As with macros, referring to a nonexistent characteristic is not an error; it merely substitutes to nothing.

18.4 Program arguments

When you invoke a program or do-file, what you type following the program or do-file name are the arguments. For instance, if you have a program called `xyz` and type

```
. xyz mpg weight
```

then `mpg` and `weight` are the program's arguments, `mpg` being the first argument and `weight` the second.

Program arguments are passed to programs via local macros:

Macro	Contents
`` `0' ``	what the user typed exactly as the user typed it, odd spacing, double quotes, and all
`` `1' ``	the first argument (first word of `` `0' ``)
`` `2' ``	the second argument (second word of `` `0' ``)
`` `3' ``	the third argument (third word of `` `0' ``)
...	...
`` `*' ``	the arguments `` `1' ``, `` `2' ``, `` `3' ``, ..., listed one after the other and with one blank in between; similar to but different from `` `0' `` because odd spacing and double quotes are removed

That is, what the user types is passed to you in three different ways:

1. It is passed in `` `0' `` exactly as the user typed it, meaning quotes, odd spacing, and all.
2. It is passed in `` `1' ``, `` `2' ``, ... broken out into arguments on the basis of blanks (but with quotes used to force binding; we will get to that).
3. It is passed in `` `*' `` as "`` `1' `2' `3' ``...", which is a crudely cleaned up version of `` `0' ``.

You will probably not use all three forms in one program.

We recommend that you ignore `` `*' ``, at least for receiving arguments; it is included so that old Stata programs will continue to work.

Operating directly with `0' takes considerable programming sophistication, although Stata's syntax command makes interpreting `0' according to standard Stata syntax easy. That will be covered in [U] **18.4.4 Parsing standard Stata syntax** below.

The easiest way to receive arguments, however, is to deal with the positional macros `1', `2', ...

At the start of this section, we imagined an xyz program invoked by typing xyz mpg weight. Then `1' would contain mpg, `2' would contain weight, and `3' would contain nothing.

Let's write a program to report the correlation between two variables. Of course, Stata already has a command that can do this—correlate—and, in fact, we will implement our program in terms of correlate. It is silly, but all we want to accomplish right now is to show how Stata passes arguments to a program.

Here is our program:

```
program xyz
        correlate `1' `2'
end
```

Once the program is defined, we can try it:

```
. use http://www.stata-press.com/data/r12/auto
(1978 Automobile Data)

. xyz mpg weight
(obs=74)

             |      mpg   weight
-------------+------------------
         mpg |   1.0000
      weight |  -0.8072   1.0000
```

See how this works? We typed xyz mpg weight, which invoked our xyz program with `1' being mpg and `2' being weight. Our program gave the command correlate `1' `2', and that expanded to correlate mpg weight.

Stylistically, this is not a good example of the use of positional arguments, but realistically, there is nothing wrong with it. The stylistic problem is that if xyz is really to report the correlation between two variables, it ought to allow standard Stata syntax, and that is not a difficult thing to do. Realistically, the program works.

Positional arguments, however, play an important role, even for programmers who care about style. When we write a subroutine—a program to be called by another program and not intended for direct human use—we often pass information by using positional arguments.

Stata forms the positional arguments `1', `2', ... by taking what the user typed following the command (or do-file), parsing it on white space with double quotes used to force binding, and stripping the quotes. The arguments are formed on the basis of words, but double-quoted strings are kept together as one argument but with the quotes removed.

Let's create a program to illustrate these concepts. Although we would not normally define programs interactively, this program is short enough that we will:

```
. program listargs
  1.  display "The 1st argument you typed is:  `1'"
  2.  display "The 2nd argument you typed is:  `2'"
  3.  display "The 3rd argument you typed is:  `3'"
  4.  display "The 4th argument you typed is:  `4'"
  5.  end
```

The display command simply types the double-quoted string following it; see [P] **display**.

Let's try our program:

```
. listargs
The 1st argument you typed is:
The 2nd argument you typed is:
The 3rd argument you typed is:
The 4th argument you typed is:
```

We type `listargs`, and the result shows us what we already know—we typed nothing after the word `listargs`. There are no arguments. Let's try it again, this time adding `this is a test`:

```
. listargs this is a test
The 1st argument you typed is:  this
The 2nd argument you typed is:  is
The 3rd argument you typed is:  a
The 4th argument you typed is:  test
```

We learn that the first argument is '`this`', the second is '`is`', and so on. Blanks always separate arguments. You can, however, override this feature by placing double quotes around what you type:

```
. listargs "this is a test"
The 1st argument you typed is:  this is a test
The 2nd argument you typed is:
The 3rd argument you typed is:
The 4th argument you typed is:
```

This time we typed only one argument, '`this is a test`'. When we place double quotes around what we type, Stata interprets whatever we type inside the quotes to be one argument. Here `` `1' `` contains '`this is a test`' (the double quotes were removed).

We can use double quotes more than once:

```
. listargs "this is" "a test"
The 1st argument you typed is:  this is
The 2nd argument you typed is:  a test
The 3rd argument you typed is:
The 4th argument you typed is:
```

The first argument is '`this is`' and the second argument is '`a test`'.

18.4.1 Named positional arguments

Positional arguments can be named: in your code, you do not have to refer to `` `1' ``, `` `2' ``, `` `3' ``, ...; you can instead refer to more meaningful names, such as `n`, `a`, and `b`; `numb`, `alpha`, and `beta`; or whatever else you find convenient. You want to do this because programs coded in terms of `` `1' ``, `` `2' ``, ... are hard to read and therefore are more likely to contain errors.

You obtain better-named positional arguments by using the `args` command:

```
program progname
        args argnames
        ...
end
```

For instance, if your program received four positional arguments and you wanted to call them `varname`, `n`, `oldval`, and `newval`, you would code

```
program progname
        args varname n oldval newval
        ...
end
```

varname, n, oldval, and newval become new local macros, and args simply copies ‘1’, ‘2’, ‘3’, and ‘4’ to them. It does not change ‘1’, ‘2’, ‘3’, and ‘4’—you can still refer to the numbered macros if you wish—and it does not verify that your program receives the right number of arguments. If our example above were invoked with just two arguments, ‘oldval’ and ‘newval’ would contain nothing. If it were invoked with five arguments, the fifth argument would still be out there, stored in local macro ‘5’.

Let's make a command to create a dataset containing n observations on x ranging from a to b. Such a command would be useful, for instance, if we wanted to graph some complicated mathematical function and experiment with different ranges. It is convenient if we can type the range of x over which we wish to make the graph rather than concocting the range by hand. (In fact, Stata already has such a command—range—but it will be instructive to write our own.)

Before writing this program, we had better know how to proceed, so here is how you could create a dataset containing n observations with x ranging from a to b:

1. clear
   to clear whatever data are in memory.
2. set obs *n*
   to make a dataset of n observations on no variables; if n were 100, we would type set obs 100.
3. gen x = (_n-1)/(*n*-1)*(*b*-*a*)+*a*
   because the built-in variable _n is 1 in the first observation, 2 in the second, and so on; see [U] **13.4 System variables (_variables)**.

So, the first version of our program might read

```
program rng                                    // arguments are n a b
        clear
        set obs ‘1’
        generate x = (_n-1)/(_N-1)*(‘3’-‘2’)+‘2’
end
```

The above is just a direct translation of what we just said. ‘1’ corresponds to n, ‘2’ corresponds to a, and ‘3’ corresponds to b. This program, however, would be far more understandable if we changed it to read

```
program rng
        args n a b
        clear
        set obs ‘n’
        generate x = (_n-1)/(_N-1)*(‘b’-‘a’)+‘a’
end
```

18.4.2 Incrementing through positional arguments

Some programs contain k arguments, where k varies, but it does not much matter because the same thing is done to each argument. One such program is summarize: type summarize mpg to obtain summary statistics on mpg, and type summarize mpg weight to obtain first summary statistics on mpg and then summary statistics on weight.

```
program ...
        local i = 1
        while "‘‘i’’" != "" {
                logic stated in terms of ‘‘i’’
                local ++i
        }
end
```

Equivalently, if the logic that uses ``i'' contains only one reference to ``i'',

```
program ...
        local i = 1
        while "``i''" != "" {
                logic stated in terms of ``i++''
        }
end
```

Note the tricky construction ``i'', which then itself is placed in double quotes—"``i''"—for the `while` loop. To understand it, say that `i` contains 1 or, equivalently, `i' is 1. Then ``i'' is `1' is the name of the first variable. "``i''" is the name of the first variable in quotes. The `while` asks if the name of the variable is nothing and, if it is not, executes. Now `i' is 2, and "``i''" is the name of the second variable, in quotes. If that name is not "", we continue. If the name is "", we are done.

Say that you were writing a subroutine that was to receive k variables, but the code that processes each variable needs to know (while it is processing) how many variables were passed to the subroutine. You need first to count the variables (and so derive k) and then, knowing k, pass through the list again.

```
program progname
        local k = 1                     // count the number of arguments
        while "``k''" != "" {
                local ++k
        }
        local --k                       // k contains one too many
                                        // now pass through again
        local i = 1
        while `i' <= `k' {
                code in terms of ``i'' and `k'
                local ++i
        }
end
```

In the above example, we have used `while`, Stata's all-purpose looping command. Stata has two other looping commands, `foreach` and `forvalues`, and they sometimes produce code that is more readable and executes more quickly. We direct you to read [P] **foreach** and [P] **forvalues**, but at this point, there is nothing they can do that `while` cannot do. Above we coded

```
        local i = 1
        while `i' <= `k' {
                code in terms of ``i'' and `k'
                local ++i
        }
```

to produce logic that looped over the values `i' = 1 to `k'. We could have instead coded

```
        forvalues i = 1(1)`k' {
                code in terms of ``i'' and `k'
        }
```

Similarly, at the beginning of this subsection, we said that you could use the following code in terms of `while` to loop over the arguments received:

```
program ...
        local i = 1
        while "``i''" != "" {
                logic stated in terms of ``i''
                local ++i
        }
end
```

Equivalent to the above would be

```
program ...
        foreach x of local 0 {
                logic stated in terms of `x'
        }
end
```

See [P] **foreach** and [P] **forvalues**.

You can combine `args` and incrementing through an unknown number of positional arguments. Say that you were writing a subroutine that was to receive `varname`, the name of some variable; `n`, which is some sort of count; and at least one and maybe 20 variable names. Perhaps you are to sum the variables, divide by `n`, and store the result in the first variable. What the program does is irrelevant; here is how we could receive the arguments:

```
program progname
        args varname n
        local i 3
        while "``i''" != "" {
                logic stated in terms of ``i''
                local ++i
        }
end
```

18.4.3 Using macro shift

Another way to code the repeat-the-same-process problem for each argument is

```
program ...
        while "`1'" != "" {
                logic stated in terms of `1'
                macro shift
        }
end
```

`macro shift` shifts `` `1' ``, `` `2' ``, `` `3' ``, ..., one to the left: what was `` `1' `` disappears, what was `` `2' `` becomes `` `1' ``, what was `` `3' `` becomes `` `2' ``, and so on.

The outside `while` loop continues the process until macro `` `1' `` contains nothing.

`macro shift` is an older construct that we no longer advocate using. Instead, we recommend that you use the techniques described in the previous subsection, that is, references to ``` ``i'' ``` and `foreach`/`forvalues`.

There are two reasons we make this recommendation: `macro shift` destroys the positional macros `` `1' ``, `` `2' ``, which must then be reset using `tokenize` should you wish to pass through the argument list again, and (more importantly) if the number of arguments is large (which in Stata/MP and Stata/SE is more likely), `macro shift` can be incredibly slow.

❑ Technical note

`macro shift` can do one thing that would be difficult to do by other means.

`` `*' ``, the result of listing the contents of the numbered macros one after the other with one blank between, changes with `macro shift`. Say that your program received a list of variables and that the first variable was the dependent variable and the rest were independent variables. You want to save the first variable name in `` `lhsvar' `` and all the rest in `` `rhsvars' ``. You could code

```
program progname
        local lhsvar "‘1’"
        macro shift 1
        local rhsvars "‘*’"
        ...
end
```

Now suppose that one macro contains a list of variables and you want to split the contents of the macro in two. Perhaps `‘varlist’` is the result of a `syntax` command (see [U] **18.4.4 Parsing standard Stata syntax**), and you now wish to split `‘varlist’` into `‘lhsvar’` and `‘rhsvars’`. `tokenize` will reset the numbered macros:

```
program progname
        ...
        tokenize ‘varlist’
        local lhsvar "‘1’"
        macro shift 1
        local rhsvars "‘*’"
        ...
end
```

❑

18.4.4 Parsing standard Stata syntax

Let's now switch to `‘0’` from the positional arguments `‘1’`, `‘2’`, ...

You can parse `‘0’` (what the user typed) according to standard Stata syntax with one command. Remember that standard Stata syntax is

[by *varlist*:] *command* [*varlist*] [=*exp*] [using *filename*] [*if*] [*in*] [*weight*]

[, *options*]

See [U] **11 Language syntax**.

The `syntax` command parses standard syntax. You code what amounts to the syntax diagram of your command in your program, and then `syntax` looks at `‘0’` (it knows to look there) and compares what the user typed with what you are willing to accept. Then one of two things happens: either `syntax` stores the pieces in an easily processable way or, if what the user typed does not match what you specified, `syntax` issues the appropriate error message and stops your program.

Consider a program that is to take two or more variable names along with an optional `if` *exp* and `in` *range*. The program would read

```
program ...
        syntax varlist(min=2) [if] [in]
        ...
end
```

You will have to read [P] **syntax** to learn how to specify the syntactical elements, but the command is certainly readable, and it will not be long until you are guessing correctly about how to fill it in. And yes, the square brackets really do indicate optional elements, and you just use them with `syntax` in the natural way.

The one `syntax` command you code encompasses the parsing process. Here, if what the user typed matches "two or more variables and an optional if and in", `syntax` defines new local macros:

`‘varlist’`	the two or more variable names
`‘if’`	the `if` *exp* specified by the user (or nothing)
`‘in’`	the `in` *range* specified by the user (or nothing)

To see that this works, experiment with the following program:

```
program tryit
        syntax varlist(min=2) [if] [in]
        display "varlist now contains |‘varlist’|"
        display ‘"if now contains |‘if’|"’
        display "in now contains |‘in’|"
end
```

Below we experiment:

```
. tryit mpg weight
varlist now contains |mpg weight|
if now contains ||
in now contains ||
. tryit mpg weight displ if foreign==1
varlist now contains |mpg weight displ|
if now contains |if foreign==1|
in now contains ||
. tryit mpg wei in 1/10
varlist now contains |mpg weight|
if now contains ||
in now contains |in 1/10|
. tryit mpg
too few variables specified
r(102);
```

In our third try we abbreviated the weight variable as `wei`, yet, after parsing, `syntax` unabbreviated the variable for us.

If this program were next going to step through the variables in the varlist, the positional macros `‘1’`, `‘2’`, ... could be reset by coding

```
tokenize ‘varlist’
```

See [P] **tokenize**. `tokenize ‘varlist’` resets `‘1’` to be the first word of `‘varlist’`, `‘2’` to be the second word, and so on.

18.4.5 Parsing immediate commands

Immediate commands are described in [U] **19 Immediate commands**—they take numbers as arguments. By convention, when you name immediate commands, you should make the last letter of the name *i*. Assume that `mycmdi` takes as arguments two numbers, the first of which must be a positive integer, and allows the options `alpha` and `beta`. The basic structure is

```
program mycmdi
        gettoken n 0 : 0, parse(" ,")          /* get first number */
        gettoken x 0 : 0, parse(" ,")          /* get second number */
        confirm integer number ‘n’             /* verify first is integer */
        confirm number ‘x’                     /* verify second is number */
        if ‘n’<=0 error 2001                   /* check that n is positive */
        place any other checks here
        syntax [, Alpha Beta]                  /* parse remaining syntax */
        make calculation and display output
end
```

See [P] **gettoken**.

18.4.6 Parsing nonstandard syntax

If you wish to interpret nonstandard syntax and positional arguments are not adequate for you, you know that you face a formidable programming task. The key to the solution is the `gettoken` command.

`gettoken` can pull one token from the front of a macro according to the parsing characters you specify and, optionally, define another macro or redefine the initial macro to contain the remaining (unparsed) characters. That is,

Say that `‘0’` contains	“`this is what the user typed`”
After `gettoken`,	
new macro `‘token’` could contain	“`this`”
and `‘0’` could still contain	“`this is what the user typed`”
or	
new macro `‘token’` could contain	“`this`”
and new macro `‘rest’` could contain	“` is what the user typed`”
and `‘0’` could still contain	“`this is what the user typed`”
or	
new macro `‘token’` could contain	“`this`”
and `‘0’` could contain	“` is what the user typed`”

A simplified syntax of `gettoken` is

`gettoken` *emname1* [*emname2*] `:` *emname3* [`,` `parse(`*pchars*`)` `quotes`

`match(`*lmacname*`)` `bind`]

where *emname1*, *emname2*, *emname3*, and *lmacname* are the names of local macros. (Stata provides a way to work with global macros, but in practice that is seldom necessary; see [P] **gettoken**.)

`gettoken` pulls the first token from *emname3* and stores it in *emname1*, and if *emname2* is specified, stores the remaining characters from *emname3* in *emname2*. Any of *emname1*, *emname2*, and *emname3* may be the same macro. Typically, `gettoken` is coded

`gettoken` *emname1* `: 0` [`,` *options*]

`gettoken` *emname1* `0 : 0` [`,` *options*]

because `‘0’` is the macro containing what the user typed. The first coding is used for token lookahead, should that be necessary, and the second is used for committing to taking the token.

`gettoken`'s options are

`parse("`*string*`")` — for specifying parsing characters
the default is `parse(" ")`, meaning to parse on white space
it is common to specify `parse(‘"" "’)`, meaning to parse on white space and double quote
(`‘"" "’` is the string double-quote-space in compound double quotes)

`quotes` — to specify that outer double quotes *not* be stripped

`match(`*lmacname*`)` — to bind on parentheses and square brackets
lmacname will be set to contain “`(`”, “`[`”, or nothing, depending on whether *emname1* was bound on parentheses or brackets or if `match()` turned out to be irrelevant
emname1 will have the outside parentheses or brackets removed

`gettoken` binds on double quotes whenever a (simple or compound) double quote is encountered at the beginning of *emname3*. Specifying `parse(‘"" "’)` ensures that double-quoted strings are isolated.

`quote` specifies that double quotes not be removed from the source in defining the token. For instance, in parsing “`"this is" a test`”, the next token is “`this is`” if `quote` is not specified and is “`"this is"`” if `quote` is specified.

`match()` specifies that parentheses and square brackets be matched in defining tokens. The outside level of parentheses or brackets is stripped. In parsing “`(2+3)/2`”, the next token is “`2+3`” if `match()` is specified. In practice, `match()` might be used with expressions, but it is more likely to be used to isolate bound varlists and time-series varlists.

18.5 Scalars and matrices

In addition to macros, scalars and matrices are provided for programmers; see [U] **14 Matrix expressions**, [P] **scalar** and [P] **matrix**.

As far as scalar calculations go, you can use macros or scalars. Remember, macros can hold numbers. Stata's scalars are, however, slightly faster and are a little more accurate than macros. The speed issue is so slight as to be nearly immeasurable. Macros are accurate to a minimum of 12 decimal digits, and scalars are accurate to roughly 16 decimal digits. Which you use makes little difference except in iterative calculations.

Stata also has a serious matrix programming language called Mata, which is the subject of another manual. Mata can be used to write subroutines that are called by Stata programs. See the *Mata Reference Manual*, and in particular, [M-1] **ado**.

18.6 Temporarily destroying the data in memory

It is sometimes necessary to modify the data in memory to accomplish a particular task. A well-behaved program, however, ensures that the user's data are always restored. The `preserve` command makes this easy:

code before the data need changing
`preserve`
code that changes data freely

When you give the `preserve` command, Stata makes a copy of the user's data on disk. When your program terminates—no matter how—Stata restores the data and erases the temporary file; see [P] **preserve**.

18.7 Temporary objects

If you write a substantial program, it will invariably require the use of temporary variables in the data, or temporary scalars, matrices, or files. Temporary objects are necessary while the program is making its calculations, and once the program completes they are discarded.

Stata provides three commands to create temporary objects: `tempvar` creates names for variables in the dataset, `tempname` creates names for scalars and matrices, and `tempfile` creates names for files. All are described in [P] **macro**, and all have the same syntax:

{ `tempvar` | `tempname` | `tempfile` } *macname* [*macname* ...]

The commands create local macros containing names you may use.

18.7.1 Temporary variables

Say that, in making a calculation, you need to add variables sum_y and sum_z to the data. You might be tempted to code

```
...
gen sum_y = ...
gen sum_z = ...
...
```

but that would be poor because the dataset might already have variables named sum_y and sum_z in it and you will have to remember to drop the variables before your program concludes. Better is

```
...
tempvar sum_y
gen `sum_y' = ...
tempvar sum_z
gen `sum_z' = ...
...
```

or

```
...
tempvar sum_y sum_z
gen `sum_y' = ...
gen `sum_z' = ...
...
```

It is not necessary to explicitly drop `sum_y' and `sum_z' when you are finished, although you may if you wish. Stata will automatically drop any variables with names assigned by tempvar. After issuing the tempvar command, you must refer to the names with the enclosing quotes, which signifies macro expansion. Thus, after typing tempvar sum_y—the one case where you do not put single quotes around the name—refer thereafter to the variable `sum_y', with quotes. tempvar does not create temporary variables. Instead tempvar creates names that may later be used to create new variables that will be temporary, and tempvar stores that name in the local macro whose name you provide.

A full description of tempvar can be found in [P] **macro**.

18.7.2 Temporary scalars and matrices

tempname works just like tempvar. For instance, a piece of your code might read

```
tempname YXX XXinv
matrix accum `YXX' = price weight mpg
matrix `XXinv' = invsym(`YXX'[2..., 2...])
tempname b
matrix `b' = `XXinv'*`YXX'[1..., 1]
```

The above code solves for the coefficients of a regression on price on weight and mpg; see [U] **14 Matrix expressions** and [P] **matrix** for more information on the matrix commands.

As with temporary variables, temporary scalars and matrices are automatically dropped at the conclusion of your program.

18.7.3 Temporary files

In cases where you ordinarily might think you need temporary files, you may not because of Stata's ability to preserve and automatically restore the data in memory; see [U] **18.6 Temporarily destroying the data in memory** above.

For more complicated programs, Stata does provide temporary files. A code fragment might read

```
preserve                          /* save original data */
tempfile males females
keep if sex==1
save "`males'"
restore, preserve                 /* get back original data */
keep if sex==0
save "`females'"
```

As with temporary variables, scalars, and matrices, it is not necessary to delete the temporary files when you are through with them; Stata automatically erases them when your program ends.

18.8 Accessing results calculated by other programs

Stata commands that report results also save the results where they can be subsequently used by other commands or programs. This is documented in the *Saved results* section of the particular command in the reference manuals. Commands save results in one of three places:

1. r-class commands, such as `summarize`, save their results in `r()`; most commands are r-class.
2. e-class commands, such as `regress`, save their results in `e()`; e-class commands are Stata's model estimation commands.
3. s-class commands (there are no good examples) save their results in `s()`; this is a rarely used class that programmers sometimes find useful to help parse input.

Commands that do not save results are called n-class commands. More correctly, these commands require that you state where the result is to be saved, as in `generate` *newvar* `=` . . . .

▷ Example 1

You wish to write a program to calculate the standard error of the mean, which is given by the formula $\sqrt{s^2/n}$, where s^2 is the calculated variance. (You could obtain this statistic by using the `ci` command, but we will pretend that is not true.) You look at [R] **summarize** and learn that the mean is stored in `r(mean)`, the variance in `r(Var)`, and the number of observations in `r(N)`. With that knowledge, you write the following program:

```
program meanse
        quietly summarize `1'
        display "      mean = " r(mean)
        display "SE of mean = " sqrt(r(Var)/r(N))
end
```

The result of executing this program is

```
. meanse mpg
      mean = 21.297297
SE of mean = .67255109
```

◁

If you run an r-class command and type `return list` or run an e-class command and type `ereturn list`, Stata will summarize what was saved:

```
. use http://www.stata-press.com/data/r12/auto
(1978 Automobile Data)
. regress mpg weight displ
  (output omitted)
. ereturn list
scalars:
                  e(N) =  74
               e(df_m) =  2
               e(df_r) =  71
                  e(F) =  66.78504752026517
                 e(r2) =  .6529306984682528
               e(rmse) =  3.45606176570828
                e(mss) =  1595.409691543724
                e(rss) =  848.0497679157351
               e(r2_a) =  .643154098425105
                 e(ll) =  -195.2397979466294
               e(ll_0) =  -234.3943376482347
               e(rank) =  3
macros:
            e(cmdline) : "regress mpg weight displ"
              e(title) : "Linear regression"
         e(marginsok) : "XB default"
                e(vce) : "ols"
             e(depvar) : "mpg"
                e(cmd) : "regress"
         e(properties) : "b V"
            e(predict) : "regres_p"
              e(model) : "ols"
          e(estat_cmd) : "regress_estat"
matrices:
                  e(b) :  1 x 3
                  e(V) :  3 x 3
functions:
             e(sample)
. summarize mpg if foreign
    Variable |       Obs        Mean    Std. Dev.       Min        Max
-------------+--------------------------------------------------------
         mpg |        22    24.77273    6.611187         14         41
. return list
scalars:
                  r(N) =  22
              r(sum_w) =  22
               r(mean) =  24.77272727272727
                r(Var) =  43.70779220779221
                 r(sd) =  6.611186898567625
                r(min) =  14
                r(max) =  41
                r(sum) =  545
```

In the example above, we ran `regress` followed by `summarize`. As a result, `e(N)` records the number of observations used by `regress` (equal to 74), and `r(N)` records the number of observations used by `summarize` (equal to 22). `r(N)` and `e(N)` are not the same.

If we now ran another r-class command—say, `tabulate`—the contents of `r()` would change, but those in `e()` would remain unchanged. You might, therefore, think that if we then ran another e-class command, say, `probit`, the contents of `e()` would change, but `r()` would remain unchanged.

Although it is true that e() results remain in place until the next e-class command is executed, do not depend on r() remaining unchanged. If an e-class or n-class command were to use an r-class command as a subroutine, that would cause r() to change. Anyway, most commands are r-class, so the contents of r() change often.

❑ Technical note

It is, therefore, of great importance that you access results stored in r() immediately after the command that sets them. If you need the mean and variance of the variable '1' for subsequent calculation, do *not* code

```
summarize '1'
...
... r(mean) ... r(Var) ...
```

Instead, code

```
summarize '1'
local mean = r(mean)
local var = r(Var)
...
... 'mean' ... 'var' ...
```

or

```
tempname mean var
summarize '1'
scalar 'mean' = r(mean)
scalar 'var' = r(Var)
...
... 'mean' ... 'var' ...
```

❑

Saved results, whether in r() or e(), come in three types: scalars, macros, and matrices. If you look back at the ereturn list and return list output, you will see that regress saves examples of all three, whereas summarize saves just scalars. (regress also saves the "function" e(sample), as do all the other e-class commands; see [U] **20.6 Specifying the estimation subsample**.)

Regardless of the type of e(*name*) or r(*name*), you can just refer to e(*name*) or r(*name*). That was the rule we gave in [U] **13.6 Accessing results from Stata commands**, and that rule is sufficient for most uses. There is, however, another way to refer to saved results. Rather than referring to r(*name*) and e(*name*), you can embed the reference in macro-substitution characters '' to produce 'r(*name*)' and 'e(*name*)'. The result is the same as macro substitution; the saved result is evaluated, and then the evaluation is substituted:

```
. display "You can refer to " e(cmd) " or to 'e(cmd)'"
You can refer to regress or to regress
```

This means, for instance, that typing 'e(cmd)' is the same as typing regress because e(cmd) contains "regress":

```
. 'e(cmd)'

      Source |       SS       df       MS              Number of obs =      74
-------------+------------------------------           F(  2,    71) =   66.79
       Model |  1595.40969     2  797.704846           Prob > F      =  0.0000
  (remaining output omitted)
```

In the ereturn list, e(cmd) was listed as a macro, and when you place a macro's name in single quotes, the macro's contents are substituted, so this is hardly a surprise.

What is surprising is that you can do this with scalar and even matrix saved results. `e(N)` is a scalar equal to 74 and may be used as such in any expression such as "`display e(mss)/e(N)`" or "`local meanss = e(mss)/e(N)`". `` `e(N)' `` substitutes to the string "74" and may be used in any context whatsoever, such as "`` local val`e(N)' = e(N) ``" (which would create a macro named `val74`). The rules for referring to saved results are

1. You may refer to `r(`*name*`)` or `e(`*name*`)` without single quotes in any expression and only in an expression. (Referring to s-class `s(`*name*`)` without single quotes is not allowed.)

   1.1 If *name* does not exist, missing value (`.`) is returned; it is not an error to refer to a nonexistent saved result.

   1.2 If *name* is a scalar, the full double-precision value of *name* is returned.

   1.3 If *name* is a macro, it is examined to determine whether its contents can be interpreted as a number. If so, the number is returned; otherwise, the first 80 characters of *name* are returned.

   1.4 If *name* is a matrix, the full *matrix* is returned.

2. You may refer to `` `r( ``*name*`)'`, `` `e( ``*name*`)'`, or `` `s( ``*name*`)'`—note the presence of quotes indicating macro substitution—in any context whatsoever.

   2.1 If *name* does not exist, nothing is substituted; it is not an error to refer to a nonexistent saved result. The resulting line is the same as if you had never typed `` `r( ``*name*`)'`, `` `e( ``*name*`)'`, or `` `s( ``*name*`)'`.

   2.2 If *name* is a scalar, a string representation of the number accurate to no less than 12 digits of precision is substituted.

   2.3 If *name* is a macro, the full contents are substituted.

   2.4 If *name* is a matrix, the word `matrix` is substituted.

In general, you should refer to scalar and matrix saved results without quotes—`r(`*name*`)` and `e(`*name*`)`—and to macro saved results with quotes—`` `r( ``*name*`)'`, `` `e( ``*name*`)'`, and `` `s( ``*name*`)'`—but it is sometimes convenient to switch. Say that returned result `r(example)` contains the number of periods patients are observed, and assume that `r(example)` was saved as a macro and not as a scalar. You could still refer to `r(example)` without the quotes in an expression context and obtain the expected result. It would have made more sense for you to have stored `r(example)` as a scalar, but really it would not matter, and the user would not even have to know how the saved result was stored.

Switching the other way is sometimes useful, too. Say that returned result `r(N)` is a scalar that contains the number of observations used. You now want to use some other command that has an option `n(`*#*`)` that specifies the number of observations used. You could not type `n(r(N))` because the syntax diagram says that the `n()` option expects its argument to be a literal number. Instead, you could type `` n(`r(N)') ``.

18.9 Accessing results calculated by estimation commands

Estimation results are saved in `e()`, and you access them in the same way you access any saved result; see [U] **18.8 Accessing results calculated by other programs** above. In summary,

1. Estimation commands—`regress`, `logistic`, etc.—save results in `e()`.

2. Estimation commands save their name in `e(cmd)`. For instance, `regress` saves "`regress`" and `poisson` saves "`poisson`" in `e(cmd)`.

3. Estimation commands save the command they executed in `e(cmdline)`. For instance, if you typed `reg mpg displ`, saved in `e(cmdline)` would be "`reg mpg displ`".
4. Estimation commands save the number of observations used in `e(N)`, and they identify the estimation subsample by setting `e(sample)`. You could type, for instance, `summarize if e(sample)` to obtain summary statistics on the observations used by the estimator.
5. Estimation commands save the entire coefficient vector and variance–covariance matrix of the estimators in `e(b)` and `e(V)`. These are matrices, and they may be manipulated like any other matrix:

```
. matrix list e(b)
e(b)[1,3]
         weight        displ        _cons
y1   -.00656711    .00528078    40.084522
. matrix y = e(b)*e(V)*e(b)'
. matrix list y
symmetric y[1,1]
           y1
y1  6556.982
```

6. Estimation commands set `_b[`*name*`]` and `_se[`*name*`]` as convenient ways to use coefficients and their standard errors in expressions; see [U] **13.5 Accessing coefficients and standard errors**.
7. Estimation commands may set other `e()` scalars, macros, or matrices containing more information. This is documented in the *Saved results* section of the particular command in the command reference.

▷ Example 2

If you are writing a command for use after `regress`, early in your code you should include the following:

```
if "`e(cmd)'" != "regress" {
        error 301
}
```

This is how you verify that the estimation results that are stored have been set by `regress` and not by some other estimation command. Error 301 is Stata's "last estimates not found" error.

◁

18.10 Saving results

If your program calculates something, it should save the results of the calculation so that other programs can access them. In this way your program not only can be used interactively but also can be used as a subroutine for other commands.

Saving results is easy:

1. On the `program` line, specify the `rclass`, `eclass`, or `sclass` option according to whether you intend to return results in `r()`, `e()`, or `s()`.
2. Code

`return scalar` *name* `=` *exp*	(same syntax as `scalar` without the `return`)
`return local` *name* ...	(same syntax as `local` without the `return`)
`return matrix` *name matname*	(moves *matname* to `r(`*name*`)`)

   to save results in `r()`.

3. Code

`ereturn` *name* = *exp*	(same syntax as `scalar` without the `ereturn`)
`ereturn local` *name* ...	(same syntax as `local` without the `ereturn`)
`ereturn matrix` *name matname*	(moves *matname* to `e(`*name*`)`)

   to save results in `e()`. You do not save the coefficient vector and variance matrix `e(b)` and `e(V)` in this way; instead you use `ereturn post`.

4. Code

`sreturn local` *name* ...	(same syntax as `local` without the `sreturn`)

   to save results in `s()`. (The s-class has only macros.)

A program must be exclusively r-class, e-class, or s-class.

18.10.1 Saving results in r()

In [U] **18.8 Accessing results calculated by other programs**, we showed an example that reported the mean and standard error of the mean. A better version would save in `r()` the results of its calculations and would read

```
program meanse, rclass
        quietly summarize `1'
        local mean = r(mean)
        local sem  = sqrt(r(Var)/r(N))
        display "      mean = " `mean'
        display "SE of mean = " `sem'
        return scalar mean = `mean'
        return scalar se = `sem'
end
```

Running `meanse` now sets `r(mean)` and `r(se)`:

```
. meanse mpg
      mean = 21.297297
SE of mean = .67255109

. return list

scalars:
          r(se)      =  .6725510870764975
          r(mean)    =  21.2972972972973
```

In this modification, we added the `rclass` option to the `program` statement, and we added two `return` commands to the end of the program.

Although we placed the `return` statements at the end of the program, they may be placed at the point of calculation if that is more convenient. A more concise version of this program would read

```
program meanse, rclass
        quietly summarize `1'
        return scalar mean = r(mean)
        return scalar se = sqrt(r(Var)/r(N))
        display "      mean = " return(mean)
        display "SE of mean = " return(se)
end
```

The `return()` function is just like the `r()` function, except that `return()` refers to the results that this program *will* return rather than to the saved results that currently *are* returned (which here are due to `summarize`). That is, when you code the `return` command, the result is not immediately posted to `r()`. Rather, Stata holds onto the result in `return()` until your program concludes, and then it copies the contents of `return()` to `r()`. While your program is active, you may use the `return()` function to access results you have already "returned". (`return()` works just like `r()` works after your program returns, meaning that you may code `‘return()’` to perform macro substitution.)

18.10.2 Saving results in e()

Saving in `e()` is in most ways similar to saving in `r()`: you add the `eclass` option to the `program` statement, and then you use `ereturn` ... just as you used `return` ... to store results. There are, however, some significant differences:

1. Unlike `r()`, estimation results are saved in `e()` the instant you issue an `ereturn scalar`, `ereturn local`, or `ereturn matrix` command. Estimation results can consume considerable memory, and Stata does not want to have multiple copies of the results floating around. That means you must be more organized and post your results at the end of your program.
2. In your code when you have your estimates and are ready to begin posting, you will first clear the previous estimates, set the coefficient vector `e(b)` and corresponding variance matrix `e(V)`, and set the estimation-sample function `e(sample)`. How you do this depends on how you obtained your estimates:

    2.1 If you obtained your estimates by using Stata's likelihood maximizer `ml`, this is automatically handled for you; skip to step 3.

    2.2 If you obtained estimates by "stealing" an existing estimator, `e(b)`, `e(V)`, and `e(sample)` already exist, and you will not want to clear them; skip to step 3.

    2.3 If you write your own code from start to finish, you use the `ereturn post` command; see [P] **ereturn**. You will code something like "`ereturn post ‘b’ ‘V’, esample(‘touse’)`", where `‘b’` is the name of the coefficient vector, `‘V’` is the name of the corresponding variance matrix, and `‘touse’` is the name of a variable containing 1 if the observation was used and 0 if it was ignored. `ereturn post` clears the previous estimates and moves the coefficient vector, variance matrix, and variable into `e(b)`, `e(V)`, and `e(sample)`.

    2.4 A variation on (2.3) is when you use an existing estimator to produce the estimates but do not want all the other `e()` results stored by the estimator. Then you code

    ```
    tempvar touse
    tempname b V
    matrix ‘b’ = e(b)
    matrix ‘V’ = e(V)
    qui gen byte ‘touse’ = e(sample)
    ereturn post ‘b’ ‘V’, esample(‘touse’)
    ```

3. You now save anything else in `e()` that you wish by using the `ereturn scalar`, `ereturn local`, or `ereturn matrix` command.
4. Save `e(cmdline)` by coding

    ```
    ereturn local cmdline ‘"‘0’"’
    ```

    This is not required, but it is considered good style.

5. You code `ereturn local cmd "`*cmdname*`"`. Stata does not consider estimation results complete until this command is posted, and Stata considers the results to be complete when this is posted, so you must remember to do this and to do this last. If you set `e(cmd)` too early and the user pressed *Break*, Stata would consider your estimates complete when they are not.

Say that you wish to write the estimation command with syntax

`myest` *depvar* var_1 var_2 [`if` *exp*] [`in` *range*]`,` *optset1 optset2*

where *optset1* affects how results are displayed and *optset2* affects the estimation results themselves. One important characteristic of estimation commands is that, when typed without arguments, they redisplay the previous estimation results. The outline is

```
program myest, eclass
        local options "optset1"
        if replay() {
                if "`e(cmd)'"!="myest" {
                        error 301              /* last estimates not found       */
                }
                syntax [, `options']
        }
        else {
                syntax varlist [if] [in] [, `options' optset2]
                marksample touse
Code contains either this,
                tempnames b V
                commands for performing estimation
                assume produces `b' and `V'
                ereturn post `b' `V', esample(`touse')
                ereturn local depvar "`depv'"
or this,
                ml model ... if `touse' ...
and regardless, concludes,
                perhaps other ereturn commands appear here
                ereturn local cmdline `"`0'"'
                ereturn local cmd "myest"
        }
                                               /* (re)display results ...        */
code typically reads
        code to output header above coefficient table
        ereturn display                       /* displays coefficient table     */
or
        ml display                            /* displays header and coef. table */
end
```

Here is a list of the commonly saved `e()` results. Of course, you may create any `e()` results that you wish.

`e(N)` (scalar)
Number of observations.

`e(df_m)` (scalar)
Model degrees of freedom.

`e(df_r)` (scalar)
"Denominator" degrees of freedom if estimates are nonasymptotic.

`e(r2_p)` (scalar)
Value of the pseudo-R^2 if it is calculated. (If a "real" R^2 is calculated as it would be in linear regression, it is stored in (scalar) `e(r2)`.)

e(F) (scalar)
Test of the model against the constant-only model, if relevant, and if results are nonasymptotic.

e(ll) (scalar)
Log-likelihood value, if relevant.

e(ll_0) (scalar)
Log-likelihood value for constant-only model, if relevant.

e(N_clust) (scalar)
Number of clusters, if any.

e(chi2) (scalar)
Test of the model against the constant-only model, if relevant, and if results are asymptotic.

e(rank) (scalar)
Rank of e(V).

e(cmd) (macro)
Name of the estimation command.

e(cmdline) (macro)
Command as typed.

e(depvar) (macro)
Names of the dependent variables.

e(wtype) and e(wexp) (macros)
If weighted estimation was performed, e(wtype) contains the weight type (fweight, pweight, etc.) and e(wexp) contains the weighting expression.

e(title) (macro)
Title in estimation output.

e(clustvar) (macro)
Name of the cluster variable, if any.

e(vcetype) (macro)
Text to appear above standard errors in estimation output; typically Robust, Bootstrap, Jackknife, or "".

e(vce) (macro)
vcetype specified in vce().

e(chi2type) (macro)
LR or Wald or other depending on how e(chi2) was performed.

e(properties) (macro)
Typically contains b V.

e(predict) (macro)
Name of the command that predict is to use; if this is blank, predict uses the default _predict.

e(b) and e(V) (matrices)
The coefficient vector and corresponding variance matrix. Saved when you coded ereturn post.

e(sample) (function)
This function was defined by ereturn post's esample() option if you specified it. You specified a variable containing 1 if you used an observation and 0 otherwise. ereturn post stole the variable and created e(sample) from it.

18.10.3 Saving results in s()

s() is a strange class because, whereas the other classes allow scalars, macros, and matrices, s() allows only macros.

s() is seldom used and is for subroutines that you might write to assist in parsing the user's input prior to evaluating any user-supplied expressions.

Here is the problem s solves: say that you create a nonstandard syntax for some command so that you have to parse through it yourself. The syntax is so complicated that you want to create subroutines to take pieces of it and then return information to your main routine. Assume that your syntax contains expressions that the user might type. Now say that one of the expressions the user types is, for example, r(mean)/sqrt(r(Var))—perhaps the user is using results left behind by summarize.

If, in your parsing step, you call subroutines that return results in r(), you will wipe out r(mean) and r(Var) before you ever get around to seeing them, much less evaluating them. So, you must be careful to leave r() intact until your parsing is complete; you must use no r-class commands, and any subroutines you write must not touch r(). You must use s-class subroutines because s-class routines return results in s() rather than r(). S-class provides macros only because that is all you need to solve parsing problems.

To create an s-class routine, specify the sclass option on the program line and then use sreturn local to return results.

S-class results are posted to s() at the instant you issue the sreturn() command, so you must organize your results. Also, s() is never automatically cleared, so occasionally coding sreturn clear at appropriate points in your code is a good idea. Few programs need s-class subroutines.

18.11 Ado-files

Ado-files were introduced in [U] **17 Ado-files**.

When a user types '*gobbledygook*', Stata first asks itself if *gobbledygook* is one of its built-in commands. If so, the command is executed. Otherwise, it asks itself if *gobbledygook* is a defined program. If so, the program is executed. Otherwise, Stata looks in various directories for *gobbledygook*.ado. If there is no such file, the process ends with the "unrecognized command" error.

If Stata finds the file, it quietly issues to itself the command 'run *gobbledygook*.ado' (specifying the path explicitly). If that runs without error, Stata asks itself again if *gobbledygook* is a defined program. If not, Stata issues the "unrecognized command" error. (Here somebody wrote a bad ado-file.) If the program is defined, as it should be, Stata executes it.

Thus you can arrange for programs you write to be loaded automatically. For instance, if you were to create hello.ado containing

begin hello.ado

```
program hello
        display "hi there"
end
```

end hello.ado

and store the file in your current directory or your personal directory (see [U] **17.5.2 Where is my personal ado-directory?**), you could type hello and be greeted by a reassuring

```
. hello
hi there
```

You could, at that point, think of hello as just another part of Stata.

There are two places to put your personal ado-files. One is the current directory, and that is a good choice when the ado-file is unique to a project. You will want to use it only when you are in that directory. The other place is your *personal ado-directory*, which is probably something like `C:\ado\personal` if you use Windows, `~/ado/personal` if you use Unix, and `~/ado/personal` if you use a Mac. We are guessing.

To find your personal ado-directory, enter Stata and type

```
. personal
```

❑ Technical note

Stata looks in various directories for ado-files, defined by the c-class value `c(adopath)`, which contains

```
UPDATES;BASE;SITE;.;PERSONAL;PLUS;OLDPLACE
```

The words in capital letters are codenames for directories, and the mapping from codenames to directories can be obtained by typing the `sysdir` command. Here is what `sysdir` shows on one particular Windows computer:

```
. sysdir
   STATA:  C:\Program Files\Stata12\
 UPDATES:  C:\Program Files\Stata12\ado\updates\
    BASE:  C:\Program Files\Stata12\ado\base\
    SITE:  C:\Program Files\Stata12\ado\site\
    PLUS:  C:\ado\plus\
PERSONAL:  C:\ado\personal\
OLDPLACE:  C:\ado\
```

Even if you use Windows, your mapping might be different because it all depends on where you installed Stata. That is the point of the codenames. They make it possible to refer to directories according to their logical purposes rather than their physical location.

The c-class value `c(adopath)` is the search path, so in looking for an ado-file, Stata first looks in `UPDATES` then in `BASE`, and so on, until it finds the file. Actually, Stata not only looks in `UPDATES` but also takes the first letter of the ado-file it is looking for and looks in the lettered subdirectory. For files with the extension `.style`, Stata will look in a subdirectory named `style` rather than a lettered subdirectory. Say that Stata was looking for `gobbledygook.ado`. Stata would look up `UPDATES` (`C:\Program Files\Stata12\ado\updates` in our example) and, if the file were not found there, it would look in the g subdirectory of `UPDATES` (`C:\Program Files\Stata12\ado\updates\g`) before looking in `BASE`, whereupon it would follow the same rules. If Stata were looking for `gobbledygook.style`, Stata would look up `UPDATES` (`C:\Program Files\Stata12\ado\updates` in our example) and, if the file were not found there, it would look in the `style` subdirectory of `UPDATES` (`C:\Program Files\Stata12\ado\updates\style`) before looking in `BASE`, whereupon it would follow the same rules.

Why the extra complication? We distribute hundreds of ado-files, help files, and other file types with Stata, and some operating systems have difficulty dealing with so many files in the same directory. All operating systems experience at least a performance degradation. To prevent this, the ado-directory we ship is split 28 ways (letters *a–z*, underscore, and `style`). Thus the Stata command `ci`, which is implemented as an ado-file, can be found in the subdirectory `c` of `BASE`.

If you write ado-files, you can structure your personal ado-directory this way, too, but there is no reason to do so until you have more than, say, 250 files in one directory.

❑

❑ Technical note

After finding and running *gobbledygook*.ado, Stata calculates the total size of all programs that it has automatically loaded. If this exceeds adosize (see [P] **sysdir**), Stata begins discarding the oldest automatically loaded programs until the total is less than adosize. Oldest here is measured by the time last used, not the time loaded. This discarding saves memory and does not affect you, because any program that was automatically loaded could be automatically loaded again if needed.

It does, however, affect performance. Loading the program takes time, and you will again have to wait if you use one of the previously loaded-and-discarded programs. Increasing adosize reduces this possibility, but at the cost of memory. The set adosize command allows you to change this parameter; see [P] **sysdir**. The default value of adosize is 1,000. A value of 1,000 for adosize means that up to 1,000 K can be allocated to autoloaded programs. Experimentation has shown that this is a good number—increasing it does not improve performance much.

❑

18.11.1 Version

We recommend that the first line following program in your ado-file declare the Stata release under which you wrote the program; hello.ado would read better as

begin hello.ado

```
program hello
        version 12
        display "hi there"
end
```

end hello.ado

We introduced the concept of version in [U] **16.1.1 Version**. In regular do-files, we recommend that the version line appear as the first line of the do-file. For ado-files, the line appears after the program because loading the ado-file is one step and executing the program is another. It is when Stata executes the program defined in the ado-file that we want to stipulate the interpretation of the commands.

The inclusion of the version line is of more importance in ado-files than in do-files because ado-files have longer lives than do-files, so it is more likely that you will use an ado-file with a later release and ado-files tend to use more of Stata's features, increasing the probability that any change to Stata will affect them.

18.11.2 Comments and long lines in ado-files

Comments in ado-files are handled the same way as in do-files: you enclose the text in /* comment */ brackets, or you begin the line with an asterisk (*), or you interrupt the line with //; see [U] **16.1.2 Comments and blank lines in do-files**.

Logical lines longer than physical lines are also handled as they are in do-files: either you change the delimiter to a semicolon (;) or you comment out the new line by using /// at the end of the previous physical line.

18.11.3 Debugging ado-files

Debugging ado-files is a little tricky because it is Stata and not you that controls when the ado-file is loaded.

Assume that you wanted to change hello to say "Hi, Mary". You open hello.ado in the Do-file Editor and change it to read

begin hello.ado
```
program hello
        version 12
        display "hi, Mary"
end
```
end hello.ado

After saving it, you try it:

```
. hello
hi there
```

Stata ran the old copy of hello—the copy it still has in its memory. Stata wants to be fast about executing ado-files, so when it loads one, it keeps it around a while—waiting for memory to get short—before clearing it from its memory. Naturally, Stata can drop hello anytime because it can always reload it from disk.

You changed the copy on disk, but Stata still has the old copy loaded into memory. You type discard to tell Stata to forget these automatically loaded things and to force itself to get new copies of the ado-files from disk:

```
. discard
. hello
hi, Mary
```

You had to type discard only because you changed the ado-file while Stata was running. Had you exited Stata and returned later to use hello, the discard would not have been necessary because Stata forgets things between sessions anyway.

18.11.4 Local subroutines

An ado-file can contain more than one program, and if it does, the other programs defined in the ado-file are assumed to be subroutines of the main program. For example,

begin decoy.ado
```
program decoy
        ...
        duck ...
        ...
end
program  duck
        ...
end
```
end decoy.ado

duck is considered a local subroutine of decoy. Even after decoy.ado was loaded, if you typed duck, you would be told "unrecognized command". To emphasize what *local* means, assume that you have also written an ado-file named duck.ado:

begin duck.ado
```
program duck
        ...
end
```
end duck.ado

Even so, when `decoy` called `duck`, it would be the program `duck` defined in `decoy.ado` that was called. To further emphasize what *local* means, assume that `decoy.ado` contains

begin decoy.ado
```
program decoy
        ...
        manic ...
        ...
        duck ...
        ...
end
program duck
        ...
end
```
end decoy.ado

and that `manic.ado` contained

begin manic.ado
```
program manic
        ...
        duck ...
        ...
end
```
end manic.ado

Here is what would happen when you executed `decoy`:

1. `decoy` in `decoy.ado` would begin execution. `decoy` calls `manic`.
2. `manic` in `manic.ado` would begin execution. `manic` calls `duck`.
3. `duck` in `duck.ado` (yes) would begin execution. `duck` would do whatever and return.
4. `manic` regains control and eventually returns.
5. `decoy` is back in control. `decoy` calls `duck`.
6. `duck` in `decoy.ado` would execute, complete, and return.
7. `decoy` would regain control and return.

When `manic` called `duck`, it was the global ado-file `duck.ado` that was executed, yet when `decoy` called `duck`, it was the local program `duck` that was executed.

Stata does not find this confusing and neither should you.

18.11.5 Development of a sample ado-command

Below we demonstrate how to create a new Stata command. We will program an influence measure for use with linear regression. It is an interesting statistic in its own right, but even if you are not interested in linear regression and influence measures, the focus here is on programming, not on the particular statistic chosen.

Belsley, Kuh, and Welsch (1980, 24) present a measure of influence in linear regression defined as

$$\frac{\text{Var}\left(\widehat{y}_i^{(i)}\right)}{\text{Var}(\widehat{y}_i)}$$

which is the ratio of the variance of the ith fitted value based on regression estimates obtained by omitting the ith observation to the variance of the ith fitted value estimated from the full dataset. This ratio is estimated using

$$\text{FVARATIO}_i \equiv \frac{n-k}{n-(k+1)}\left\{1-\frac{d_i^2}{1-h_{ii}}\right\}(1-h_{ii})^{-1}$$

where n is the sample size; k is the number of estimated coefficients; $d_i^2 = e_i^2/\mathbf{e}'\mathbf{e}$ and e_i is the ith residual; and h_{ii} is the ith diagonal element of the hat matrix. The ingredients of this formula are all available through Stata, so, after estimating the regression parameters, we can easily calculate FVARATIO_i. For instance, we might type

```
. regress mpg weight displ
. predict hii if e(sample), hat
. predict ei if e(sample), resid
. quietly count if e(sample)
. scalar nreg = r(N)
. gen eTe = sum(ei*ei)
. gen di2 = (ei*ei)/eTe[_N]
. gen FVi = (nreg - 3) / (nreg - 4) * (1 - di2/(1-hii)) / (1-hii)
```

The number 3 in the formula for `FVi` represents k, the number of estimated parameters (which is an intercept plus coefficients on `weight` and `displ`), and the number 4 represents $k+1$.

❑ Technical note

Do you understand why this works? `predict` can create h_{ii} and e_i, but the trick is in getting $\mathbf{e}'\mathbf{e}$—the sum of the squared e_is. Stata's `sum()` function creates a running sum. The first observation of `eTe` thus contains e_1^2; the second, $e_1^2+e_2^2$; the third, $e_1^2+e_2^2+e_3^2$; and so on. The last observation, then, contains $\sum_{i=1}^N e_i^2$, which is $\mathbf{e}'\mathbf{e}$. (We specified `if e(sample)` on our `predict` commands to restrict calculations to the estimation subsample, so `hii` and `eii` might have missing values, but that does not matter because `sum()` treats missing values as contributing zero to the sum.) We use Stata's explicit subscripting feature and then refer to `eTe[_N]`, the last observation. (See [U] **13.3 Functions** and [U] **13.7 Explicit subscripting**.) After that, we plug into the formula to obtain the result.

❑

Assuming that we often wanted this influence measure, it would be easier and less prone to error if we canned this calculation in a program. Our first draft of the program reflects exactly what we would have typed interactively:

begin fvaratio.ado, version 1

```
program fvaratio
        version 12
        predict hii if e(sample), hat
        predict ei if e(sample), resid
        quietly count if e(sample)
        scalar nreg = r(N)
        gen eTe = sum(ei*ei)
        gen di2 = (ei*ei)/eTe[_N]
        gen FVi = (nreg - 3) / (nreg - 4) * (1 - di2/(1-hii)) / (1-hii)
        drop hii ei eTe di2
end
```

end fvaratio.ado, version 1

All we have done is to enter what we would have typed into a file, bracketing it with `program fvaratio` and `end`. Because our command is to be called `fvaratio`, the file must be named `fvaratio.ado` and must be stored in either the current directory or our personal ado-directory (see [U] **17.5.2 Where is my personal ado-directory?**).

Now when we type `fvaratio`, Stata will be able to find it, load it, and execute it. In addition to copying the interactive lines into a program, we added the line '`drop hii` ...' to eliminate the working variables we had to create along the way.

So, now we can interactively type

```
. regress mpg weight displ
. fvaratio
```

and add the new variable `FVi` to our data.

Our program is not general. It is suitable for use after fitting a regression model on two, and only two, independent variables because we coded a `3` in the formula for k. Stata statistical commands such as `regress` store information about the problem and answer in `e()`. Looking in *Saved results* in [R] **regress**, we find that `e(df_m)` contains the model degrees of freedom, which is $k - 1$, assuming that the model has an intercept. Also, the sample size of the dataset used in the regression is stored in `e(N)`, eliminating our need to count the observations and define a scalar containing this count. Thus the second draft of our program reads

begin fvaratio.ado, version 2

```
program fvaratio
        version 12
        predict hii if e(sample), hat
        predict ei if e(sample), resid
        gen eTe = sum(ei*ei)
        gen di2 = (ei*ei)/eTe[_N]
        gen FVi = (e(N)-(e(df_m)+1)) / (e(N)-(e(df_m)+2)) *     /// changed this
                  (1 - di2/(1-hii)) / (1-hii)                   // version
        drop hii ei eTe di2
end
```

end fvaratio.ado, version 2

In the formula for `FVi`, we substituted `(e(df_m)+1)` for the literal number `3`, `(e(df_m)+2)` for the literal number `4`, and `e(N)` for the sample size.

Back to the substance of our problem, `regress` also saves the residual sum of squares in `e(rss)`, so calculating `eTe` is not really necessary:

begin fvaratio.ado, version 3

```
program fvaratio
        version 12
        predict hii if e(sample), hat
        predict ei if e(sample), resid
        gen di2 = (ei*ei)/e(rss)                          // changed this version
        gen FVi = (e(N)-(e(df_m)+1)) / (e(N)-(e(df_m)+2)) *       ///
                  (1 - di2/(1-hii)) / (1-hii)
        drop hii ei di2
end
```

end fvaratio.ado, version 3

Our program is now shorter and faster, and it is completely general. This program is probably good enough for most users; if you were implementing this solely for your own occasional use, you could stop right here. The program does, however, have the following deficiencies:

1. When we use it with data with missing values, the answer is correct, but we see messages about the number of missing values generated. (These messages appear when the program is generating the working variables.)
2. We cannot control the name of the variable being produced—it is always called `FVi`. Moreover, when `FVi` already exists (say, from a previous regression), we get an error message that `FVi` already exists. We then have to `drop` the old `FVi` and type `fvaratio` again.
3. If we have created any variables named `hii`, `ei`, or `di2`, we also get an error that the variable already exists, and the program refuses to run.

Fixing these problems is not difficult. The fix for problem 1 is easy; we embed the entire program in a `quietly` block:

begin fvaratio.ado, version 4

```
program fvaratio
        version 12
        quietly {                                          // new this version
                predict hii if e(sample), hat
                predict ei if e(sample), resid
                gen di2 = (ei*ei)/e(rss)
                gen FVi = (e(N)-(e(df_m)+1)) / (e(N)-(e(df_m)+2)) *       ///
                          (1 - di2/(1-hii)) / (1-hii)
                drop hii ei di2
        }                                                  // new this version
end
```

end fvaratio.ado, version 4

The output for the commands between the `quietly {` and `}` is now suppressed—the result is the same as if we had put `quietly` in front of each command.

Solving problem 2—that the resulting variable is always called `FVi`—requires use of the `syntax` command. Let's put that off and deal with problem 3—that the working variables have nice names like `hii`, `ei`, and `di2`, and so prevent users from using those names in their data.

One solution would be to change the nice names to unlikely names. We could change `hii` to `MyHiiVaR`, which would not guarantee the prevention of a conflict but would certainly make it unlikely. It would also make our program difficult to read, an important consideration should we want to change it in the future. There is a better solution. Stata's `tempvar` command (see [U] **18.7.1 Temporary variables**) places names into local macros that are guaranteed to be unique:

begin fvaratio.ado, version 5

```
program fvaratio
        version 12
        tempvar hii ei di2                                  // new this version
        quietly {
                predict `hii' if e(sample), hat   // changed, as are other lines
                predict `ei' if e(sample), resid
                gen `di2' = (`ei'*`ei')/e(rss)
                gen FVi = (e(N)-(e(df_m)+1)) / (e(N)-(e(df_m)+2)) *      ///
                          (1 - `di2'/(1-`hii')) / (1-`hii')
        }
end
```

end fvaratio.ado, version 5

At the beginning of our program, we declare the temporary variables. (We can do it outside or inside the `quietly`—it makes no difference—and we do not have to do it at the beginning or even all at once; we could declare them as we need them, but at the beginning is prettiest.) When we refer to a temporary variable, we do not refer directly to it (such as by typing `hii`); we refer to it indirectly by typing open and close single quotes around the name (`` `hii' ``). And at the end of our program, we no longer bother to `drop` the temporary variables—temporary variables are dropped automatically by Stata when a program concludes.

❑ Technical note

Why do we type single quotes around the names? `tempvar` creates local macros containing the real temporary variable names. `hii` in our program is now a local macro, and `` `hii' `` refers to the contents of the local macro, which is the variable's actual name.

❑

We now have an excellent program—its only fault is that we cannot specify the name of the new variable to be created. Here is the solution to that problem:

begin fvaratio.ado, version 6

```
program fvaratio
        version 12
        syntax newvarname                                   // new this version
        tempvar hii ei di2
        quietly {
                predict `hii' if e(sample), hat
                predict `ei' if e(sample), resid
                gen `di2' = (`ei'*`ei')/e(rss)
                gen `typlist' `varlist' = ///              changed this version
                        (e(N)-(e(df_m)+1)) / (e(N)-(e(df_m)+2)) *      ///
                        (1 - `di2'/(1-`hii')) / (1-`hii')
        }
end
```

end fvaratio.ado, version 6

It took a change to one line and the addition of another to obtain the solution. This magic all happens because of `syntax` (see [U] **18.4.4 Parsing standard Stata syntax** above).

'`syntax newvarname`' specifies that one new variable name must be specified (had we typed '`syntax [newvarname]`', the new varname would have been optional; had we typed '`syntax newvarlist`', the user would have been required to specify at least one new variable and allowed to specify more). In any case, `syntax` compares what the user types to what is allowed. If what the user types does not match what we have declared, `syntax` will issue the appropriate error message and stop our program. If it does match, our program will continue, and what the user typed will be broken out and stored in local macros for us. For a `newvarname`, the new name typed by the user is placed in the local macro `varlist`, and the type of the variable (`float`, `double`, ...) is placed in `typlist` (even if the user did not specify a storage type, in which case the type is the current default storage type).

This is now an excellent program. There are, however, two more improvements we could make. First, we have demonstrated that, by the use of '`syntax newvarname`', we can allow the user to define not only the name of the created variable but also the storage type. However, when it comes to the creation of intermediate variables, such as '`hii`' and '`di2`', it is good programming practice to keep as much precision as possible. We want our final answer to be precise as possible, regardless of how we ultimately decide to store it. Any calculation that uses a previously generated variable would benefit if the previously generated variable were stored in double precision. Below we modify our program appropriately:

——— begin fvaratio.ado, version 7 ———

```
program fvaratio
        version 12
        syntax newvarname
        tempvar hii ei di2
        quietly {
                predict double `hii' if e(sample), hat        // changed, as are
                predict double `ei' if e(sample), resid       // other lines
                gen double `di2' = (`ei'*`ei')/e(rss)
                gen `typlist' `varlist' = ///
                        (e(N)-(e(df_m)+1)) / (e(N)-(e(df_m)+2)) *      ///
                        (1 - `di2'/(1-`hii')) / (1-`hii')
        }
end
```

——— end fvaratio.ado, version 7 ———

As for the second improvement we could make, `fvaratio` is intended to be used sometime after `regress`. How do we know the user is not misusing our program and executing it after, say, `logistic`? `e(cmd)` will tell us the name of the last estimation command; see [U] **18.9 Accessing results calculated by estimation commands** and [U] **18.10.2 Saving results in e()** above. We should change our program to read

——— begin fvaratio.ado, version 8 ———

```
program fvaratio
        version 12
        if "`e(cmd)'"!="regress" {                         // new this version
                error 301
        }
        syntax newvarname
        tempvar hii ei di2
        quietly {
                predict double `hii' if e(sample), hat
                predict double `ei' if e(sample), resid
                gen double `di2' = (`ei'*`ei')/e(rss)
                gen `typlist' `varlist' = ///
                        (e(N)-(e(df_m)+1)) / (e(N)-(e(df_m)+2)) *      ///
                        (1 - `di2'/(1-`hii')) / (1-`hii')
        }
end
```

——— end fvaratio.ado, version 8 ———

The `error` command issues one of Stata's prerecorded error messages and stops our program. Error 301 is "last estimates not found"; see [P] **error**. (Try typing `error 301` at the command line.)

In any case, this is a perfect program.

❑ Technical note

You do not have to go to all the trouble we did to program the FVARATIO measure of influence or any other statistic that appeals to you. Whereas version 1 was not really an acceptable solution—it was too specialized—version 2 was acceptable. Version 3 was better, and version 4 better yet, but the improvements were of less and less importance.

Putting aside the details of Stata's language, you should understand that final versions of programs do not just happen—they are the results of drafts that have been refined. How much refinement depends on how often and who will be using the program. In this sense, the "official" ado-files that come with Stata are poor examples. They have been subject to substantial refinement because they will be used by strangers with no knowledge of how the code works. When writing programs for yourself, you may want to stop refining at an earlier draft.

❑

18.11.6 Writing online help

When you write an ado-file, you should also write a help file to go with it. This file is a standard text file, named *command*`.sthlp`, that you place in the same directory as your ado-file *command*`.ado`. This way, when users type `help` followed by the name of your new command (or pull down **Help**), they will see something better than "help for ... not found".

You can obtain examples of help files by examining the `.sthlp` files in the official ado-directory; type "`sysdir`" and look in the lettered subdirectories of the directory defined as `BASE`:

```
. sysdir
   STATA:  C:\Program Files\Stata12\
 UPDATES:  C:\Program Files\Stata12\ado\updates\
    BASE:  C:\Program Files\Stata12\ado\base\
    SITE:  C:\Program Files\Stata12\ado\site\
    PLUS:  C:\ado\plus\
PERSONAL:  C:\ado\personal\
OLDPLACE:  C:\ado\
```

Here you would find examples of `.sthlp` files in the `a`, `b`, ... subdirectories of `C:\Program Files\Stata12\ado\base`.

Help files are physically written on the disk in text format, but their contents are Stata Markup and Control Language (SMCL). For the most part, you can ignore that. If the file contains a line that reads

```
Also see help for the finishup command
```

it will display in just that way. However, SMCL contains many special directives, so that if the line in the file were to read

```
Also see {hi:help} for the {help finishup} command
```

what would be displayed would be

Also see **help** for the **finishup** command

and moreover, **finishup** would appear as a hypertext link, meaning that if users clicked on it, they would see help on `finishup`.

You can read about the details of SMCL in [P] **smcl**. The following is a SMCL help file:

begin examplehelpfile.sthlp

```
{smcl}
{* *! version 1.2.0  02jun2011}{...}
{vieweralsosee "[R] help" "help help "}{...}
{viewerjumpto "Syntax" "examplehelpfile##syntax"}{...}
{viewerjumpto "Description" "examplehelpfile##description"}{...}
{viewerjumpto "Options" "examplehelpfile##options"}{...}
{viewerjumpto "Remarks" "examplehelpfile##remarks"}{...}
{viewerjumpto "Examples" "examplehelpfile##examples"}{...}
{title:Title}

{phang}
{bf:whatever} {hline 2}  Calculate whatever statistic


{marker syntax}{...}
{title:Syntax}

{p 8 17 2}
{cmdab:wh:atever}
[{varlist}]
{ifin}
{weight}
[{cmd:,}
{it:options}]

{synoptset 20 tabbed}{...}
{synopthdr}
{synoptline}
{syntab:Main}
{synopt:{opt d:etail}}display additional statistics{p_end}
{synopt:{opt mean:only}}suppress the display; calculate only the mean;
                     programmer's option{p_end}
{synopt:{opt f:ormat}}use variable's display format{p_end}
{synopt:{opt sep:arator(#)}}draw separator line after every {it:#} variables;
                     default is {cmd:separator(5)}{p_end}
{synopt:{opth g:enerate(newvar)}}create variable name {it:newvar}{p_end}
{synoptline}
{p2colreset}{...}
{p 4 6 2}
{cmd:by} is allowed; see {manhelp by D}.{p_end}
{p 4 4 2}
{cmd:fweight}s are allowed; see {help weight}.


{marker description}{...}
{title:Description}

{pstd}
{cmd:whatever} calculates the whatever statistic for the variables in
{varlist} when the data are not stratified.


{marker options}{...}
{title:Options}

{dlgtab:Main}

{phang}
{opt detail} displays detailed output of the calculation.

{phang}
{opt meanonly} restricts the calculation to be based on only the
means.  The default is to use a trimmed mean.
```

```
{phang}
{opt format} requests that the summary statistics be displayed using the display
formats associated with the variables, rather than the default {cmd:g} display
format; see {bf:[U] 12.5 Formats: Controlling how data are displayed}.

{phang}
{opt separator(#)} specifies how often to insert separation lines
into the output.  The default is {cmd:separator(5)}, meaning that a
line is drawn after every 5 variables.  {cmd:separator(10)} would
draw a line after every 10 variables. {cmd:separator(0)} suppresses
the separation line.

{phang}
{opth generate(newvar)} creates {it:newvar} containing the whatever values.


{marker remarks}{...}
{title:Remarks}

{pstd}
For detailed information on the whatever statistic, see {bf:[R] intro}.


{marker examples}{...}
{title:Examples}

{phang}{cmd:. whatever mpg weight}{p_end}

{phang}{cmd:. whatever mpg weight, meanonly}{p_end}
```
end examplehelpfile.sthlp

If you were to select **Help > Stata Command**, and type `examplehelpfile` and click on **OK**, or if you were to type `help examplehelpfile`, this is what you would see:

Title

whatever — Calculate whatever statistic

Syntax

whatever [*varlist*] [*if*] [*in*] [*weight*] [, *options*]

options	description
Main	
detail	display additional statistics
meanonly	suppress the display; calculate only the mean; programmer's option
format	use variable's display format
separator(#)	draw separator line after every # variables; default is **separator(5)**
generate(*newvar*)	create variable name *newvar*

by is allowed; see **[D] by**.
fweights are allowed; see **weight**.

Description

whatever calculates the whatever statistic for the variables in *varlist* when the data are not stratified.

```
Options

    Main
detail displays detailed output of the calculation.

meanonly restricts the calculation to be based on only the means.
    The default is to use a trimmed mean.

format requests that the summary statistics be displayed using the display
    formats associated with the variables, rather than the default g display
    format; see [U] 12.5 Formats: controlling how data are displayed.

separator(#) specifies how often to insert separation lines into the output.
    The default is separator(5), meaning that a line is drawn after every 5
    variables.  separator(10) would draw a line after every 10 variables.
    separator(0) suppresses the separation line.

generate(newvar) creates newvar containing the whatever values.


Remarks

    For detailed information on the whatever statistic, see [R] intro.


Examples

    . whatever mpg weight

    . whatever mpg weight, meanonly
```

Users will find it easier to understand your programs if you document them the same way that we document ours. We offer the following guidelines:

1. The first line must be

   ```
   {smcl}
   ```

   This notifies Stata that the help file is in SMCL format.

2. The second line should be

   ```
   {* *! version #.#.# date}{...}
   ```

   The * indicates a comment and the {...} will suppress the blank line. Whenever you edit the help file, update the version number and the date found in the comment line.

3. The next several lines denote what will be displayed in the quick access toolbar with the three pulldown menus: Dialog, Also See, and Jump To.

   ```
   {vieweralsosee "[R] help" "help help "}{...}
   {viewerjumpto "Syntax" "examplehelpfile##syntax"}{...}
   {viewerjumpto "Description" "examplehelpfile##description"}{...}
   {viewerjumpto "Options" "examplehelpfile##options"}{...}
   {viewerjumpto "Remarks" "examplehelpfile##remarks"}{...}
   {viewerjumpto "Examples" "examplehelpfile##examples"}{...}
   ```

4. Then place the title.

   ```
   {title:Title}

   {phang}
   {bf:yourcmd} {hline 2} Your title
   ```

5. Include two blank lines, and place the Syntax title, syntax diagram, and options table:

    {title:Syntax}

    {p 8 17 2}
    syntax line

    {p 8 17 2}
    second syntax line, if necessary

    {synoptset 20 tabbed}{...}
    {synopthdr}
    {synoptline}
    {syntab:*tab*}
    {synopt:{*option*}}*brief description of option*{p_end}
    {synoptline}
    {p2colreset}{...}

    {p 4 6 2}
    clarifying text, if required

6. Include two blank lines, and place the Description title and text:

    {title:Description}

    {pstd}
    description text

   Briefly describe what the command does. Do not burden the user with details yet. Assume that the user is at the point of asking whether this is what he or she is looking for.

7. If your command allows options, include two blank lines, and place the Options title and descriptions:

    {title:Options}

    {phang}
    {opt *optionname*} *option description*

    {pmore}
    continued option description, if necessary

    {phang}
    {opt *optionname*} *second option description*

   Options should be included in the order in which they appear in the option table. Option paragraphs are reverse indented, with the option name on the far left, where it is easily spotted. If an option requires more than one paragraph, subsequent paragraphs are set using {pmore}. One blank line separates one option from another.

8. Optionally include two blank lines, and place the Remarks title and text:

    {title:Remarks}

    {pstd}
    text

   Include whatever long discussion you feel necessary. Stata's official online help files often omit this because the discussions appear in the manual. Stata's official help files for features added between releases (obtained from the *Stata Journal*, the Stata website, etc.), however, include this section because the appropriate *Stata Journal* may not be as accessible as the manuals.

9. Optionally include two blank lines, and place the Examples title and text:

    {title:Examples}

    {phang}
    {cmd:. *first example*}

    {phang}
    {cmd:. *second example*}

    Nothing communicates better than providing something beyond theoretical discussion. Examples rarely need much explanation.

10. Optionally include two blank lines, and place the Author title and text:

    {title:Author}

    {pstd}
    Name, affiliation, etc.

    Exercise caution. If you include a telephone number, expect your phone to ring. An email address may be more appropriate.

11. Optionally include two blank lines, and place the References title and text:

    {title:References}

    {pstd}
    Author. year.
    Title. Location: Publisher.

We also warn that it is easy to use too much `{hi:highlighting}`. Use it sparingly. In text, use `{cmd:...}` to show what would be shown in `typewriter` typeface it the documentation were printed in this manual.

❑ Technical note

Sometimes it is more convenient to describe two or more related commands in the same `.sthlp` file. Thus `xyz.sthlp` might document both the `xyz` and `abc` commands. To arrange that typing `help abc` displays `xyz.sthlp`, create the file `abc.sthlp`, containing

```
begin abc.sthlp
.h xyz
end abc.sthlp
```

When a `.sthlp` file contains one line of the form '`.h` *refname*', Stata interprets that as an instruction to display `help` for *refname*.

❑

❑ Technical note

If you write a collection of programs, you need to somehow index the programs so that users (and you) can find the command they want. We do that with our `contents.sthlp` entry. You should create a similar kind of entry. We suggest that you call your private entry `user.sthlp` in your personal ado-directory; see [U] **17.5.2 Where is my personal ado-directory?**. This way, to review what you have added, you can type `help user`.

We suggest that Unix users at large sites also add `site.sthlp` to the `SITE` directory (typically `/usr/local/ado`, but type `sysdir` to be sure). Then you can type `help site` for a list of the commands available sitewide.

❑

18.11.7 Programming dialog boxes

You cannot only write new Stata commands and help files, but you can also create your own interface, or dialog box, for a command you have written. Stata provides a dialog box programming language to allow you to create your own dialog boxes. In fact, most of the dialog boxes you see in Stata's interface have been created using this language.

This is not for the faint of heart, but if you want to create your own dialog box for a command, type

```
. help dialog programming
```

The help file contains all the details on creating and programming dialog boxes.

18.12 A compendium of useful commands for programmers

You can use any Stata command in your programs and ado-files. Also, some commands are intended solely for use by Stata programmers. You should see the section under the *Programming* heading in the subject table of contents at the beginning of the *Quick Reference and Index*.

18.13 References

Baum, C. F. 2009. *An Introduction to Stata Programming*. College Station, TX: Stata Press.

Belsley, D. A., E. Kuh, and R. E. Welsch. 1980. *Regression Diagnostics: Identifying Influential Data and Sources of Collinearity*. New York: Wiley.

Gould, W. W. 2001. Statistical software certification. *Stata Journal* 1: 29–50.

Herrin, J. 2009. Stata tip 77: (Re)using macros in multiple do-files. *Stata Journal* 9: 497–498.

19 Immediate commands

Contents

19.1 Overview

An *immediate* command is a command that obtains data not from the data stored in memory but from numbers typed as arguments. Immediate commands, in effect, turn Stata into a glorified hand calculator.

There are many instances when you may not have the data, but you do know something about the data, and what you know is adequate to perform statistical tests. For instance, you do not have to have individual-level data to obtain the standard error of the mean, and thereby a confidence interval, if you know the mean, standard deviation, and number of observations. In other instances, you may actually have the data, and you could enter the data and perform the test, but it would be easier if you could just ask for the statistic based on a summary. For instance, you flip a coin 10 times, and it comes up heads twice. You could enter a 10-observation dataset with two ones (standing for heads) and eight zeros (meaning tails).

Immediate commands are meant to solve those problems. Immediate commands have the following properties:

1. They never disturb the data in memory. You can perform an immediate calculation as an aside without changing your data.
2. The syntax for these commands is the same, the command name followed by numbers, which are the summary statistics from which the statistic is calculated. The numbers are almost always summary statistics, and the order in which they are specified is in some sense "natural".
3. Immediate commands all end in the letter *i*, although the converse is not true. Usually, if there is an immediate command, there is a nonimmediate form also, that is, a form that works on the data in memory. For every statistical command in Stata, we have included an immediate form if it is reasonable to assume that you might know the requisite summary statistics without having the underlying data and if typing those statistics is not absurdly burdensome.
4. Immediate commands are documented along with their nonimmediate counterparts. Thus, if you want to obtain a confidence interval, whether it be from summary data with an immediate command or using the data in memory, use the table of contents or index to discover that [R] **ci** discusses confidence intervals. There, you learn that `ci` calculates confidence intervals by using the data in memory and that `cii` does the same with the data specified immediately following the command.

19.1.1 Examples

▷ Example 1

Let's take the example of confidence intervals. Professional papers often publish the mean, standard deviation, and number of observations for variables used in the analysis. Those statistics are sufficient for calculating a confidence interval. If we know that the mean mileage rating of cars in some sample is 24, that the standard deviation is 6, and that there are 97 cars in the sample, we can calculate

```
. cii 97 24 6
    Variable |        Obs        Mean    Std. Err.       [95% Conf. Interval]
-------------+---------------------------------------------------------------
             |         97          24    .6092077        22.79073    25.20927
```

We learn that the mean's standard error is 0.61 and its 95% confidence interval is $[22.8, 25.2]$. To obtain this, we typed cii (the immediate form of the ci command) followed by the number of observations, the mean, and the standard deviation. We knew the order in which to specify the numbers because we had read [R] **ci**.

We could use the immediate form of the ttest command to test the hypothesis that the true mean is 22:

```
. ttesti 97 24 6 22
One-sample t test
------------------------------------------------------------------------------
         |     Obs        Mean    Std. Err.   Std. Dev.   [95% Conf. Interval]
---------+--------------------------------------------------------------------
       x |      97          24    .6092077           6    22.79073    25.20927
------------------------------------------------------------------------------
    mean = mean(x)                                                t =   3.2830
Ho: mean = 22                                    degrees of freedom =       96

    Ha: mean < 22               Ha: mean != 22                 Ha: mean > 22
 Pr(T < t) = 0.9993         Pr(|T| > |t|) = 0.0014          Pr(T > t) = 0.0007
```

The first three numbers were as we specified in the cii command. ttesti requires a fourth number, which is the constant against which the mean is being tested; see [R] **ttest**.

◁

▷ Example 2

We mentioned flipping a coin 10 times and having it come up heads twice. The 99% confidence interval can also be obtained from ci:

```
. cii 10 2, level(99)
                                                         — Binomial Exact —
    Variable |        Obs        Mean    Std. Err.       [99% Conf. Interval]
-------------+---------------------------------------------------------------
             |         10          .2    .1264911        .0108505    .6482012
```

In the previous example, we specified cii with three numbers following it; in this example, we specify 2. Immediate commands often determine what to do by the number of arguments following the command. With two arguments, ci assumes that we are specifying the number of trials and successes from a binomial experiment; see [R] **ci**.

The immediate form of the bitest command performs exact hypothesis testing:

```
. bitesti 10 2 .5
        N   Observed k   Expected k   Assumed p   Observed p
------------------------------------------------------------
       10          2            5      0.50000      0.20000

  Pr(k >= 2)            = 0.989258  (one-sided test)
  Pr(k <= 2)            = 0.054688  (one-sided test)
  Pr(k <= 2 or k >= 8) = 0.109375  (two-sided test)
```

For a full explanation of this output, see [R] **bitest**.

◁

▷ Example 3

Stata's tabulate command makes tables and calculates various measures of association. The immediate form, tabi, does the same, but we specify the contents of the table following the command:

```
. tabi 5 10 \ 2 14
           |          col
       row |         1          2 |     Total
-----------+----------------------+----------
         1 |         5         10 |        15 
         2 |         2         14 |        16 
-----------+----------------------+----------
     Total |         7         24 |        31 
           Fisher's exact =                 0.220
   1-sided Fisher's exact =                 0.170
```

The tabi command is slightly different from most immediate commands because it uses '\' to indicate where one row ends and another begins.

◁

19.1.2 A list of the immediate commands

Command	Reference	Description
bitesti	[R] **bitest**	Binomial probability test
cci csi iri mcci	[ST] **epitab**	Tables for epidemiologists
cii	[R] **ci**	Confidence intervals for means, proportions, and counts
prtesti	[R] **prtest**	One- and two-sample tests of proportions
sampsi	[R] **sampsi**	Sample size and power for means and proportions
sdtesti	[R] **sdtest**	Variance comparison tests
symmi	[R] **symmetry**	Symmetry and marginal homogeneity tests
tabi	[R] **tabulate twoway**	Two-way tables of frequencies
ttesti	[R] **ttest**	Mean comparison tests
twoway pci	[G-2] **graph twoway pci**	Paired-coordinate plot with spikes or lines
twoway pcarrowi	[G-2] **graph twoway pcarrowi**	Paired-coordinate plot with arrows
twoway scatteri	[G-2] **graph twoway scatteri**	Twoway scatterplot

19.2 The display command

display is not really an immediate command, but it can be used as a hand calculator.

```
. display 2+5
7

. display sqrt(2+sqrt(3^2-4*2*-2))/(2*3)
.44095855
```

See [R] **display**.

20 Estimation and postestimation commands

Contents

20.1 All estimation commands work the same way

All Stata commands that fit statistical models—commands such as `regress`, `logit`, `sureg`, and so on—work the same way. Most single-equation estimation commands have the syntax

command varlist [*if*] [*in*] [*weight*] [, *options*]

and most multiple-equation estimation commands have the syntax

command (*varlist*) (*varlist*) ... (*varlist*) [*if*] [*in*] [*weight*] [, *options*]

Adopt a loose definition of single and multiple equation in interpreting this. For instance, `heckman` is a two-equation system, mathematically speaking, yet we categorize it, syntactically, with single-equation commands because most researchers think of it as a linear regression with an adjustment for the censoring. The important thing is that most estimation commands have one or the other of these two syntaxes.

In single-equation commands, the first variable in the *varlist* is the dependent variable, and the remaining variables are the independent variables, with some exceptions. For instance, `xtmixed` allows special variable prefixes to identify random factors.

Prefix commands may be specified in front of an estimation command to modify what it does. The syntax is

prefix: *command* ...

where the prefix commands are

Prefix command	Description	Manual entry
`by`	repeat command on subsets of data	[D] **by**
`statsby`	collect results across subsets of data	[D] **statsby**
`rolling`	time-series rolling estimation	[TS] **rolling**
*`svy`	estimation for complex survey data	[SVY] **svy**
*`mi estimate`	multiply imputed data and multiple imputation	[MI] **mi estimate**
*`nestreg`	nested model statistics	[R] **nestreg**
*`stepwise`	stepwise estimation	[R] **stepwise**
*`xi`	interaction expansion	[R] **xi**
*`fracpoly`	fractional polynomials	[R] **fracpoly**
*`mfp`	multiple fractional polynomials	[R] **mfp**

*Available for some but not all estimation commands

Two other prefix commands—`bootstrap` and `jackknife`—also work with estimation commands—see [R] **bootstrap** and [R] **jackknife**—but usually it is easier to specify the estimation-command option `vce(bootstrap)` or `vce(jackknife)`.

Also, all estimation commands—whether single or multiple equation—share the following features:

1. You can use the standard features of Stata's syntax— `if` *exp* and `in` *range*—to specify the estimation subsample; you do not have to make a special dataset.

2. You can retype the estimation command without arguments to redisplay the most recent estimation results. For instance, after fitting a model with `regress`, you can see the estimates again by typing `regress` by itself. You do not have to do this immediately—any number of commands can occur between the estimation and the replaying, and, in fact, you can even replay the last estimates after the data have changed or you have dropped the data altogether. Stata never forgets (unless you type `discard`; see [P] **discard**).

3. You can specify the `level()` option at the time of estimation, or when you redisplay results if that makes sense, to specify the width of the confidence intervals for the coefficients. The default is `level(95)`, meaning 95% confidence intervals. You can reset the default with `set level`; see [R] **level**.

4. You can use the postestimation command `margins` to display model results in terms of marginal effects (dy/dx or even $df(y)/dx$), which can be displayed as either derivatives or elasticities; see [R] **margins**.

5. You can use the postestimation command `margins` to obtain tables of estimated marginal means, adjusted predictions, and predictive margins; see [U] **20.16 Obtaining conditional and average marginal effects** and see [R] **margins**.

6. You can use the postestimation command `pwcompare` to obtain pairwise comparisons across levels of factor variables. You can compare estimated cell means, marginal means, intercepts, marginal intercepts, slopes, or marginal slopes—collectively called margins. See [U] **20.17 Obtaining pairwise comparisons**, [R] **margins**, and [R] **margins, pwcompare**.

7. You can use the postestimation command `contrast` to obtain contrasts, which is to say, to compare levels of factor variables and their interactions. This command can also produce ANOVA-style tests of main effects, interactions effects, simple effects, and nested effects; and it can be used after most estimation commands. See [U] **20.18 Obtaining contrasts, tests of interactions, and main effects**, [R] **contrast**, and [R] **margins, contrast**.

8. You can use the postestimation command `marginsplot` to graph any of the results produced by `margins`, and because `margins` can replicate any result produced by `pwcompare` and `contrast`, you can graph any result produced by them, too. See [R] **marginsplot**.

9. You can use the postestimation command `estat` to obtain common statistics associated with the model. Which statistics are available are documented in the postestimation section following the documentation of the estimation command, for instance, in [R] **regress postestimation** following [R] **regress**.

   You can always use the postestimation command `estat vce` to obtain the variance–covariance matrix of the estimators (VCE), presented as either a correlation matrix or a covariance matrix. (You can also obtain the estimated coefficients and covariance matrix as vectors and matrices and manipulate them with Stata's matrix capabilities; see [U] **14.5 Accessing matrices created by Stata commands**.)

10. You can use the postestimation command `predict` to obtain predictions, residuals, influence statistics, and the like, either for the data on which you just estimated or for some other data. You can use postestimation command `predictnl` to obtain point estimates, standard errors, etc., for customized predictions. See [R] **predict** and [R] **predictnl**.

11. You can refer to the values of coefficients and standard errors in expressions (such as with `generate`) by using standard notation; see [U] **13.5 Accessing coefficients and standard errors**. You can refer in expressions to the values of other estimation-related statistics by using `e(`*resultname*`)`. For instance, all commands define `e(N)` recording the number of observations in the estimation subsample. After estimation, type `ereturn list` to see a list of all that is available. See the *Saved results* section in the estimation command's documentation for their definition.

    An especially useful `e()` result is `e(sample)`: it returns 1 if an observation was used in the estimation and 0 otherwise, so you can add `if e(sample)` to the end of other commands to restrict them to the estimation subsample. You could type, for instance, `summarize if e(sample)`.

12. You can use the postestimation command `test` to perform tests on the estimated parameters (Wald tests of linear hypotheses), `testnl` to perform Wald tests of nonlinear hypotheses, and `lrtest` to perform likelihood-ratio tests. You can use the postestimation command `lincom` to obtain point estimates and confidence intervals for linear combinations of the estimated parameters and the postestimation command `nlcom` to obtain nonlinear combinations.
13. You can specify the `coeflegend` option at the time of estimation or when you redisplay results to see how to type your coefficients in postestimation commands such as `test` and `lincom` (see [R] **test** and [R] **lincom**) and in expressions.
14. You can use the `statsby` prefix command (see [D] **statsby**) to fit models over each category in a categorical variable and collect the results in a Stata dataset.
15. You can use the postestimation command `estimates` to store estimation results by name for later retrieval or for displaying/comparing multiple models by using `estimates`, or for saving them in a file; see [R] **estimates**.
16. You can use the postestimation command `_estimates` to hold estimates, perform other estimation commands, and then restore the prior estimates. This is of particular interest to programmers. See [P] **_estimates**.
17. You can use the postestimation command `suest` to obtain the joint parameter vector and variance–covariance matrix for coefficients from two different models by using seemingly unrelated estimation. This is especially useful for testing the equality, say, of coefficients across models; see [R] **suest**.
18. You can use the postestimation command `hausman` to perform Hausman model-specification tests by using `hausman`; see [R] **hausman**.
19. With some exceptions, you can specify the `vce(robust)` option at the time of estimation to obtain the Huber/White/robust alternate estimate of variance, or you can specify the `vce(cluster` *clustvar*`)` option to relax the assumption of independence of the observations. See [R] ***vce_option***.

    Most estimation commands also allow a `vce(`*vcetype*`)` option to specify other alternative variance estimators—which ones are allowed are documented with the estimator—and usually `vce(opg)`, `vce(bootstrap)`, and `vce(jackknife)` are available.

20.2 Standard syntax

You can combine Stata's `if` *exp* and `in` *range* with any estimation command. Estimation commands also allow `by` *varlist*`:`, where it would be sensible.

▷ Example 1

We have data on 74 automobiles that record the mileage rating (`mpg`), weight (`weight`), and whether the car is domestic or foreign produced (`foreign`). We can fit a linear regression model of `mpg` on `weight` and the square of `weight`, using just the foreign-made automobiles, by typing

```
. use http://www.stata-press.com/data/r12/auto
(1978 Automobile Data)

. regress mpg weight c.weight#c.weight if foreign

      Source |       SS       df       MS              Number of obs =      22
-------------+------------------------------           F(  2,    19) =    8.31
       Model |  428.256889     2  214.128444           Prob > F      =  0.0026
    Residual |  489.606747    19  25.7687762           R-squared     =  0.4666
-------------+------------------------------           Adj R-squared =  0.4104
       Total |  917.863636    21  43.7077922           Root MSE      =  5.0763

------------------------------------------------------------------------------
         mpg |      Coef.   Std. Err.      t    P>|t|     [95% Conf. Interval]
-------------+----------------------------------------------------------------
      weight |  -.0132182   .0275711    -0.48   0.637    -.0709252    .0444888
             |
    c.weight#|
    c.weight |   5.50e-07   5.41e-06     0.10   0.920    -.0000108    .0000119
             |
       _cons |   52.33775    34.1539     1.53   0.142    -19.14719    123.8227
------------------------------------------------------------------------------
```

We use the factor-variable notation `c.weight#c.weight` to add the square of `weight` to our regression; see [U] **11.4.3 Factor variables**.

We can run separate regressions for the domestic and foreign-produced automobiles with the `by` *varlist*`:` prefix:

```
. use http://www.stata-press.com/data/r12/auto
(1978 Automobile Data)

. by foreign: regress mpg weight c.weight#c.weight

-------------------------------------------------------------------------------
-> foreign = Domestic

      Source |       SS       df       MS              Number of obs =      52
-------------+------------------------------           F(  2,    49) =   91.64
       Model |  905.395466     2  452.697733           Prob > F      =  0.0000
    Residual |  242.046842    49  4.93973146           R-squared     =  0.7891
-------------+------------------------------           Adj R-squared =  0.7804
       Total |  1147.44231    51  22.4988688           Root MSE      =  2.2226

------------------------------------------------------------------------------
         mpg |      Coef.   Std. Err.      t    P>|t|     [95% Conf. Interval]
-------------+----------------------------------------------------------------
      weight |  -.0131718   .0032307    -4.08   0.000    -.0196642   -.0066794
             |
    c.weight#|
    c.weight |   1.11e-06   4.95e-07     2.25   0.029     1.19e-07    2.11e-06
             |
       _cons |   50.74551   5.162014     9.83   0.000     40.37205    61.11896
------------------------------------------------------------------------------

-------------------------------------------------------------------------------
-> foreign = Foreign

      Source |       SS       df       MS              Number of obs =      22
-------------+------------------------------           F(  2,    19) =    8.31
       Model |  428.256889     2  214.128444           Prob > F      =  0.0026
    Residual |  489.606747    19  25.7687762           R-squared     =  0.4666
-------------+------------------------------           Adj R-squared =  0.4104
       Total |  917.863636    21  43.7077922           Root MSE      =  5.0763

------------------------------------------------------------------------------
         mpg |      Coef.   Std. Err.      t    P>|t|     [95% Conf. Interval]
-------------+----------------------------------------------------------------
      weight |  -.0132182   .0275711    -0.48   0.637    -.0709252    .0444888
             |
    c.weight#|
    c.weight |   5.50e-07   5.41e-06     0.10   0.920    -.0000108    .0000119
             |
       _cons |   52.33775    34.1539     1.53   0.142    -19.14719    123.8227
------------------------------------------------------------------------------
```

Although all estimation commands allow `if` *exp* and `in` *range*, only some allow the `by` *varlist*: prefix. For `by()`, the duration of Stata's memory is limited: it remembers the *last* set of estimates only. This means that, if we were to use any of the other features described below, they would use the last regression estimated, which right now is `mpg` on `weight` and square of `weight` for the `Foreign` subsample.

We can instead collect the statistics from each of the by-groups using the [D] **statsby** prefix.

```
. statsby, by(foreign): regress mpg weight c.weight#c.weight
(running regress on estimation sample)

      command:  regress mpg weight c.weight#c.weight
           by:  foreign

Statsby groups
----+--- 1 ---+--- 2 ---+--- 3 ---+--- 4 ---+--- 5
..
```

`statsby` runs the regression first on Domestic cars then on Foreign cars and saves the coefficients by overwriting our dataset. Do not worry; if the dataset has not been previously saved, `statsby` will refuse to run unless we also specify the `clear` option.

Here is what we now have in memory.

```
. list

     +--------------------------------------------------+
     |  foreign   _b_weight      _stat_2       _b_cons  |
     |--------------------------------------------------|
  1. | Domestic   -.0131718     1.11e-06      50.74551  |
  2. |  Foreign   -.0132182     5.50e-07      52.33775  |
     +--------------------------------------------------+
```

These are the coefficients from the two regressions above. `statsby` does not know how to name the coefficient for `c.weight#c.weight` so it labels the coefficient with the generic name `_stat_2`. We can also save the standard errors and other statistics from the regressions; see [D] **statsby**. ◁

20.3 Replaying prior results

When you type an estimation command without arguments, it redisplays prior results.

▷ Example 2

To perform a regression of `mpg` on the variables `weight` and `displacement`, we could type

```
. regress mpg weight displacement
      Source |       SS       df       MS              Number of obs =      74
-------------+------------------------------           F(  2,    71) =   66.79
       Model |  1595.40969     2  797.704846           Prob > F      =  0.0000
    Residual |  848.049768    71  11.9443629           R-squared     =  0.6529
-------------+------------------------------           Adj R-squared =  0.6432
       Total |  2443.45946    73  33.4720474           Root MSE      =  3.4561

------------------------------------------------------------------------------
         mpg |      Coef.   Std. Err.      t    P>|t|     [95% Conf. Interval]
-------------+----------------------------------------------------------------
      weight |  -.0065671   .0011662    -5.63   0.000    -.0088925   -.0042417
displacement |   .0052808   .0098696     0.54   0.594    -.0143986    .0249602
       _cons |   40.08452    2.02011    19.84   0.000     36.05654    44.11251
------------------------------------------------------------------------------
```

We now go on to do other things, summarizing data, listing observations, performing hypothesis tests, or anything else. If we decide that we want to see the last set of estimates again, we type the estimation command without arguments.

```
. regress
      Source |       SS       df       MS              Number of obs =      74
-------------+------------------------------           F(  2,    71) =   66.79
       Model |  1595.40969     2  797.704846           Prob > F      =  0.0000
    Residual |  848.049768    71  11.9443629           R-squared     =  0.6529
-------------+------------------------------           Adj R-squared =  0.6432
       Total |  2443.45946    73  33.4720474           Root MSE      =  3.4561

------------------------------------------------------------------------------
         mpg |      Coef.   Std. Err.      t    P>|t|     [95% Conf. Interval]
-------------+----------------------------------------------------------------
      weight |  -.0065671   .0011662    -5.63   0.000    -.0088925   -.0042417
displacement |   .0052808   .0098696     0.54   0.594    -.0143986    .0249602
       _cons |   40.08452    2.02011    19.84   0.000     36.05654    44.11251
------------------------------------------------------------------------------
```

This feature works with *every* estimation command, so we could just as well have done it with, say, `stcox` or `logit`.

◁

20.4 Cataloging estimation results

Stata keeps only the results of the most recently fit model in active memory. You can use Stata's `estimates` command, however, to temporarily store estimation results for displaying, comparing, cross-model testing, etc., during the same session. You can also save estimation results to disk, but that will be the subject of the next section. You may temporarily store up to 300 sets of estimation results.

▷ Example 3

Continuing with our automobile data, we fit four models and `estimates store` them. We fit the models `quietly` to keep the output to a minimum.

```
. quietly regress mpg weight displ
. estimates store r_base
. quietly regress mpg weight displ foreign
. estimates store r_alt
. quietly qreg mpg weight displ
. estimates store q_base
. quietly qreg mpg weight displ foreign
. estimates store q_alt
```

We saved the four models under the names `r_base`, `r_alt`, `q_base`, and `q_alt` but, if we forget, we can ask to see a directory of what is stored:

```
. estimates dir
```

model	command	depvar	npar	title
r_base	regress	mpg	3	
r_alt	regress	mpg	4	
q_base	qreg	mpg	3	
q_alt	qreg	mpg	4	

We can ask that any of the previous models be replayed:

```
. estimates replay r_base

Model r_base

      Source |       SS       df       MS              Number of obs =      74
-------------+------------------------------           F(  2,    71) =   66.79
       Model |  1595.40969     2  797.704846           Prob > F      =  0.0000
    Residual |  848.049768    71  11.9443629           R-squared     =  0.6529
-------------+------------------------------           Adj R-squared =  0.6432
       Total |  2443.45946    73  33.4720474           Root MSE      =  3.4561

         mpg |      Coef.   Std. Err.      t    P>|t|     [95% Conf. Interval]
-------------+----------------------------------------------------------------
      weight |  -.0065671   .0011662    -5.63   0.000    -.0088925   -.0042417
displacement |   .0052808   .0098696     0.54   0.594    -.0143986    .0249602
       _cons |   40.08452    2.02011    19.84   0.000     36.05654    44.11251
```

Or we can ask to see all the models in a combined table:

```
. estimates table _all
```

Variable	r_base	r_alt	q_base	q_alt
weight	-.00656711	-.00677449	-.00581173	-.00595056
displacement	.00528078	.00192865	.00428411	.00018552
foreign		-1.6006312		-2.1326004
_cons	40.084522	41.847949	37.559865	39.213348

estimates displayed just the coefficients, but we could ask for other statistics. Finally, we can also select one of the stored estimates to be made active, meaning things are just as if we had just fit the model:

```
. estimates restore r_alt
(results r_alt are active now)

. regress

      Source |       SS       df       MS              Number of obs =      74
-------------+------------------------------           F(  3,    70) =   45.88
       Model |  1619.71935     3  539.906448           Prob > F      =  0.0000
    Residual |  823.740114    70  11.7677159           R-squared     =  0.6629
-------------+------------------------------           Adj R-squared =  0.6484
       Total |  2443.45946    73  33.4720474           Root MSE      =  3.4304

------------------------------------------------------------------------------
         mpg |      Coef.   Std. Err.      t    P>|t|     [95% Conf. Interval]
-------------+----------------------------------------------------------------
      weight |  -.0067745   .0011665    -5.81   0.000    -.0091011   -.0044479
displacement |   .0019286   .0100701     0.19   0.849    -.0181556    .0220129
     foreign |  -1.600631   1.113648    -1.44   0.155    -3.821732    .6204699
       _cons |   41.84795   2.350704    17.80   0.000     37.15962    46.53628
------------------------------------------------------------------------------
```

◁

You can do a lot more with estimates; see [R] **estimates**. In particular, estimates makes it easy to perform cross-model tests, such as the Hausman specification test.

20.5 Saving estimation results

estimates can also save estimation results into a file.

```
. estimates save alt
  file alt.ster saved
```

That saved the active estimation results, the ones we just estimated or, in our case, the ones we just restored. Later, even in another Stata session, we could reload our estimates:

```
. estimates use alt
. regress
      Source |       SS       df       MS              Number of obs =      74
-------------+------------------------------           F(  3,    70) =   45.88
       Model |  1619.71935     3  539.906448           Prob > F      =  0.0000
    Residual |  823.740114    70  11.7677159           R-squared     =  0.6629
-------------+------------------------------           Adj R-squared =  0.6484
       Total |  2443.45946    73  33.4720474           Root MSE      =  3.4304

------------------------------------------------------------------------------
         mpg |      Coef.   Std. Err.      t    P>|t|     [95% Conf. Interval]
-------------+----------------------------------------------------------------
      weight |  -.0067745   .0011665    -5.81   0.000    -.0091011   -.0044479
displacement |   .0019286   .0100701     0.19   0.849    -.0181556    .0220129
     foreign |  -1.600631   1.113648    -1.44   0.155    -3.821732    .6204699
       _cons |   41.84795   2.350704    17.80   0.000     37.15962    46.53628
------------------------------------------------------------------------------
```

There is one important difference between storing results in memory and saving them in a file: `e(sample)` is lost. We have not discussed `e(sample)` yet, but it allows us to identify the observations among those currently in memory that were used in the estimation. For instance, after estimation, we could type

```
. summarize mpg weight displ foreign if e(sample)
  (output omitted)
```

and see the summary statistics of the relevant data. We could do that after an `estimates restore`, too. But we cannot do it after an `estimates use`. Part of the reason is that we might not even have the relevant data in memory. Even if we do, however, here is what will happen:

```
. summarize mpg weight displ foreign if e(sample)
    Variable |       Obs        Mean    Std. Dev.       Min        Max
-------------+--------------------------------------------------------
         mpg |         0
      weight |         0
displacement |         0
     foreign |         0
```

Stata will just assume that none of the data in memory played a role in obtaining the estimation results.

There is more worth knowing. You could, for instance, type `estimates describe` to see the command line that produced the estimates. See [R] **estimates**.

20.6 Specifying the estimation subsample

You specify the estimation subsample—the sample to be used in estimation—by specifying the `if` *exp* and/or `in` *range* modifiers with the estimation command.

Once an estimation command has been run or previous estimates restored, Stata remembers the estimation subsample, and you can use the modifier `if e(sample)` on the end of other Stata commands. The term *estimation subsample* refers to the set of observations used to produce the active estimation results. That might turn out to be all the observations (as it was in the above example) or some of the observations:

```
. regress mpg weight 5.rep78 if foreign
      Source |       SS       df       MS              Number of obs =      21
-------------+------------------------------           F(  2,    18) =   10.21
       Model |  423.317154     2  211.658577           Prob > F      =  0.0011
    Residual |   372.96856    18  20.7204756           R-squared     =  0.5316
-------------+------------------------------           Adj R-squared =  0.4796
       Total |  796.285714    20  39.8142857           Root MSE      =   4.552

------------------------------------------------------------------------------
         mpg |      Coef.   Std. Err.      t    P>|t|     [95% Conf. Interval]
-------------+----------------------------------------------------------------
      weight |  -.0131402   .0029684    -4.43   0.000    -.0193765   -.0069038
     5.rep78 |   5.052676    2.13492     2.37   0.029     .5673764    9.537977
       _cons |   52.86088   6.540147     8.08   0.000     39.12054    66.60122
------------------------------------------------------------------------------

. summarize mpg weight 5.rep78 if e(sample)
    Variable |       Obs        Mean    Std. Dev.       Min        Max
-------------+--------------------------------------------------------
         mpg |        21    25.28571    6.309856         17         41
      weight |        21    2263.333    364.7099       1760       3170
     5.rep78 |        21    .4285714    .5070926          0          1
```

Twenty-one observations were used in the above regression and we subsequently obtained the means for those same 21 observations by typing `summarize ... if e(sample)`. There are two reasons observations were dropped: we specified `if foreign` when we ran the regression, and there were observations for which `5.rep78` was missing. The reason does not matter; `e(sample)` is true if the observation was used and false otherwise.

You can use `if e(sample)` on the end of any Stata command that allows an `if` *exp*.

Here Stata has a shorthand command that produces the same results as `summarize ... if e(sample)`:

```
. estat summarize, label
Estimation sample regress                           Number of obs =     21

--------------------------------------------------------------------------
    Variable |      Mean   Std. Dev.        Min        Max  Label
-------------+------------------------------------------------------------
         mpg |  25.28571   6.309856          17         41  Mileage (mpg)
      weight |  2263.333   364.7099        1760       3170  Weight (lbs.)
     5.rep78 |  .4285714   .5070926           0          1  Repair Record 1978
--------------------------------------------------------------------------
```

See [R] **estat**.

20.7 Specifying the width of confidence intervals

You can specify the width of the confidence intervals for the coefficients using the `level()` option at estimation or when you play back the results.

▷ Example 4

To obtain narrower, 90% confidence intervals when we fit the model, we type

```
. regress mpg weight displ, level(90)
      Source |       SS       df       MS              Number of obs =      74
-------------+------------------------------           F(  2,    71) =   66.79
       Model |  1595.40969     2  797.704846           Prob > F      =  0.0000
    Residual |  848.049768    71  11.9443629           R-squared     =  0.6529
-------------+------------------------------           Adj R-squared =  0.6432
       Total |  2443.45946    73  33.4720474           Root MSE      =  3.4561

------------------------------------------------------------------------------
         mpg |      Coef.   Std. Err.      t    P>|t|     [90% Conf. Interval]
-------------+----------------------------------------------------------------
      weight |  -.0065671   .0011662    -5.63   0.000    -.0085108   -.0046234
displacement |   .0052808   .0098696     0.54   0.594    -.0111679    .0217294
       _cons |   40.08452    2.02011    19.84   0.000     36.71781    43.45124
------------------------------------------------------------------------------
```

If we subsequently typed **regress**, without arguments, 95% confidence intervals would be reported. If we initially fit the model with 95% confidence intervals, we could later type **regress, level(90)** to redisplay results with 90% confidence intervals.

Also, we could type **set level 90** to make 90% intervals our default; see [R] **level**.

Stata allows noninteger confidence intervals between 10.00 and 99.99, with a maximum of two digits following the decimal point. For instance, we could type

```
. regress mpg weight displ, level(92.5)
      Source |       SS       df       MS              Number of obs =      74
-------------+------------------------------           F(  2,    71) =   66.79
       Model |  1595.40969     2  797.704846           Prob > F      =  0.0000
    Residual |  848.049768    71  11.9443629           R-squared     =  0.6529
-------------+------------------------------           Adj R-squared =  0.6432
       Total |  2443.45946    73  33.4720474           Root MSE      =  3.4561

------------------------------------------------------------------------------
         mpg |      Coef.   Std. Err.      t    P>|t|   [92.5% Conf. Interval]
-------------+----------------------------------------------------------------
      weight |  -.0065671   .0011662    -5.63   0.000    -.0086745   -.0044597
displacement |   .0052808   .0098696     0.54   0.594    -.0125535     .023115
       _cons |   40.08452    2.02011    19.84   0.000     36.43419    43.73485
------------------------------------------------------------------------------
```

◁

20.8 Formatting the coefficient table

You can change the formatting of the coefficient table with the `sformat()`, `pformat()`, and `cformat()` options. The `sformat()` option changes the output format of test statistics; `pformat()` changes p-values; and `cformat()` changes coefficients, standard errors, and confidence limits. We can reduce the number of decimal places by specifying `%f` fixed-width formats:

```
. regress mpg weight displ, cformat(%6.3f) sformat(%4.1f) pformat(%4.2f)

      Source |       SS       df       MS              Number of obs =      74
-------------+------------------------------           F(  2,    71) =   66.79
       Model |  1595.40969     2  797.704846           Prob > F      =  0.0000
    Residual |  848.049768    71  11.9443629           R-squared     =  0.6529
-------------+------------------------------           Adj R-squared =  0.6432
       Total |  2443.45946    73  33.4720474           Root MSE      =  3.4561

------------------------------------------------------------------------------
         mpg |      Coef.   Std. Err.      t    P>|t|     [95% Conf. Interval]
-------------+----------------------------------------------------------------
      weight |     -0.007      0.001    -5.6   0.00       -0.009      -0.004
displacement |      0.005      0.010     0.5   0.59       -0.014       0.025
       _cons |     40.085      2.020    19.8   0.00       36.057      44.113
------------------------------------------------------------------------------
```

The option cformat(%6.3f), for example, fixes a width of six characters with three digits to the right of the decimal point. For more information on formats, see [U] **12.5.1 Numeric formats**.

The formatting options may also be specified when replaying results, so you can try different formats without refitting the model:

```
. regress, cformat(%7.4f)

      Source |       SS       df       MS              Number of obs =      74
-------------+------------------------------           F(  2,    71) =   66.79
       Model |  1595.40969     2  797.704846           Prob > F      =  0.0000
    Residual |  848.049768    71  11.9443629           R-squared     =  0.6529
-------------+------------------------------           Adj R-squared =  0.6432
       Total |  2443.45946    73  33.4720474           Root MSE      =  3.4561

------------------------------------------------------------------------------
         mpg |      Coef.   Std. Err.      t    P>|t|     [95% Conf. Interval]
-------------+----------------------------------------------------------------
      weight |    -0.0066     0.0012   -5.63   0.000     -0.0089     -0.0042
displacement |     0.0053     0.0099    0.54   0.594     -0.0144      0.0250
       _cons |    40.0845     2.0201   19.84   0.000     36.0565     44.1125
------------------------------------------------------------------------------
```

20.9 Obtaining the variance–covariance matrix

Typing estat vce displays the variance–covariance matrix of the estimators in active memory.

▷ Example 5

In example 2, we typed regress mpg weight displacement. The full variance–covariance matrix of the estimators can be displayed at any time after estimation:

```
. estat vce

Covariance matrix of coefficients of regress model

        e(V) |     weight  displace~t       _cons
-------------+------------------------------------
      weight |  1.360e-06
displacement |  -.0000103   .00009741
       _cons | -.00207455   .01188356   4.0808455
```

Typing `estat vce` with the `corr` option presents this matrix as a correlation matrix:

```
. estat vce, corr

Correlation matrix of coefficients of regress model

        e(V) |   weight  displa~t     _cons
-------------+------------------------------
      weight |   1.0000
displacement |  -0.8949    1.0000
       _cons |  -0.8806    0.5960    1.0000
```

See [R] **estat**.

Also, Stata's matrix commands understand that `e(V)` refers to the matrix:

```
. matrix list e(V)

symmetric e(V)[3,3]
                   weight  displacement         _cons
      weight    1.360e-06
displacement    -.0000103     .00009741
       _cons   -.00207455     .01188356     4.0808455

. mat Vinv = invsym(e(V))

. mat list Vinv

symmetric Vinv[3,3]
                   weight  displacement         _cons
      weight     60175851
displacement    4081161.2     292709.46
       _cons    18706.732     1222.3339     6.1953911
```

See [U] **14.5 Accessing matrices created by Stata commands**.

◁

20.10 Obtaining predicted values

Our discussion below, although cast in terms of predicted values, applies equally to the other statistics generated by `predict`; see [R] **predict**.

When Stata fits a model, whether it is regression or anything else, it internally saves the results, including the estimated coefficients and the variable names. The `predict` command allows you to use that information.

▷ Example 6

Let's perform a linear regression of mpg on weight and the square of weight:

```
. regress mpg weight c.weight#c.weight
      Source |       SS       df       MS              Number of obs =      74
-------------+------------------------------           F(  2,    71) =   72.80
       Model |  1642.52197     2  821.260986           Prob > F      =  0.0000
    Residual |  800.937487    71  11.2808097           R-squared     =  0.6722
-------------+------------------------------           Adj R-squared =  0.6630
       Total |  2443.45946    73  33.4720474           Root MSE      =  3.3587

------------------------------------------------------------------------------
         mpg |      Coef.   Std. Err.      t    P>|t|     [95% Conf. Interval]
-------------+----------------------------------------------------------------
      weight |  -.0141581   .0038835    -3.65   0.001    -.0219016   -.0064145
             |
    c.weight#|
    c.weight |   1.32e-06   6.26e-07     2.12   0.038     7.67e-08    2.57e-06
             |
       _cons |   51.18308   5.767884     8.87   0.000     39.68225    62.68392
------------------------------------------------------------------------------
```

After the regression, predict is defined to be

$$-.0141581\texttt{weight} + 1.32 \cdot 10^{-6}\texttt{weight}^2 + 51.18308$$

(Actually, it is more precise because the coefficients are internally stored at much higher precision than shown in the output.) Thus we can create a new variable—call it fitted—equal to the prediction by typing predict fitted and then use scatter to display the fitted and actual values separately for domestic and foreign automobiles:

```
. predict fitted
(option xb assumed; fitted values)
. scatter mpg fitted weight, by(foreign, total) c(. l) m(o i) sort
```

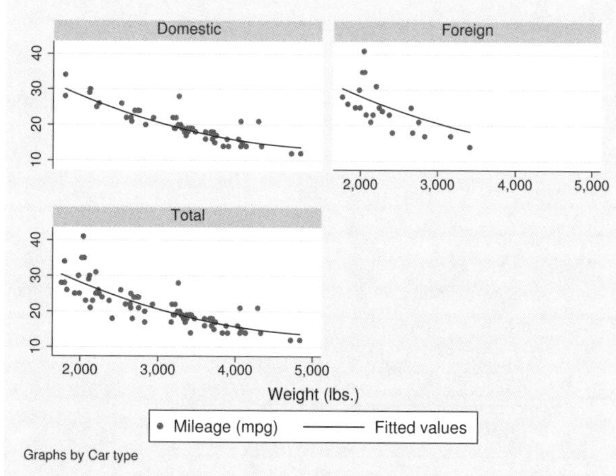

predict can calculate much more than just predicted values. For predict after linear regression, predict can calculate residuals, standardized residuals, Studentized residuals, influence statistics, etc. In any case, we specify what is to be calculated via an option, so if we wanted the residuals stored in new variable r, we would type

```
. predict r, resid
```

The options that may be specified following `predict` vary according to the estimation command previously used; the `predict` options are documented along with the estimation command. For instance, to discover all the things `predict` can do following `regress`, see [R] **regress**.

◁

20.10.1 Using predict

The use of `predict` is not limited to linear regression. `predict` can be used after any estimation command.

▷ Example 7

You fit a logistic regression model of whether a car is manufactured outside the United States on the basis of its weight and mileage rating using either the `logistic` or the `logit` command; see [R] **logistic** and [R] **logit**. We will use `logit`.

```
. use http://www.stata-press.com/data/r12/auto, clear
. logit foreign weight mpg
Iteration 0:   log likelihood =  -45.03321
Iteration 1:   log likelihood = -29.898968
Iteration 2:   log likelihood = -27.495771
Iteration 3:   log likelihood = -27.184006
Iteration 4:   log likelihood = -27.175166
Iteration 5:   log likelihood = -27.175156
Logistic regression                               Number of obs   =         74
                                                  LR chi2(2)      =      35.72
                                                  Prob > chi2     =     0.0000
Log likelihood = -27.175156                       Pseudo R2       =     0.3966

------------------------------------------------------------------------------
     foreign |      Coef.   Std. Err.      z    P>|z|     [95% Conf. Interval]
-------------+----------------------------------------------------------------
      weight |  -.0039067   .0010116    -3.86   0.000    -.0058894    -.001924
         mpg |  -.1685869   .0919174    -1.83   0.067    -.3487418     .011568
       _cons |   13.70837   4.518707     3.03   0.002     4.851864    22.56487
------------------------------------------------------------------------------
```

After `logit`, `predict` without options calculates the probability of a positive outcome (we learned that by looking at [R] **logit**). To obtain the predicted probabilities that each car is manufactured outside the United States, we type

```
. predict probhat
(option pr assumed; Pr(foreign))
. summarize probhat
    Variable |       Obs        Mean    Std. Dev.       Min        Max
-------------+--------------------------------------------------------
     probhat |        74    .2972973    .3052979    .000729   .8980594
. list make mpg weight foreign probhat in 1/5

     +------------------------------------------------------+
     | make            mpg   weight    foreign     probhat  |
     |------------------------------------------------------|
  1. | AMC Concord      22    2,930   Domestic    .1904363  |
  2. | AMC Pacer        17    3,350   Domestic    .0957767  |
  3. | AMC Spirit       22    2,640   Domestic    .4220815  |
  4. | Buick Century    20    3,250   Domestic    .0862625  |
  5. | Buick Electra    15    4,080   Domestic    .0084948  |
     +------------------------------------------------------+
```

◁

20.10.2 Making in-sample predictions

predict does not retrieve a vector of prerecorded values—it calculates the predictions on the basis of the recorded coefficients and the data currently in memory. In the above examples, when we typed things like

```
. predict probhat
```

predict filled in the prediction everywhere that it could be calculated.

Sometimes we have more data in memory than were used by the estimation command, either because we explicitly ignored some of the observations by specifying an if *exp* with the estimation command or because there are missing values. In such cases, if we want to restrict the calculation to the estimation subsample, we would do that in the usual way by adding if e(sample) to the end of the command:

```
. predict probhat if e(sample)
```

20.10.3 Making out-of-sample predictions

Because predict makes its calculations on the basis of the recorded coefficients and the data in memory, predict can do more than calculate predicted values for the data on which the estimation took place—it can make out-of-sample predictions, as well.

If you fit your model on a subset of the observations, you could then predict the outcome for all the observations:

```
. logit foreign weight mpg if rep78 > 3
. predict pall
```

If you do not specify if e(sample) at the end of the predict command, predict calculates the predictions for all observations possible.

In fact, because predict works from the active estimation results, you can use predict with *any* dataset that contains the necessary variables.

▷ Example 8

Continuing with our previous logit example, assume that we have a second dataset containing the mpg and weight of a different sample of cars. We have just fit your model and now continue:

```
. use otherdat, clear
(Different cars)
. predict probhat                          Stata remembers the previous model
(option pr assumed; Pr(foreign))
. summarize probhat foreign

    Variable |       Obs        Mean    Std. Dev.       Min        Max
-------------+--------------------------------------------------------
     probhat |        12    .2505068    .3187104   .0084948   .8920776
     foreign |        12    .1666667    .3892495          0          1
```

◁

▷ Example 9

There are many ways to obtain out-of-sample predictions. Above, we estimated on one dataset and then used another. If our first dataset had contained both sets of cars, marked, say, by the variable `difcars` being 0 if from the first sample and 1 if from the second, we could type

```
. logit foreign weight mpg if difcars==0
same output as above appears
. predict probhat
(option pr assumed; Pr(foreign))
. summarize probhat foreign if difcars==1
same output as directly above appears
```

If we just had a few additional cars, we could even input them after estimation. Assume that our data once again contain only the first sample of cars, and assume that we are interested in an additional sample of only two rather than 12 cars; we could type

```
. use http://www.stata-press.com/data/r12/auto
. keep make mpg weight foreign
. logit foreign weight mpg
same output as above appears
. input
                      make        mpg     weight    foreign
 75. "Merc. Zephyr" 20 2830 0                  we type in our new data
 76. "VW Dasher" 23 2160 1
 77. end
. predict probhat                              obtain all the predictions
(option pr assumed; Pr(foreign))
. list in -2/1
```

	make	mpg	weight	foreign	probhat
75.	Merc. Zephyr	20	2830	Domestic	.3275397
76.	VW Dasher	23	2160	Foreign	.8009743

◁

20.10.4 Obtaining standard errors, tests, and confidence intervals for predictions

When you use `predict`, you create, for each observation in the prediction sample, a statistic that is a function of the data and the estimated model parameters. You also could have generated your own customized predictions by using `generate`. In either case, to get standard errors, Wald tests, and confidence intervals for your predictions, use `predictnl`. For example, if we wanted the standard errors for our predicted probabilities, we could type

```
. drop probhat
. predictnl probhat = predict(), se(phat_se)
. list in 1/5
```

	make	mpg	weight	foreign	probhat	phat_se
1.	AMC Concord	22	2,930	Domestic	.1904363	.0658386
2.	AMC Pacer	17	3,350	Domestic	.0957767	.0536296
3.	AMC Spirit	22	2,640	Domestic	.4220815	.0892845
4.	Buick Century	20	3,250	Domestic	.0862625	.0461927
5.	Buick Electra	15	4,080	Domestic	.0084948	.0093079

Comparing this output to our previous listing of the first five predicted probabilities, you will notice that the output is identical, except that we now have an additional variable, `phat_se`, which contains the estimated standard error for each predicted probability.

We first had to `drop probhat` because `predictnl` will regenerate it. Note also the use of `predict()` within `predictnl`—it specified that we wanted to generate a point estimate (and standard error) for the default prediction after `logit`; see [R] **predictnl** for more details.

20.11 Accessing estimated coefficients

You can access coefficients and standard errors after estimation by referring to `_b[`*name*`]` and `_se[`*name*`]`; see [U] **13.5 Accessing coefficients and standard errors**.

▷ Example 10

Let's return to linear regression. We are doing a study of earnings of men and women at a particular company. In addition to each person's earnings, we have information on their educational attainment and tenure with the company. We type the following:

```
. regress lnearn ed tenure i.female female#(c.ed c.tenure)
  (output omitted)
```

If you are not familiar with the # notation, see [U] **11.4.3 Factor variables**.

We now wish to predict everyone's income as if they were male and then compare these as-if earnings with the actual earnings:

```
. generate asif = _b[_cons] + _b[ed]*ed + _b[tenure]*tenure
```

◁

▷ Example 11

We are analyzing the mileage of automobiles and are using a slightly more sophisticated model than any we have used so far. As we have previously, we will fit a linear regression model of `mpg` on `weight` and the square of `weight`, but we also add the interaction of `foreign` with `weight`, the car's gear ratio (`gear_ratio`), and `foreign` interacted with `gear_ratio`. We will use factor-variable notation to create the squared term and the interactions; see [U] **11.4.3 Factor variables**.

```
. use http://www.stata-press.com/data/r12/auto, clear
(1978 Automobile Data)

. regress mpg weight c.weight#c.weight i.foreign#c.weight gear_ratio
> i.foreign#c.gear_ratio

      Source |       SS       df       MS              Number of obs =      74
-------------+------------------------------           F(  5,    68) =   33.44
       Model |  1737.05293     5  347.410585           Prob > F      =  0.0000
    Residual |  706.406534    68  10.3883314           R-squared     =  0.7109
-------------+------------------------------           Adj R-squared =  0.6896
       Total |  2443.45946    73  33.4720474           Root MSE      =  3.2231

------------------------------------------------------------------------------
         mpg |      Coef.   Std. Err.      t    P>|t|     [95% Conf. Interval]
-------------+----------------------------------------------------------------
      weight |  -.0118517   .0045136    -2.63   0.011    -.0208584   -.002845
             |
   c.weight#|
    c.weight |   9.81e-07   7.04e-07     1.39   0.168    -4.25e-07   2.39e-06
             |
    foreign#|
    c.weight |
           1 |  -.0032241   .0015577    -2.07   0.042    -.0063326  -.0001157
             |
  gear_ratio |   1.159741   1.553418     0.75   0.458    -1.940057   4.259539
             |
    foreign#|
c.gear_ratio |
           1 |   1.597462   1.205313     1.33   0.189    -.8077036   4.002627
             |
       _cons |   44.61644   8.387943     5.32   0.000     27.87856   61.35432
------------------------------------------------------------------------------
```

If you are not experienced in both regression technology and automobile technology, you may find it difficult to interpret this regression. Putting aside issues of statistical significance, we find that mileage decreases with a car's weight but increases with the square of weight; decreases even more rapidly with weight for foreign cars; increases with higher gear ratio; and increases even more rapidly with higher gear ratio in foreign cars.

Thus do foreign cars yield better or worse gas mileage? Results are mixed. As the foreign cars' weight increases, they do more poorly in relation to domestic cars, but they do better at higher gear ratios. One way to compare the results is to predict what mileage foreign cars would have *if they were manufactured domestically*. The regression provides all the information necessary for making that calculation; mileage for domestic cars is estimated to be

$$-.012\,\texttt{weight} + 9.81 \cdot 10^{-7}\,\texttt{weight}^2 + 1.160\,\texttt{gear\_ratio} + 44.6$$

We can use that equation to predict the mileage of foreign cars and then compare it with the true outcome. The `_b[]` function simplifies reference to the estimated coefficients. We can type

```
. gen asif=_b[weight]*weight + _b[c.weight#c.weight]*c.weight#c.weight +
          _b[gear_ratio]*gear_ratio + _b[_cons]
```

`_b[weight]` refers to the estimated coefficient on `weight`, `_b[c.weight#c.weight]` to the estimated coefficient on `c.weight#c.weight`, and so on.

We might now ask how the actual mileage of a Honda compares with the asif prediction:

```
. list make asif mpg if strpos(make,"Honda")

     +------------------------------+
     | make              asif   mpg |
     |------------------------------|
 61. | Honda Accord   26.52597   25 |
 62. | Honda Civic    30.62202   28 |
     +------------------------------+
```

Notice the way we constructed our if clause to select Hondas. strpos() is the string function that returns the location in the first string where the second string is found or, if the second string does not occur in the first, zero. Thus any recorded make that contains the string "Honda" anywhere in it would be listed; see [D] **functions**.

We find that both Honda models yield slightly lower gas mileage than the asif domestic car–based prediction. (We do not endorse this model as a complete model of the determinants of mileage, nor do we single out the Honda for any special scorn. In fact, please note that the observed values are within the root mean squared error of the average prediction.)

We might wish to compare the overall average mpg and the asif prediction over all foreign cars in the data:

```
. summarize mpg asif if foreign
    Variable |       Obs        Mean    Std. Dev.       Min        Max
-------------+--------------------------------------------------------
         mpg |        22    24.77273    6.611187         14         41
        asif |        22    26.67124    3.142912   19.70466   30.62202
```

We find that, on average, foreign cars yield slightly lower mileage than our asif prediction. This might lead us to ask if any foreign cars do better than the asif prediction:

```
. list make asif mpg if foreign & mpg>asif, sep(0)

     +---------------------------------+
     | make                asif    mpg |
     |---------------------------------|
 55. | BMW 320i        24.31697     25 |
 57. | Datsun 210      28.96818     35 |
 63. | Mazda GLC       29.32015     30 |
 66. | Subaru          28.85993     35 |
 68. | Toyota Corolla  27.01144     31 |
 71. | VW Diesel       28.90355     41 |
     +---------------------------------+
```

We find six such automobiles.

◁

20.12 Performing hypothesis tests on the coefficients

20.12.1 Linear tests

After estimation, `test` is used to perform tests of linear hypotheses on the basis of the variance–covariance matrix of the estimators (Wald tests).

▷ Example 12

(`test` has many syntaxes and features, so do not use this example as an excuse for not reading [R] **test**.) Using the automobile data, we perform the following regression:

```
. use http://www.stata-press.com/data/r12/auto, clear
(1978 Automobile Data)

. generate weightsq=weight^2

. regress mpg weight weightsq foreign

      Source |       SS       df       MS              Number of obs =      74
-------------+------------------------------           F(  3,    70) =   52.25
       Model |  1689.15372     3   563.05124           Prob > F      =  0.0000
    Residual |   754.30574    70  10.7757963           R-squared     =  0.6913
-------------+------------------------------           Adj R-squared =  0.6781
       Total |  2443.45946    73  33.4720474           Root MSE      =  3.2827

------------------------------------------------------------------------------
         mpg |      Coef.   Std. Err.      t    P>|t|     [95% Conf. Interval]
-------------+----------------------------------------------------------------
      weight |  -.0165729   .0039692    -4.18   0.000    -.0244892   -.0086567
    weightsq |   1.59e-06   6.25e-07     2.55   0.013     3.45e-07    2.84e-06
     foreign |    -2.2035   1.059246    -2.08   0.041      -4.3161   -.0909002
       _cons |   56.53884   6.197383     9.12   0.000     44.17855    68.89913
------------------------------------------------------------------------------
```

We can use the `test` command to calculate the joint significance of `weight` and `weightsq`:

```
. test weight weightsq

 ( 1)  weight = 0
 ( 2)  weightsq = 0

       F(  2,    70) =   60.83
            Prob > F =    0.0000
```

We are not limited to testing whether the coefficients are zero. We can test whether the coefficient on `foreign` is -2 by typing

```
. test foreign = -2

 ( 1)  foreign = -2

       F(  1,    70) =    0.04
            Prob > F =    0.8482
```

We can even test more complicated hypotheses because `test` can perform basic algebra. Here is an absurd hypothesis:

```
. test 2*(weight+weightsq)=-3*(foreign-(weight-weightsq))

 ( 1)  - weight + 5.0 weightsq + 3.0 foreign = 0

       F(  1,    70) =    4.31
            Prob > F =    0.0416
```

test simplified the algebra of our hypothesis and then presented the test results. We discover that the hypothesis may be absurd but we cannot reject it at the 1% or even 4% level. We can also use test's accumulate option to combine this test with another test:

```
. test foreign+weight=0, accum

 ( 1)  - weight + 5.0 weightsq + 3.0 foreign = 0
 ( 2)  weight + foreign = 0

       F(  2,    70) =    9.12
            Prob > F =    0.0003
```

There are limitations. test can test only linear hypotheses. If we attempt to test a nonlinear hypothesis, test will tell us that it is not possible:

```
. test weight/foreign=0
not possible with test
r(131);
```

Testing nonlinear hypotheses is discussed in [U] **20.12.4 Nonlinear Wald tests** below. ◁

20.12.2 Using test

test bases its results on the estimated variance–covariance matrix of the estimators (that is, performs a Wald test), so it can be used after any estimation command. For maximum likelihood estimation, you will have to decide whether you want to perform tests on the basis of the information matrix instead of constraining the equation, reestimating it, and then calculating the likelihood-ratio test (see [U] **20.12.3 Likelihood-ratio tests**). Because test bases its results on the information matrix, its results have the same standing as the asymptotic z statistic presented in the coefficient table.

▷ Example 13

Let's examine the repair records of the cars in our automobile data as rated by *Consumer Reports*:

```
. tabulate rep78 foreign

    Repair │
    Record │       Car type
      1978 │  Domestic    Foreign │     Total
───────────┼──────────────────────┼──────────
         1 │         2          0 │         2
         2 │         8          0 │         8
         3 │        27          3 │        30
         4 │         9          9 │        18
         5 │         2          9 │        11
───────────┼──────────────────────┼──────────
     Total │        48         21 │        69
```

The values are coded 1–5, corresponding to well below average to well above average. We will fit this variable by using a maximum-likelihood ordered logit model (the nolog option suppresses the iteration log, saving us some paper):

```
. ologit rep78 price foreign weight weightsq displ, nolog
Ordered logistic regression                       Number of obs   =         69
                                                  LR chi2(5)      =      33.12
                                                  Prob > chi2     =     0.0000
Log likelihood = -77.133082                       Pseudo R2       =     0.1767

------------------------------------------------------------------------------
       rep78 |      Coef.   Std. Err.      z    P>|z|     [95% Conf. Interval]
-------------+----------------------------------------------------------------
       price |   -.000034   .0001188    -0.29   0.775    -.0002669     .000199
     foreign |   2.685648   .9320398     2.88   0.004     .8588833    4.512412
      weight |  -.0037447   .0025609    -1.46   0.144    -.0087639    .0012745
    weightsq |   7.87e-07   4.50e-07     1.75   0.080    -9.43e-08    1.67e-06
displacement |  -.0108919   .0076805    -1.42   0.156    -.0259455    .0041617
-------------+----------------------------------------------------------------
       /cut1 |  -9.417196   4.298201                     -17.84151   -.9928766
       /cut2 |  -7.581864    4.23409                     -15.88053    .7168002
       /cut3 |   -4.82209   4.147679                     -12.95139    3.307212
       /cut4 |   -2.79344   4.156219                     -10.93948    5.352599
------------------------------------------------------------------------------
```

We now wonder whether all our variables other than `foreign` are jointly significant. We test the hypothesis just as we would after linear regression:

```
. test weight weightsq displ price
 ( 1)  [rep78]weight = 0
 ( 2)  [rep78]weightsq = 0
 ( 3)  [rep78]displacement = 0
 ( 4)  [rep78]price = 0
           chi2(  4) =    3.63
         Prob > chi2 =    0.4590
```

To compare this with the results performed by a likelihood-ratio test, see [U] **20.12.3 Likelihood-ratio tests**. Here results differ little.

◁

20.12.3 Likelihood-ratio tests

After maximum likelihood estimation, you can obtain likelihood-ratio tests by fitting both the unconstrained and constrained models, storing the results using `estimates store`, and then running `lrtest`. See [R] **lrtest** for the full details.

▷ Example 14

In [U] **20.12.2 Using test** above, we fit an ordered logit on `rep78` and then tested the significance of all the explanatory variables except `foreign`.

To obtain the likelihood-ratio test, sometime after fitting the full model, we type `estimates store` *full_model_name*, where *full_model_name* is just a label that we assign to these results.

```
. ologit rep78 price foreign weight weightsq displ
  (output omitted)
. estimates store myfullmodel
```

This command saves the current model results with the name `myfullmodel`.

Next we fit the constrained model. After that, typing '`lrtest myfullmodel .`' compares the current model with the model we saved:

```
. ologit rep78 foreign

Iteration 0:   log likelihood = -93.692061
Iteration 1:   log likelihood = -79.696089
Iteration 2:   log likelihood = -79.044933
Iteration 3:   log likelihood = -79.029267
Iteration 4:   log likelihood = -79.029243

Ordered logistic regression                       Number of obs   =         69
                                                  LR chi2(1)      =      29.33
                                                  Prob > chi2     =     0.0000
Log likelihood = -79.029243                       Pseudo R2       =     0.1565

------------------------------------------------------------------------------
       rep78 |      Coef.   Std. Err.      z    P>|z|     [95% Conf. Interval]
-------------+----------------------------------------------------------------
     foreign |    2.98155   .6203637     4.81   0.000      1.76566    4.197441
-------------+----------------------------------------------------------------
       /cut1 |  -3.158382   .7224269                     -4.574313   -1.742452
       /cut2 |  -1.362642   .3557343                     -2.059868   -.6654154
       /cut3 |   1.232161   .3431227                      .5596533     1.90467
       /cut4 |   3.246209   .5556646                      2.157127    4.335292
------------------------------------------------------------------------------

. lrtest myfullmodel .

Likelihood-ratio test                                 LR chi2(4)  =       3.79
(Assumption: . nested in myfullmodel)                 Prob > chi2 =     0.4348
```

When we tested the same constraint with `test` (which performed a Wald test), we obtained a χ^2 of 3.63 and a significance level of 0.4590. We used `.` (the dot) to specify the results in active memory, although we also could have stored them with `estimates store` and referred to them by name instead. Also, the order in which you specify the two models to `lrtest` doesn't matter; `lrtest` is smart enough to know the full model from the constrained model.

◁

Two other postestimation commands work in the same way as `lrtest`, meaning that they accept names of stored estimation results as their input: `hausman` for performing Hausman specification tests and `suest` for seemingly unrelated estimation. We do not cover these commands here; see [R] **hausman** and [R] **suest** for more details.

20.12.4 Nonlinear Wald tests

`testnl` can be used to test nonlinear hypotheses about the parameters of the active estimation results. `testnl`, like `test`, bases its results on the variance–covariance matrix of the estimators (that is, performs a Wald test), so it can be used after any estimation command; see [R] **testnl**.

▷ Example 15

We fit the model

```
. regress price mpg weight foreign
```
(output omitted)

and then type

```
. testnl (38*_b[mpg]^2 = _b[foreign]) (_b[mpg]/_b[weight]=4)
  (1)  38*_b[mpg]^2 = _b[foreign]
  (2)  _b[mpg]/_b[weight]=4
               F(2, 70) =        0.02
               Prob > F =        0.9806
```

We performed this test on linear regression estimates, but tests of this type could be performed after any estimation command.

◁

20.13 Obtaining linear combinations of coefficients

lincom computes point estimates, standard errors, t or z statistics, p-values, and confidence intervals for a linear combination of coefficients after any estimation command. Results can optionally be displayed as odds ratios, incidence-rate ratios, or relative-risk ratios.

▷ Example 16

We fit a linear regression:

```
. use http://www.stata-press.com/data/r12/regress, clear
. regress y x1 x2 x3
      Source |       SS       df       MS              Number of obs =     148
-------------+------------------------------           F(  3,   144) =   96.12
       Model |  3259.3561      3  1086.45203           Prob > F      =  0.0000
    Residual |  1627.56282   144  11.3025196           R-squared     =  0.6670
-------------+------------------------------           Adj R-squared =  0.6600
       Total |  4886.91892   147  33.2443464           Root MSE      =  3.3619

------------------------------------------------------------------------------
           y |      Coef.   Std. Err.      t    P>|t|     [95% Conf. Interval]
-------------+----------------------------------------------------------------
          x1 |   1.457113    1.07461     1.36   0.177    -.6669339    3.581161
          x2 |   2.221682   .8610358     2.58   0.011     .5197797    3.923583
          x3 |   -.006139   .0005543   -11.08   0.000    -.0072345   -.0050435
       _cons |   36.10135   4.382693     8.24   0.000     27.43863    44.76407
------------------------------------------------------------------------------
```

Suppose that we want to see the difference of the coefficients of x2 and x1. We type

```
. lincom x2 - x1
 ( 1) - x1 + x2 = 0

------------------------------------------------------------------------------
           y |      Coef.   Std. Err.      t    P>|t|     [95% Conf. Interval]
-------------+----------------------------------------------------------------
         (1) |   .7645682   .9950282     0.77   0.444     -1.20218    2.731316
------------------------------------------------------------------------------
```

◁

lincom is handy for computing the odds ratio of one covariate group relative to another.

▷ Example 17

We estimate the parameters of a logistic model of low birthweight:

```
. use http://www.stata-press.com/data/r12/lbw3
(Hosmer & Lemeshow data)

. logit low age lwd i.race smoke ptd ht ui

Iteration 0:   log likelihood =   -117.336
Iteration 1:   log likelihood =   -99.3982
Iteration 2:   log likelihood = -98.780418
Iteration 3:   log likelihood = -98.777998
Iteration 4:   log likelihood = -98.777998

Logistic regression                               Number of obs   =        189
                                                  LR chi2(8)      =      37.12
                                                  Prob > chi2     =     0.0000
Log likelihood = -98.777998                       Pseudo R2       =     0.1582
```

low	Coef.	Std. Err.	z	P>|z|	[95% Conf.	Interval]
age	-.0464796	.0373888	-1.24	0.214	-.1197603	.0268011
lwd	.8420615	.4055338	2.08	0.038	.0472299	1.636893
race						
2	1.073456	.5150753	2.08	0.037	.0639273	2.082985
3	.815367	.4452979	1.83	0.067	-.0574008	1.688135
smoke	.8071996	.404446	2.00	0.046	.0145001	1.599899
ptd	1.281678	.4621157	2.77	0.006	.3759478	2.187408
ht	1.435227	.6482699	2.21	0.027	.1646414	2.705813
ui	.6576256	.4666192	1.41	0.159	-.2569313	1.572182
_cons	-1.216781	.9556797	-1.27	0.203	-3.089878	.656317

Level 1 of `race` designates white, level 2 designates black, and level 3 designates other.

If we want to obtain the odds ratio for black smokers relative to white nonsmokers (the reference group), we type

```
. lincom 2.race + smoke, or
 ( 1)  [low]2.race + [low]smoke = 0
```

low	Odds Ratio	Std. Err.	z	P>|z|	[95% Conf.	Interval]
(1)	6.557805	4.744692	2.60	0.009	1.588176	27.07811

`lincom` computed $\exp(\beta_{\text{race}==2} + \beta_{\text{smoke}}) = 6.56$.

◁

20.14 Obtaining nonlinear combinations of coefficients

`lincom` is limited to estimating linear combinations of coefficients, for example, `2.race + smoke`, or exponentiated linear combinations, as in the above. For general nonlinear combinations, use `nlcom`.

▷ Example 18

Continuing our previous example: suppose that we wanted the ratio of the coefficients (and standard errors, Wald test, confidence interval, etc.) of blacks and races other than white and black:

```
. nlcom _b[2.race]/_b[3.race]
       _nl_1:  _b[2.race]/_b[3.race]
```

low	Coef.	Std. Err.	z	P>|z|	[95% Conf.	Interval]
_nl_1	1.316531	.7359262	1.79	0.074	-.1258574	2.75892

The Wald test given is that of the null hypothesis that the nonlinear combination is zero versus the two-sided alternative—this is probably not informative for a ratio. If we would instead like to test whether this ratio is one, we can rerun `nlcom`, this time subtracting one from our ratio estimate.

```
. nlcom _b[2.race]/_b[3.race] - 1
       _nl_1:  _b[2.race]/_b[3.race] - 1
```

low	Coef.	Std. Err.	z	P>|z|	[95% Conf.	Interval]
_nl_1	.3165314	.7359262	0.43	0.667	-1.125857	1.75892

We can interpret this as not much evidence that the "ratio minus 1" is different from zero, meaning that we cannot reject the null hypothesis that the ratio equals one.

When using `nlcom`, we needed to refer to the model coefficients by their "proper" names, for example, `_b[2.race]`, and not by the shorthand `2.race`, such as when using `lincom`. If we had typed

```
. nlcom 2.race/3.race
```

Stata would have reported an error.

If you have difficulty determining what to type for a coefficient when using `lincom` or `nlcom` replay your results using the `coeflegend` option. Here are the results for our current estimates,

```
. logit, coeflegend
Logistic regression                               Number of obs   =        189
                                                  LR chi2(8)      =      37.12
                                                  Prob > chi2     =     0.0000
Log likelihood = -98.777998                       Pseudo R2       =     0.1582
```

low	Coef.	Legend
age	-.0464796	_b[low:age]
lwd	.8420615	_b[low:lwd]
race		
2	1.073456	_b[low:2.race]
3	.815367	_b[low:3.race]
smoke	.8071996	_b[low:smoke]
ptd	1.281678	_b[low:ptd]
ht	1.435227	_b[low:ht]
ui	.6576256	_b[low:ui]
_cons	-1.216781	_b[low:_cons]

◁

20.15 Obtaining marginal means, adjusted predictions, and predictive margins

`predict` uses the current estimation results (the coefficients and the VCE) to estimate the value of statistics for observations in the data. `lincom` and `nlcom` use the current estimation results to estimate a specific linear or nonlinear expression of the coefficients. The `margins` command combines aspects of both and estimates margins of responses.

`margins` answers the question "What does my model have to say about such-and-such a group or such-and-such a person", where such-and-such might be

- my estimation sample or another sample
- a sample with the values of some covariates fixed
- a sample evaluated at each level of a treatment
- a population represented by a complex survey sample
- someone who looks like the fifth person in my sample
- someone who looks like the mean of the covariates in my sample
- someone who looks like the median of the covariates in my sample
- someone who looks like the 25th percentile of the covariates in my sample
- someone who looks like some other function of the covariates in my sample
- a standardized population
- a balanced experimental design
- any combination of the above
- any comparison of the above

`margins` answers these questions either conditionally on fixed values of all covariates or averaged over the observations in a sample. It answers these questions about almost any predictions or any other response that you can calculate as a function of your estimated parameters—linear responses, probabilities, hazards, survival times, odds ratios, risk differences, etc. It answers these questions in terms of the response given covariate levels, or in terms of the change in the response for a change in levels (also known as marginal effects). It answers these questions providing standard errors, test statistics, and confidence intervals; and those statistics can take the covariates as given or adjust for sampling, also known as predictive margins and survey statistics.

A margin is a statistic based on a response for a fitted model calculated over a dataset in which some of or all the covariates are fixed at values different from what they really are.

Margins go by different names in different fields, and they can estimate many interesting statistics related to a fitted model. We discuss some common uses below; see [R] **margins** for more applications.

20.15.1 Obtaining estimated marginal means

A classic application of margins is to estimate the expected marginal means from a linear estimator as though the design for the covariates were balanced—assuming an equal number of observations for each unique combination of levels for the factor-variable covariates. These means have a long history in the study of ANOVA and MANOVA but are of limited use with nonexperimental data. For a discussion, see *Obtaining margins as though the data were balanced* in [R] **margins** and example 4 in [R] **anova**.

Estimated marginal means are also called "least-squares means".

Consider an analysis of variance of the change in systolic blood pressure as determined by one of four drug treatments and adjusting for the patient's disease (Afifi and Azen 1979).

```
. use http://www.stata-press.com/data/r12/systolic
(Systolic Blood Pressure Data)

. tabulate drug disease

           |        Patient's Disease
 Drug Used |         1          2          3 |     Total
-----------+---------------------------------+----------
         1 |         6          4          5 |        15
         2 |         5          4          6 |        15
         3 |         3          5          4 |        12
         4 |         5          6          5 |        16
-----------+---------------------------------+----------
     Total |        19         19         20 |        58

. anova systolic drug##disease

                       Number of obs =      58     R-squared     =  0.4560
                       Root MSE      = 10.5096     Adj R-squared =  0.3259

              Source |  Partial SS    df       MS           F     Prob > F
        -------------+----------------------------------------------------
               Model |  4259.33851    11  387.212591       3.51     0.0013
                     |
                drug |  2997.47186     3  999.157287       9.05     0.0001
             disease |  415.873046     2  207.936523       1.88     0.1637
        drug#disease |  707.266259     6   117.87771       1.07     0.3958
                     |
            Residual |  5080.81667    46  110.452536
        -------------+----------------------------------------------------
               Total |  9340.15517    57  163.862371
```

Despite having randomized on drug, we see in the tabulation that our data are not balanced—for example, 12 patients were administered drug 3, whereas 16 were administered drug 4. The diseases are also not balanced across drugs. To estimate the marginal mean for each level of drug while treating the design as though it were balanced, we type

```
. margins drug, asbalanced

Adjusted predictions                              Number of obs   =         58

Expression   : Linear prediction, predict()
at           : drug             (asbalanced)
               disease          (asbalanced)

------------------------------------------------------------------------------
             |            Delta-method
             |     Margin   Std. Err.      z    P>|z|     [95% Conf. Interval]
-------------+----------------------------------------------------------------
        drug |
          1  |   25.99444   2.751008     9.45   0.000     20.60257    31.38632
          2  |   26.55556   2.751008     9.65   0.000     21.16368    31.94743
          3  |   9.744444   3.100558     3.14   0.002     3.667462    15.82143
          4  |   13.54444   2.637123     5.14   0.000     8.375778    18.71311
------------------------------------------------------------------------------
```

Assuming everyone in the sample were treated with drug 4 and that the diseases were equally distributed across the drug treatments, the expected mean change in pressure resulting from treatment with drug 4 is 13.54. Because we are treating the data as balanced, we could also say that 13.54 is the expected mean change resulting from drug 4 for any sample where an equal number of patients has each of the three diseases.

If we want an estimate of the mean that uses the distribution of diseases observed in the sample, we would remove the `asbalanced` option:

```
. margins drug

Predictive margins                              Number of obs   =        58

Expression   : Linear prediction, predict()

------------------------------------------------------------------------------
             |            Delta-method
             |     Margin   Std. Err.      z    P>|z|     [95% Conf. Interval]
-------------+----------------------------------------------------------------
        drug |
          1  |   25.89799   2.750533     9.42   0.000     20.50704    31.28893
          2  |   26.41092   2.742762     9.63   0.000      21.0352    31.78664
          3  |   9.722989   3.099185     3.14   0.002     3.648697    15.79728
          4  |   13.55575   2.640602     5.13   0.000     8.380261    18.73123
------------------------------------------------------------------------------
```

We can now say that a pressure change of 13.56 is expected if everyone in the sample is given drug 4 and the distribution of diseases is as observed in the sample.

The second set of margins are not usually called estimated marginal means because they do not impose a balanced design when estimating the mean. They are adjusted predictions that just happen to be means because the response is linear.

Neither of these values is the average pressure change for those taking drug 4 in our sample because `margins` treats everyone in the sample as having taken drug 4. Treating everyone as though they have taken each drug is what makes the means comparable. We are essentially standardizing on the values of all the other covariates in our model (in this example, just `disease`).

To obtain the observed mean for those taking drug 4, we must tell `margins` to treat drug 4 as its sample, which we do with the `over()` option.

```
. summarize systolic if drug==4

    Variable |       Obs        Mean    Std. Dev.       Min        Max
-------------+--------------------------------------------------------
    systolic |        16        13.5    9.323805         -5         27

. margins, over(drug)

Predictive margins                              Number of obs   =        58

Expression   : Linear prediction, predict()
over         : drug

------------------------------------------------------------------------------
             |            Delta-method
             |     Margin   Std. Err.      z    P>|z|     [95% Conf. Interval]
-------------+----------------------------------------------------------------
        drug |
          1  |   26.06667   2.713577     9.61   0.000     20.74815    31.38518
          2  |   25.53333   2.713577     9.41   0.000     20.21482    30.85185
          3  |       8.75   3.033872     2.88   0.004     2.803721    14.69628
          4  |       13.5    2.62741     5.14   0.000     8.350371    18.64963
------------------------------------------------------------------------------
```

The margin in the last line of the table matches the mean from `summarize`.

For many questions, we prefer one of the first two estimates of `margins` to the last one. If we compare drugs 3 and 4 from the last results, the 8.75 and 13.5 include both the effect from the drug and the differing distribution of diseases among patients taking drug 3 and drug 4 in our sample. Our first set of margins, those from `margins drug, asbalanced`, assumed that for both drug 3 and drug 4, we had an equal number of patients with each disease. Our second set of margins, those from `margins drug`, assumed that for both drug 3 and drug 4, we wanted the observed distribution of patients from the whole sample. By assuming a common distribution of diseases across the drugs, our first two sets of margins remove the effect of disease when we compare across drugs.

20.15.2 Obtaining adjusted predictions

We will use the term adjusted predictions to refer to margins that are evaluated at fixed values for all covariates. The `margins` command has a great deal of flexibility in letting you choose what those fixed values are. Consider a model of high blood pressure as a function of sex, age group, and body mass index (BMI, a common measure of weight relative to height; variable `bmi`). We will allow the effect of age to differ for males and females by interacting the age group and sex variables. We will also allow the effect of BMI to differ across all combinations of age group and sex by specifying a full factorial model.

```
. use http://www.stata-press.com/data/r12/nhanes2

. logistic highbp sex##agegrp##c.bmi

Logistic regression                               Number of obs   =      10351
                                                  LR chi2(23)     =    1151.80
                                                  Prob > chi2     =     0.0000
Log likelihood = -3403.3345                       Pseudo R2       =     0.1447

------------------------------------------------------------------------------
      highbp | Odds Ratio   Std. Err.      z    P>|z|     [95% Conf. Interval]
-------------+----------------------------------------------------------------
       2.sex |   .8875349   1.134604    -0.09   0.926     .0724484    10.87281
             |
      agegrp |
          2  |   1.096495   1.407414     0.07   0.943     .0886015    13.56978
          3  |   11.85822   13.86645     2.11   0.034     1.198572    117.3208
          4  |   54.32365   58.08088     3.74   0.000     6.682134    441.6343
          5  |   36.58197   35.67608     3.69   0.000     5.409339    247.3945
          6  |   412.8948   484.3409     5.13   0.000     41.43321    4114.626
             |
  sex#agegrp |
        2 2  |    2.13253   3.817311     0.42   0.672     .0638574    71.21623
        2 3  |   2.877815   4.650595     0.65   0.513     .1212014    68.33104
        2 4  |   1.929862   2.926537     0.43   0.665     .0987923    37.69897
        2 5  |   2.361945    3.33027     0.61   0.542     .1489711    37.44879
        2 6  |   .2006528   .3252387    -0.99   0.322     .0083701     4.81017
             |
         bmi |   1.244033   .0370629     7.33   0.000     1.173471    1.318838
             |
   sex#c.bmi |
          2  |   .9582013   .0407832    -1.00   0.316      .881511    1.041563
             |
agegrp#c.bmi |
          2  |   1.007464   .0451481     0.17   0.868     .9227504    1.099955
          3  |   .9501658    .039221    -1.24   0.216     .8763215    1.030233
          4  |   .9181214   .0349998    -2.24   0.025     .8520232    .9893474
          5  |   .9314019   .0324088    -2.04   0.041     .8699994    .9971381
          6  |   .8421064   .0362688    -3.99   0.000     .7739384    .9162785
             |
sex#agegrp#c.bmi |
        2 2  |   .9895392   .0588217    -0.18   0.860     .8807134    1.111812
        2 3  |   .9976554   .0546945    -0.04   0.966     .8960147    1.110826
        2 4  |   1.006481   .0518212     0.13   0.900     .9098702    1.113351
        2 5  |   1.003616   .0478563     0.08   0.940     .9140686    1.101935
        2 6  |   1.116955   .0630488     1.96   0.050     .9999723    1.247624
             |
       _cons |   .0001527    .000129   -10.40   0.000     .0000292    .0007999
------------------------------------------------------------------------------
```

There are six levels for `agegrp` and two levels for `sex`. This dataset has value labels attached to those levels, so the easiest way to see what those levels mean is to let Stata tell us:

```
. table sex agegrp

──────────┬───────────────────────────────────────────────
1=male,   │                    Age Group
2=female  │  20-29  30-39  40-49  50-59  60-69     70+
──────────┼───────────────────────────────────────────────
     Male │  1,116    770    610    602  1,369     448
   Female │  1,204    852    662    689  1,491     538
──────────┴───────────────────────────────────────────────
```

We can evaluate the probability of having high blood pressure for each age group while holding the proportion of males and females and the value of `bmi` to its average by specifying the covariate `agegrp` to `margins` and including the option `atmeans`:

```
. margins agegrp, atmeans
Adjusted predictions                              Number of obs   =      10351
Model VCE    : OIM

Expression   : Pr(highbp), predict()
at           : 1.sex          =    .4748333 (mean)
               2.sex          =    .5251667 (mean)
               1.agegrp       =    .2241329 (mean)
               2.agegrp       =    .1566998 (mean)
               3.agegrp       =    .1228867 (mean)
               4.agegrp       =    .1247222 (mean)
               5.agegrp       =    .2763018 (mean)
               6.agegrp       =    .0952565 (mean)
               bmi            =     25.5376 (mean)

─────────────┬────────────────────────────────────────────────────────────────
             │            Delta-method
             │     Margin   Std. Err.      z    P>|z|     [95% Conf. Interval]
─────────────┼────────────────────────────────────────────────────────────────
      agegrp │
          1  │   .0209216   .0035683     5.86   0.000     .0139278    .0279153
          2  │   .0353277   .0051429     6.87   0.000     .0252477    .0454077
          3  │   .1038999   .0091758    11.32   0.000     .0859157    .1218842
          4  │   .1679085   .0111415    15.07   0.000     .1460715    .1897454
          5  │   .1734584     .00754    23.01   0.000     .1586803    .1882365
          6  │   .1720227   .0124465    13.82   0.000     .1476281    .1964173
─────────────┴────────────────────────────────────────────────────────────────
```

The header of the table showed us the mean values of each covariate. These are the values at which the probabilities were evaluated. The mean values for the levels of `agegrp` appear in the header even though they were not used. `agegrp` assumed the values 1, 2, 3, 4, 5, and 6, as shown in the table. The means of the levels of `agegrp` are shown because we might have asked for more margins in the table, for example, `margins sex agegrp`.

The modeled probability remains below 4% for those under 40 years of age, and then increases fairly rapidly to just under 16.8% in the 50–59 age group. Above age 59, the probability almost flattens, remaining under 17.5% in the last two age groups. It is often easier for nonstatisticians to interpret the statistics computed by `margins` than the coefficients of a fitted model.

20.15.3 Obtaining predictive margins

Rather than evaluate the probability of having high blood pressure at one fixed point (the means), as we did above, we can evaluate the probability at the covariate values for each observation in our data and average those probabilities. Here is the modeled probability averaged over our sample:

```
. margins

Predictive margins                                Number of obs   =      10351
Model VCE    : OIM

Expression   : Pr(highbp), predict()

------------------------------------------------------------------------------
             |            Delta-method
             |     Margin   Std. Err.      z    P>|z|     [95% Conf. Interval]
-------------+----------------------------------------------------------------
       _cons |    .128973   .0031014    41.59   0.000     .1228945    .1350516
------------------------------------------------------------------------------
```

If we fix the level of agegrp to 1, compute the probability for each observation, and then average those probabilities, the result is the predictive margin for level 1 of agegrp. margins, by default, computes these margins for each level of each variable specified on the command line. Let's compute the predictive margins for agegrp:

```
. margins agegrp

Predictive margins                                Number of obs   =      10351
Model VCE    : OIM

Expression   : Pr(highbp), predict()

------------------------------------------------------------------------------
             |            Delta-method
             |     Margin   Std. Err.      z    P>|z|     [95% Conf. Interval]
-------------+----------------------------------------------------------------
      agegrp |
          1  |   .0375314    .004423     8.49   0.000     .0288624    .0462003
          2  |   .0557658   .0053934    10.34   0.000     .0451949    .0663367
          3  |   .1214814   .0086634    14.02   0.000     .1045015    .1384612
          4  |    .181929   .0102836    17.69   0.000     .1617736    .2020844
          5  |   .1892586   .0069432    27.26   0.000     .1756501     .202867
          6  |   .1818377   .0118687    15.32   0.000     .1585755       .2051
------------------------------------------------------------------------------
```

One way of looking at predictive margins is that they answer the question "What would the average response (probability) be in my sample if everyone were in one age group?" Another way of looking at predictive margins is that they standardize the effect of being in an age group with the distribution of other covariate values in our sample. The margins above are comparable because only the level of agegrp is changing across the margins. They represent our sample because all the other covariates take on their values in the sample when the margins are evaluated.

The predictive margins in this table differ from the adjusted predictions we estimated in [U] **20.15.2 Obtaining adjusted predictions** because the probability is a nonlinear function of the coefficients in a logistic model; see *Example 3: Average response versus response at average* in [R] **margins** for details.

Our analysis so far has been a bit naïve. The dataset we are using is from the Second National Health and Nutrition Examination Survey (NHANES II). It has weights to make it representative of the population from which it was drawn as well as other survey characteristics—strata and primary sampling units. The data have already been svyset; see [SVY] **svyset**. We should take note of these characteristics and use the svy prefix when fitting our model.

```
. svy: logistic highbp sex##agegrp##c.bmi
  (output omitted)
```

If we were to repeat the command margins agegrp, we would see that our point estimates differ only a little, but our standard errors are generally larger.

We are not restricted to margining over a single factor variable. Let's see if the pattern of high blood pressure over age groups differs for men and women. We do that by specifying the interaction of `sex` and `agegrp` to `margins`. We add the `vce(unconditional)` option to accommodate the survey design.

```
. margins sex#agegrp, vce(unconditional)
Predictive margins                                Number of obs   =      10351
Expression   : Pr(highbp), predict()
```

	Margin	Linearized Std. Err.	t	P>|t|	[95% Conf.	Interval]
sex#agegrp						
1 1	.0597492	.0098977	6.04	0.000	.0395628	.0799357
1 2	.0743686	.0090521	8.22	0.000	.0559067	.0928304
1 3	.139588	.0126715	11.02	0.000	.1137444	.1654317
1 4	.2027485	.018315	11.07	0.000	.1653949	.2401021
1 5	.2154491	.0124251	17.34	0.000	.190108	.2407902
1 6	.1562754	.0200764	7.78	0.000	.1153294	.1972215
2 1	.0178699	.0048	3.72	0.001	.0080803	.0276595
2 2	.0324522	.0058921	5.51	0.000	.0204352	.0444692
2 3	.08906	.0129256	6.89	0.000	.062698	.1154221
2 4	.1452838	.0171334	8.48	0.000	.11034	.1802276
2 5	.1657727	.0121933	13.60	0.000	.1409042	.1906411
2 6	.1833488	.0211122	8.68	0.000	.1402901	.2264075

Each line in the table corresponds to holding both `sex` and `agegrp` to fixed values while using the observed level of `bmi` to evaluate the probability and then averaging over the observations in the sample. To calculate the results in the first line of the table, `margins` fixed `sex` = 1 and `agegrp` = 1, evaluated the probability for each observation, and then averaged the probabilities. All these margins reflect the observed distribution of `bmi` in the sample.

The first six lines represent males, and the second six lines represent females. Comparing males with females for the same age groups, males are over three times as likely to have high blood pressure in the first age group ($0.060/0.018 = 3.3$), they are over twice as likely in the second age group, and while the relative gap narrows, it is not until above age 70 that the probability for males drops below the probability for females.

Can the pattern of probabilities be affected by controlling one's `bmi`? Let's reevaluate the probabilities while holding `bmi` to two levels—20 (which is well within the normal range) and 30 (which is at the boundary between overweight and obese). We add the option `at(bmi=(20 30))` to set `bmi` first to 20 and then to 30.

```
. margins sex#agegrp, at(bmi=(20 30)) vce(unconditional)

Adjusted predictions                              Number of obs     =      10351

Expression   : Pr(highbp), predict()

1._at        : bmi             =          20

2._at        : bmi             =          30

------------------------------------------------------------------------------
             |            Linearized
             |     Margin   Std. Err.      t    P>|t|     [95% Conf. Interval]
-------------+----------------------------------------------------------------
   _at#sex#  |
     agegrp  |
      1 1 1  |   .0109371   .0032537     3.36   0.002     .0043012     .0175731
      1 1 2  |   .0156812   .0047009     3.34   0.002     .0060937     .0252687
      1 1 3  |   .0482213   .0122271     3.94   0.000     .0232839     .0731586
      1 1 4  |   .0949467   .0173769     5.46   0.000     .0595064     .1303871
      1 1 5  |    .101735   .0142451     7.14   0.000      .072682      .130788
      1 1 6  |    .108137   .0203938     5.30   0.000     .0665436     .1497305
      1 2 1  |   .0053248   .0022848     2.33   0.026     .0006648     .0099847
      1 2 2  |   .0101007   .0036738     2.75   0.010     .0026078     .0175935
      1 2 3  |   .0422142   .0118314     3.57   0.001      .018084     .0663444
      1 2 4  |   .0912119   .0201067     4.54   0.000      .050204     .1322197
      1 2 5  |   .0927126    .011551     8.03   0.000     .0691542      .116271
      1 2 6  |   .1035166   .0235733     4.39   0.000     .0554385     .1515946
      2 1 1  |    .095307    .018967     5.02   0.000     .0566236     .1339903
      2 1 2  |   .1218371   .0174947     6.96   0.000     .0861564     .1575178
      2 1 3  |   .2275182    .027168     8.37   0.000     .1721088     .2829277
      2 1 4  |   .3084116   .0306717    10.06   0.000     .2458561      .370967
      2 1 5  |   .3276512   .0235645    13.90   0.000     .2795911     .3757113
      2 1 6  |   .2001169   .0352892     5.67   0.000     .1281442     .2720897
      2 2 1  |   .0258852    .007303     3.54   0.001     .0109906     .0407799
      2 2 2  |   .0483494   .0092976     5.20   0.000     .0293868     .0673119
      2 2 3  |   .1303522   .0198102     6.58   0.000      .089949     .1707553
      2 2 4  |   .1946575   .0192303    10.12   0.000     .1554371      .233878
      2 2 5  |   .2338165   .0167405    13.97   0.000     .1996741     .2679589
      2 2 6  |   .2582884   .0256129    10.08   0.000     .2060506     .3105262
------------------------------------------------------------------------------
```

That is a lot of margins, but they are in sets of six age groups. The first six margins are men with a BMI of 20, the second six are women with a BMI of 20, the third six are men with a BMI of 30, and the last six are women with a BMI of 30. These margins tell a more complete story. The probability of high blood pressure is much lower for both men and women who maintain a BMI of 20. More interesting is that the relationship between men and women differs depending on BMI. While young men who maintain a BMI of 20 are still twice as likely as young women to have high blood pressure (0.011/0.005) and youngish men are over 50% more likely (0.016/0.010), the gap narrows substantially for men in the four older groups. The story is worse for those with a BMI of 30. Both men and women with a high BMI have a substantially increased risk of high blood pressure, with men ages 50–69 almost 10 percentage points higher than women. Before you dismiss these differences as caused by the usual attenuation of the logistic curve in the tails, recall that when we fit the model, we allowed the effect of `bmi` to be different for each combination of `sex` and `agegrp`.

You may have noticed that the header of the prior results says "Adjusted predictions" rather than "Predictive margins". That is because our model has only three covariates, and we have fixed the values of each. `margins` is no longer averaging over the data, but is instead evaluating the margins at fixed points that we have requested. It lets us know that by changing the header.

We could post the results of `margins` and form linear combinations or perform tests about any of the assertions above; see *Example 10: Testing margins—contrasts of margins* in [R] **margins**.

There is much more to know about margins and the `margins` command. Consider the headings for the *Remarks* section of [R] **margins**:

Introduction
Obtaining margins of responses
- *Example 1: A simple case after regress*
- *Example 2: A simple case after logistic*
- *Example 3: Average response versus response at average*
- *Example 4: Multiple margins from one command*
- *Example 5: Margins with interaction terms*
- *Example 6: Margins with continuous variables*
- *Example 7: Margins of continuous variables*
- *Example 8: Margins of interactions*
- *Example 9: Decomposing margins*
- *Example 10: Testing margins—contrasts of margins*
- *Example 11: Margins of a specified prediction*
- *Example 12: Margins of a specified expression*
- *Example 13: Margins with multiple outcomes (responses)*
- *Example 14: Margins with multiple equations*
- *Example 15: Margins evaluated out of sample*

Obtaining margins of derivatives of responses (a.k.a. marginal effects)
- *Do not specify marginlist when you mean over()*
- *Use at() freely, especially with continuous variables*
- *Expressing derivatives as elasticities*
- *Derivatives versus discrete differences*
- *Example 16: Average marginal effect (partial effects)*
- *Example 17: Average marginal effect of all covariates*
- *Example 18: Evaluating marginal effects over the response surface*

Obtaining margins with survey data and representative samples
- *Example 19: Inferences for populations, margins of response*
- *Example 20: Inferences for populations, marginal effects*
- *Example 21: Inferences for populations with svyset data*

Standardizing margins
Obtaining margins as though the data were balanced
- *Balancing using asbalanced*
- *Balancing by standardization*
- *Balancing nonlinear responses*
- *Treating a subset of covariates as balanced*
- *Using fvset design*
- *Balancing in the presence of empty cells*

Obtaining margins with nested designs
- *Introduction*
- *Margins with nested designs as though the data were balanced*
- *Coding of nested designs*

Special topics
- *Requirements for model specification*
- *Estimability of margins*
- *Manipulability of tests*
- *Using margins after the estimates use command*
- *Syntax of at()*
- *Estimation commands that may be used with margins*

Glossary

20.16 Obtaining conditional and average marginal effects

Marginal effects measure the change in a response given a change in a covariate, which is to say that marginal effects are derivatives. As used here, marginal effects can also be the discrete change in a response as an indicator goes from 0 to 1. Some authors reserve the term marginal effect for the continuous change and use partial effect for the discrete change. We will not make that distinction. Regardless, marginal effects are most often used to make it easier to interpret how changes in covariates affect a nonlinear response from a fitted model—a probability, a censored dependent variable, a survival time, a hazard, etc.

Marginal effects can either be evaluated at a specified point for all the covariates in our model (conditional marginal effects) or be evaluated at the observed values of the covariates in a dataset and then averaged (average marginal effects).

To Stata, marginal effects are just margins whose response happens to be the derivative of another response. Those interested in marginal effects will be interested in all or most of [R] **margins**.

20.16.1 Obtaining conditional marginal effects

We call a marginal effect conditional when we fix the values of all the covariates and then take the derivative of the response with respect to a covariate. The mean of all covariates is often used as the fixed point, and this is sometimes called the marginal effect at the means.

Consider a simple probit model of union membership for women as a function of having graduated from college (`collgrad`), living in the South (`south`), tenure on the job (`tenure`), and the interaction of `south` and `tenure`. We are interested in how being in the South affects union membership. We fit the model by using an extract from 1988 of the U.S. National Longitudinal Survey of Labor Market Experience (see [XT] **xt**).

```
. use http://www.stata-press.com/data/r12/nlsw88, clear
(NLSW, 1988 extract)

. probit union i.collgrad i.south tenure south#c.tenure

Iteration 0:   log likelihood = -1042.6816
Iteration 1:   log likelihood = -997.71809
Iteration 2:   log likelihood = -997.60984
Iteration 3:   log likelihood = -997.60983

Probit regression                                 Number of obs   =       1868
                                                  LR chi2(4)      =      90.14
                                                  Prob > chi2     =     0.0000
Log likelihood = -997.60983                       Pseudo R2       =     0.0432

------------------------------------------------------------------------------
       union |      Coef.   Std. Err.      z    P>|z|     [95% Conf. Interval]
-------------+----------------------------------------------------------------
  1.collgrad |   .2783278   .0726167     3.83   0.000     .1360018    .4206539
     1.south |  -.2534964   .1050552    -2.41   0.016    -.4594008   -.0475921
      tenure |   .0362944   .0068205     5.32   0.000     .0229264    .0496624
             |
      south#|
    c.tenure |
          1  |  -.0239785   .0119533    -2.01   0.045    -.0474065   -.0005504
             |
       _cons |  -.8497418   .0664524   -12.79   0.000    -.9799862   -.7194974
------------------------------------------------------------------------------
```

Clearly, being located in the South decreases union membership. Using the `dydx()` and `atmeans` options of `margins`, we can ask how much it decreases membership by evaluating the marginal effect of being southern at the mean of all covariates:

```
. margins, dydx(south) atmeans

Conditional marginal effects                    Number of obs   =       1868
Model VCE    : OIM

Expression   : Pr(union), predict()
dy/dx w.r.t. : 1.south
at           : 0.collgrad       =     .7521413 (mean)
               1.collgrad       =     .2478587 (mean)
               0.south          =     .5744111 (mean)
               1.south          =     .4255889 (mean)
               tenure           =     6.571065 (mean)

------------------------------------------------------------------------------
             |            Delta-method
             |      dy/dx   Std. Err.      z    P>|z|     [95% Conf. Interval]
-------------+----------------------------------------------------------------
     1.south |  -.1236055    .019431    -6.36   0.000    -.1616896   -.0855215
------------------------------------------------------------------------------
Note: dy/dx for factor levels is the discrete change from the base level.
```

At the means of all the covariates, southern women are 12 percentage points less likely to be members of a union. This marginal effect includes both the direct effect of `i.south` and the interaction `south#c.tenure`.

As `margins` reports below the table, this change in the response is for the discrete change of going from not southern (0) to southern (1).

The header of `margins` tells us where the marginal effect was estimated. This margin fixes `tenure` to be 6.6 years. There is nothing special about this point. We could also evaluate the marginal effect at the median of `tenure`:

```
. margins, dydx(south) atmeans at((medians) _continuous)

Conditional marginal effects                    Number of obs   =       1868
Model VCE    : OIM

Expression   : Pr(union), predict()
dy/dx w.r.t. : 1.south
at           : 0.collgrad       =     .7521413 (mean)
               1.collgrad       =     .2478587 (mean)
               0.south          =     .5744111 (mean)
               1.south          =     .4255889 (mean)
               tenure           =     4.666667 (median)

------------------------------------------------------------------------------
             |            Delta-method
             |      dy/dx   Std. Err.      z    P>|z|     [95% Conf. Interval]
-------------+----------------------------------------------------------------
     1.south |  -.1061338   .0201722    -5.26   0.000    -.1456706    -.066597
------------------------------------------------------------------------------
Note: dy/dx for factor levels is the discrete change from the base level.
```

With `tenure` at its median of 4.67, the marginal effect is about 2 percentage points less than it was at the mean of 6.6.

When examining conditional marginal effects, it is often useful to evaluate them at a range of values for the covariates. We can do that by asking both for values of the indicator covariate `collgrad` and for a range of values for `tenure`:

```
. margins collgrad, dydx(south) at(tenure=(0(5)25))

Conditional marginal effects                      Number of obs    =       1868
Model VCE    : OIM

Expression   : Pr(union), predict()
dy/dx w.r.t. : 1.south

1._at        : tenure          =           0

2._at        : tenure          =           5

3._at        : tenure          =          10

4._at        : tenure          =          15

5._at        : tenure          =          20

6._at        : tenure          =          25

------------------------------------------------------------------------------
             |            Delta-method
             |      dy/dx   Std. Err.      z    P>|z|     [95% Conf. Interval]
-------------+----------------------------------------------------------------
1.south      |
_at#collgrad |
        1 0  |  -.0627725   .0254161    -2.47   0.014     -.112587   -.0129579
        1 1  |  -.0791483   .0321151    -2.46   0.014    -.1420928   -.0162038
        2 0  |  -.1031957   .0189184    -5.45   0.000     -.140275   -.0661164
        2 1  |  -.1256566   .0232385    -5.41   0.000    -.1712031   -.0801101
        3 0  |  -.1496772   .0222259    -6.73   0.000    -.1932392   -.1061151
        3 1  |  -.1760137   .0266874    -6.60   0.000    -.2283202   -.1237073
        4 0  |  -.2008801    .036154    -5.56   0.000    -.2717407   -.1300196
        4 1  |     -.2282   .0419237    -5.44   0.000    -.3103689    -.146031
        5 0  |  -.2549707   .0546355    -4.67   0.000    -.3620543   -.1478872
        5 1  |  -.2799495   .0613127    -4.57   0.000    -.4001201   -.1597789
        6 0  |  -.3097656   .0747494    -4.14   0.000    -.4562717   -.1632594
        6 1  |  -.3289702   .0816342    -4.03   0.000    -.4889703   -.1689701
------------------------------------------------------------------------------
Note: dy/dx for factor levels is the discrete change from the base level.
```

We now have a more complete picture of the effect that being in the South has on union participation. For those with no tenure and without a college degree (the first line in the table), being in the South decreases union participation by only 6 percentage points. For those with 25 years of tenure and with a college degree (the last line in the table), being in the South decreases participation by almost 33 percentage points. We can read the effect for any combination of tenure and college graduation status from the other lines in the table.

20.16.2 Obtaining average marginal effects

To compute average marginal effects, the marginal effect is first computed for each observation in the dataset and then averaged. If the sample over which we compute the average marginal effect represents a population, then we have estimated the marginal effect for the population.

We continue with our example of labor union participation.

```
. use http://www.stata-press.com/data/r12/nlsw88
(NLSW, 1988 extract)

. probit union i.collgrad i.south tenure south#c.tenure
  (output omitted)
```

To estimate the average marginal effect for each of our regressors, we type

```
. margins, dydx(*)

Average marginal effects                          Number of obs   =       1868
Model VCE    : OIM

Expression   : Pr(union), predict()
dy/dx w.r.t. : 1.collgrad 1.south tenure

------------------------------------------------------------------------------
             |            Delta-method
             |      dy/dx   Std. Err.      z    P>|z|     [95% Conf. Interval]
-------------+----------------------------------------------------------------
  1.collgrad |   .0878847   .0238065     3.69   0.000     .0412248    .1345447
     1.south |   -.126164   .0191504    -6.59   0.000    -.1636981   -.0886299
      tenure |   .0083571   .0016521     5.06   0.000      .005119    .0115951
------------------------------------------------------------------------------
Note: dy/dx for factor levels is the discrete change from the base level.
```

For this sample, the average marginal effect is very close to the marginal effect at the mean that we computed earlier. That is not always true; it depends on the distribution of the other covariates. The results also tell us that on average, for populations like the one from which our sample was drawn, union participation increases 0.8 percentage points for every year of tenure on the job. College graduates are, on average, 8.8 percentage points more likely to participate.

In the examples above, we treated the covariates in the sample as fixed and known. We could have accounted for the fact that this sample was drawn from a population and the covariates represent just one sample from that population. We do that by adding the `vce(robust)` or `vce(cluster` *clustvar*`)` option when fitting the model and the `vce(unconditional)` option when estimating the margins; see *Obtaining margins with survey data and representative samples* in [R] **margins**. It makes little difference in the examples above.

20.17 Obtaining pairwise comparisons

`pwcompare` performs pairwise comparisons across the levels of factor variables. `pwcompare` can compare estimated cell means, marginal means, intercepts, marginal intercepts, slopes, or marginal slopes—collectively called margins. `pwcompare` reports comparisons as contrasts (differences) of margins along with significance tests or confidence intervals for the contrasts. The tests and confidence intervals can be adjusted for multiple comparisons.

`pwcompare` is for use after an estimation command in which you have used factor variables in specifying the model. You could not use `pwcompare` after typing

```
. regress yield fertilizer1-fertilizer5
```

but you could use `pwcompare` after typing

```
. regress yield i.fertilizer
```

Below we fit a linear regression of wheat yield on type of fertilizer, and then we compare the mean yields for each pair of fertilizers and obtain p-values and confidence intervals adjusted for multiple comparisons using Tukey's honestly significant difference.

```
. use http://www.stata-press.com/data/r12/yield
(Artificial wheat yield dataset)

. regress yield i.fertilizer

      Source |       SS       df       MS              Number of obs =     200
-------------+------------------------------           F(  4,   195) =    5.33
       Model |  1078.84207     4  269.710517           Prob > F      =  0.0004
    Residual |  9859.55334   195   50.561812           R-squared     =  0.0986
-------------+------------------------------           Adj R-squared =  0.0801
       Total |  10938.3954   199  54.9668111           Root MSE      =  7.1107

------------------------------------------------------------------------------
       yield |      Coef.   Std. Err.      t    P>|t|     [95% Conf. Interval]
-------------+----------------------------------------------------------------
  fertilizer |
          2  |    3.62272   1.589997     2.28   0.024     .4869212    6.758518
          3  |   .4906299   1.589997     0.31   0.758    -2.645169    3.626428
          4  |   4.922803   1.589997     3.10   0.002     1.787005    8.058602
          5  |  -1.238328   1.589997    -0.78   0.437    -4.374127     1.89747
             |
       _cons |   41.36243   1.124298    36.79   0.000     39.14509    43.57977
------------------------------------------------------------------------------

. pwcompare fertilizer, effects mcompare(tukey)

Pairwise comparisons of marginal linear predictions

Margins      : asbalanced

---------------------------
             |    Number of
             |  Comparisons
-------------+-------------
  fertilizer |           10
---------------------------

------------------------------------------------------------------------------
             |                              Tukey                Tukey
             |   Contrast   Std. Err.      t    P>|t|     [95% Conf. Interval]
-------------+----------------------------------------------------------------
  fertilizer |
     2 vs 1  |    3.62272   1.589997     2.28   0.156    -.7552913    8.000731
     3 vs 1  |   .4906299   1.589997     0.31   0.998    -3.887381    4.868641
     4 vs 1  |   4.922803   1.589997     3.10   0.019     .5447922    9.300815
     5 vs 1  |  -1.238328   1.589997    -0.78   0.936    -5.616339    3.139683
     3 vs 2  |   -3.13209   1.589997    -1.97   0.285    -7.510101    1.245921
     4 vs 2  |   1.300083   1.589997     0.82   0.925    -3.077928    5.678095
     5 vs 2  |  -4.861048   1.589997    -3.06   0.021    -9.239059   -.4830368
     4 vs 3  |   4.432173   1.589997     2.79   0.046     .0541623    8.810185
     5 vs 3  |  -1.728958   1.589997    -1.09   0.813    -6.106969    2.649053
     5 vs 4  |  -6.161132   1.589997    -3.87   0.001    -10.53914    -1.78312
------------------------------------------------------------------------------
```

See [R] **pwcompare** and [R] **margins, pwcompare**.

20.18 Obtaining contrasts, tests of interactions, and main effects

`contrast` estimates and tests contrasts—comparisons of levels of factor variables. It also performs joint tests of these contrasts and can produce ANOVA-style tests of main effects, interaction effects, simple effects, and nested effects. It can be used after most estimation commands.

`contrast` provides a set of "contrast operators" such as `r.`, `ar.`, and `p.`. These operators are prefixed onto variable names—for example, `r.`*varname*—to specify the contrasts to be performed. The operators can be used with the `contrast` and `margins` commands.

Below we fit a regression of cholesterol level on age group category.

```
. regress chol i.agegrp
```

The reported coefficients on `i.agegrp` will themselves be contrasts, namely, contrasts on the reference category. After estimation, if we wanted to compare the cell mean of each age group with that of the previous group, we would perform a reverse-adjacent contrast by typing

```
. contrast ar.agegrp
```

That is exactly what we will do:

```
. use http://www.stata-press.com/data/r12/cholesterol
(Artificial cholesterol data)

. regress chol i.agegrp

      Source |       SS       df       MS              Number of obs =      75
-------------+------------------------------           F(  4,    70) =   35.02
       Model |  14943.3997     4  3735.84993           Prob > F      =  0.0000
    Residual |  7468.21971    70  106.688853           R-squared     =  0.6668
-------------+------------------------------           Adj R-squared =  0.6477
       Total |  22411.6194    74  302.859722           Root MSE      =  10.329

------------------------------------------------------------------------------
        chol |      Coef.   Std. Err.      t    P>|t|     [95% Conf. Interval]
-------------+----------------------------------------------------------------
      agegrp |
          2  |   8.203575   3.771628     2.18   0.033     .6812991    15.72585
          3  |   21.54105   3.771628     5.71   0.000     14.01878    29.06333
          4  |   30.15067   3.771628     7.99   0.000      22.6284    37.67295
          5  |   38.76221   3.771628    10.28   0.000     31.23993    46.28448
             |
       _cons |   180.5198   2.666944    67.69   0.000     175.2007    185.8388
------------------------------------------------------------------------------

. contrast ar.agegrp

Contrasts of marginal linear predictions

Margins      : asbalanced

------------------------------------------------
             |         df           F        P>F
-------------+----------------------------------
      agegrp |
    (2 vs 1) |          1        4.73     0.0330
    (3 vs 2) |          1       12.51     0.0007
    (4 vs 3) |          1        5.21     0.0255
    (5 vs 4) |          1        5.21     0.0255
       Joint |          4       35.02     0.0000
             |
    Residual |         70
------------------------------------------------

--------------------------------------------------------------
             |   Contrast   Std. Err.     [95% Conf. Interval]
-------------+------------------------------------------------
      agegrp |
    (2 vs 1) |   8.203575   3.771628      .6812991    15.72585
    (3 vs 2) |   13.33748   3.771628      5.815204    20.85976
    (4 vs 3) |    8.60962   3.771628      1.087345     16.1319
    (5 vs 4) |   8.611533   3.771628      1.089257    16.13381
--------------------------------------------------------------
```

We could use orthogonal polynomial contrasts to test whether there is a linear, quadratic, or even higher-order trend in the estimated cell means.

```
. contrast p.agegrp, noeffects
Contrasts of marginal linear predictions
Margins      : asbalanced
```

	df	F	P>F
agegrp			
(linear)	1	139.11	0.0000
(quadratic)	1	0.15	0.6962
(cubic)	1	0.37	0.5448
(quartic)	1	0.43	0.5153
Joint	4	35.02	0.0000
Residual	70		

You are not limited to using `contrast` in one-way models. Had we fit

```
. regress chol agegrp##race
```

we could `contrast` to obtain tests of the main effects and interaction effects.

```
. contrast agegrp##race
```

These results would be the same as would be reported by `anova`. We mention this because you can use `contrast` after any estimation command that allows factor variables and works with `margins`. You could type

```
. logistic highbp agegrp##race
. contrast agegrp##race
```

See [R] **contrast** and [R] **margins, contrast**.

20.19 Graphing margins, marginal effects, and contrasts

Using `marginsplot`, you can graph any of the results produced by `margins`, and because `margins` can replicate any of the results produced by `pwcompare` and `contrast`, you can graph any of the results produced by them, too.

In [U] **20.15.3 Obtaining predictive margins**, we did the following:

```
. use http://www.stata-press.com/data/r12/nhanes2
. svy: logistic highbp sex##agegrp##c.bmi
. margins sex#agegrp, vce(unconditional)
```

We can now graph those results by typing

```
. marginsplot, xdimension(agegrp)
```

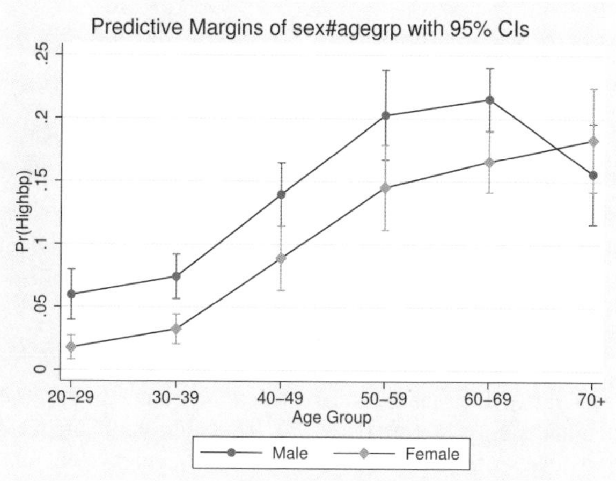

See [R] **marginsplot**.

20.20 Obtaining robust variance estimates

Estimates of variance refer to estimated standard errors or, more completely, the estimated variance–covariance matrix of the estimators of which the standard errors are a subset, being the square root of the diagonal elements. Call this matrix the variance. All estimation commands produce an estimate of variance and, using that, produce confidence intervals and significance tests.

In addition to the conventional estimator of variance, there is another estimator that has been called by various names because it has been derived independently in different ways by different authors. Two popular names associated with the calculation are Huber and White, but it is also known as the sandwich estimator of variance (because of how the calculation formula physically appears) and the robust estimator of variance (because of claims made about it). Also, this estimator has an independent and long tradition in the survey literature.

The conventional estimator of variance is derived by starting with a model. Let's start with the regression model

$$y_i = \mathbf{x}_i\boldsymbol{\beta} + \epsilon_i, \qquad \epsilon_i \sim N(0, \sigma^2)$$

although it is not important for the discussion that we are using regression. Under the model-based approach, we assume that the model is true and thereby derive an estimator for $\boldsymbol{\beta}$ and its variance.

The estimator of the standard error of $\widehat{\boldsymbol{\beta}}$ we develop is based on the assumption that the model is true in every detail. y_i is not exactly equal to $\mathbf{x}_i\boldsymbol{\beta}$ (so that we would only need to solve an equation to obtain precisely that value of $\boldsymbol{\beta}$) because the observed y_i has noise ϵ_i added to it, the noise is Gaussian, and it has constant variance. That noise leads to the uncertainty about $\boldsymbol{\beta}$, and it is from the characteristics of that noise that we are able to calculate a sampling distribution for $\widehat{\boldsymbol{\beta}}$.

The key thought here is that the standard error of $\widehat{\boldsymbol{\beta}}$ arises because of ϵ and is valid only because the model is absolutely, without question, true; we just do not happen to know the particular values of $\boldsymbol{\beta}$ and σ^2 that make the model true. The implication is that, in an infinite-sized sample, the estimator $\widehat{\boldsymbol{\beta}}$ for $\boldsymbol{\beta}$ would converge to the true value of $\boldsymbol{\beta}$ and that its variance would go to 0.

Now here is another interpretation of the estimation problem: We are going to fit the model

$$y_i = \mathbf{x}_i\mathbf{b} + e_i$$

and, to obtain estimates of $\mathbf{b}$, we are going to use the calculation formula

$$\widehat{\mathbf{b}} = (\mathbf{X}'\mathbf{X})^{-1}\mathbf{X}'\mathbf{y}$$

We have made no claims that the model is true or any claims about e_i or its distribution. We shifted our notation from β and ϵ_i to $\mathbf{b}$ and e_i to emphasize this. All we have stated are the physical actions we intend to carry out on the data. Interestingly, it is possible to calculate a standard error for $\widehat{\mathbf{b}}$ here. At least, it is possible if you will agree with us on what the standard error measures are.

We are going to define the standard error as measuring the standard error of the calculated $\widehat{\mathbf{b}}$ if we were to repeat the data collection followed by estimation over and over again.

This is a different concept of the standard error from the conventional, model-based ideas, but it is related. Both measure uncertainty about $\mathbf{b}$ (or β). The regression model–based derivation states from where the variation arises and so can make grander statements about the applicability of the measured standard error. The weaker second interpretation makes fewer assumptions and so produces a standard error suitable for one purpose.

There is a subtle difference in interpretation of these identically calculated point estimates. $\widehat{\beta}$ is the estimate of β under the assumption that the model is true. $\widehat{\mathbf{b}}$ is the estimate of $\mathbf{b}$, which is merely what the estimator would converge to if we collected more and more data.

Is the estimate of $\mathbf{b}$ unbiased? If we mean, "Does $\mathbf{b} = \beta$?" that depends on whether the model is true. $\widehat{\mathbf{b}}$ is, however, an unbiased estimate of $\mathbf{b}$, which admittedly is not saying much.

What if $\mathbf{x}$ and e are correlated? Don't we have a problem then? We may have an interpretation problem—$\mathbf{b}$ may not measure what we want to measure, namely, β—but we measure $\widehat{\mathbf{b}}$ to be such and such and expect, if the experiment and estimation were repeated, that we would observe results in the range we have reported.

So, we have two different understandings of what the parameters mean and how the variance in their estimators arises. However, both interpretations must confront the issue of how to make valid statistical inference about the coefficient estimates when the data do not come from either a simple random sample or the distribution of $(\mathbf{x}_i, \epsilon_i)$ is not independent and identically distributed (i.i.d.). In essence, we need an estimator of the standard errors that is robust to this deviation from the standard case.

Hence, the name *the robust estimate of variance*; its associated authors are Huber (1967) and White (1980, 1982) (who developed it independently), although many others have extended its development, including Gail, Tan, and Piantadosi (1988); Kent (1982); Royall (1986); and Lin and Wei (1989). In the survey literature, this same estimator has been developed; see Kish and Frankel (1974), Fuller (1975), and Binder (1983).

Many of Stata's estimation commands can produce this alternative estimate of variance, and, if they can, they have a `vce(robust)` option. Without `vce(robust)`, we get one measure of variance:

```
. use http://www.stata-press.com/data/r12/auto7
(1978 Automobile Data)
. regress mpg weight foreign
      Source |       SS       df       MS              Number of obs =      74
-------------+------------------------------           F(  2,    71) =   69.75
       Model |  1619.2877     2  809.643849           Prob > F      =  0.0000
    Residual |  824.171761    71   11.608053           R-squared     =  0.6627
-------------+------------------------------           Adj R-squared =  0.6532
       Total |  2443.45946    73  33.4720474           Root MSE      =  3.4071

------------------------------------------------------------------------------
         mpg |      Coef.   Std. Err.      t    P>|t|     [95% Conf. Interval]
-------------+----------------------------------------------------------------
      weight |  -.0065879   .0006371   -10.34   0.000    -.0078583   -.0053175
     foreign |  -1.650029   1.075994    -1.53   0.130      -3.7955    .4954422
       _cons |    41.6797   2.165547    19.25   0.000     37.36172    45.99768
------------------------------------------------------------------------------
```

With vce(robust), we get another:

```
. regress mpg weight foreign, vce(robust)
Linear regression                                      Number of obs =      74
                                                       F(  2,    71) =   73.81
                                                       Prob > F      =  0.0000
                                                       R-squared     =  0.6627
                                                       Root MSE      =  3.4071

------------------------------------------------------------------------------
             |               Robust
         mpg |      Coef.   Std. Err.      t    P>|t|     [95% Conf. Interval]
-------------+----------------------------------------------------------------
      weight |  -.0065879   .0005462   -12.06   0.000     -.007677   -.0054988
     foreign |  -1.650029   1.132566    -1.46   0.150    -3.908301    .6082424
       _cons |    41.6797   1.797553    23.19   0.000     38.09548    45.26392
------------------------------------------------------------------------------
```

Either way, the point estimates are the same. (See [R] **regress** for an example where specifying vce(robust) produces strikingly different standard errors.)

How do we interpret these results? Let's consider the model-based interpretation. Suppose that

$$y_i = \mathbf{x}_i\boldsymbol{\beta} + \epsilon_i,$$

where $(\mathbf{x}_i, \epsilon_i)$ are independently and identically distributed (i.i.d.) with variance σ^2. For the model-based interpretation, we also must assume that $\mathbf{x}_i$ and ϵ_i are uncorrelated. With these assumptions and a few technical regularity conditions, our first regression gives us consistent parameter estimates and standard errors that we can use for valid statistical inference about the coefficients. Now suppose that we weaken our assumptions so that $(\mathbf{x}_i, \epsilon_i)$ are independently and—but not necessarily—identically distributed. Our parameter estimates are still consistent, but the standard errors from the first regression can no longer be used to make valid inference. We need estimates of the standard errors that are robust to the fact that the error term is not identically distributed. The standard errors in our second regression are just what we need. We can use them to make valid statistical inference about our coefficients, even though our data are not identically distributed.

Now consider a non–model-based interpretation. If our data come from a survey design that ensures that $(\mathbf{x}_i, e_i)$ are i.i.d., then we can use the nonrobust standard errors for valid statistical inference about the population parameters $\mathbf{b}$. For this interpretation, we do not need to assume that $\mathbf{x}_i$ and e_i are uncorrelated. If they are uncorrelated, the population parameters $\mathbf{b}$ and the model parameters $\boldsymbol{\beta}$ are the same. However, if they are correlated, then the population parameters $\mathbf{b}$ that we are estimating are not the same as the model-based β. So, what we are estimating is different, but we still need standard errors that allow us to make valid statistical inference. So, if the process that we used to collect the data caused $(\mathbf{x}_i, e_i)$ to be independently but not identically distributed, then we need to use the robust standard errors to make valid statistical inference about the population parameters $\mathbf{b}$.

The robust estimator of variance has one feature that the conventional estimator does not have: the ability to relax the assumption of independence of the observations. That is, if you specify the `vce(cluster` *clustvar*`)` option, it can produce "correct" standard errors (in the measurement sense), even if the observations are correlated.

For the automobile data, it is difficult to believe that the models of the various manufacturers are truly independent. Manufacturers, after all, use common technology, engines, and drive trains across their model lines. The VW Dasher in the above regression has a measured residual of −2.80. Having been told that, do you really believe that the residual for the VW Rabbit is as likely to be above 0 as below? (The residual is −2.32.) Similarly, the measured residual for the Chevrolet Malibu is 1.27. Does that provide information about the expected value of the residual of the Chevrolet Monte Carlo (which turns out to be 1.53)?

We need to be careful about picking examples from data; we have not told you about the Datsun 210 and 510 (residuals +8.28 and −1.01) or the Cadillac Eldorado and Seville (residuals −1.99 and +7.58), but you should, at least, question the assumption of independence. It may be believable that the measured `mpg` given the weight of one manufacturer's vehicles is independent of other manufacturers' vehicles, but it is at least questionable whether a manufacturer's vehicles are independent of one another.

In commands with the `vce(robust)` option, another option—`vce(cluster` *clustvar*`)`—relaxes the independence assumption and requires only that the observations be independent across the clusters:

```
. regress mpg weight foreign, vce(cluster manufacturer)
Linear regression                                      Number of obs =      74
                                                       F(  2,     22) =   90.93
                                                       Prob > F      =  0.0000
                                                       R-squared     =  0.6627
                                                       Root MSE      =  3.4071

                              (Std. Err. adjusted for 23 clusters in manufacturer)
------------------------------------------------------------------------------
             |               Robust
         mpg |      Coef.   Std. Err.      t    P>|t|     [95% Conf. Interval]
-------------+----------------------------------------------------------------
      weight |  -.0065879   .0005339   -12.34   0.000    -.0076952   -.0054806
     foreign |  -1.650029   1.039033    -1.59   0.127    -3.804852    .5047939
       _cons |    41.6797   1.844559    22.60   0.000     37.85432    45.50508
------------------------------------------------------------------------------
```

It turns out that, in these data, whether or not we specify `vce(cluster` *clustvar*`)` makes little difference. The VW and Chevrolet examples above were not representative; had they been, the confidence intervals would have widened. (In the above, `manuf` is a variable that takes on values such as "Chev." or "VW", recording the manufacturer of the vehicle. This variable was created from variable `make`, which contains values such as "Chev. Malibu" or "VW Rabbit", by extracting the first word.)

As a demonstration of how well clustering can work, in [R] **regress** we fit a random-effects model with `regress, vce(robust)` and then compared the results with ordinary least squares and the GLS random-effects estimator. Here we will simply summarize the results.

We start with a dataset on 4,711 women aged 14–46 years. Subjects appear an average of 6.056 times in the data; there are a total of 28,534 observations. The model we use is log wage on age,

age-squared, and job tenure. The focus of the example is the estimated coefficient on tenure. We obtain the following results:

Estimator	Point estimate	Confidence interval
(Inappropriate) least squares	.039	[.038, .041]
Robust clusterered	.039	[.036, .042]
GLS random effects	.026	[.025, .027]

Notice how well the robust clustered estimate does compared with the GLS random-effects model. We then run a Hausman specification test, obtaining $\chi^2(3) = 336.62$, which casts grave doubt on the assumptions justifying the use of the GLS estimator and hence on the GLS results. At this point, we will simply quote our comments:

> Meanwhile, our robust regression results still stand, as long as we are careful about the interpretation. The correct interpretation is that, if the data collection were repeated (on women sampled the same way as in the original sample), and if we were to refit the model, 95% of the time we would expect the estimated coefficient on tenure to be in the range $[.036, .042]$.
>
> Even with robust regression, we must be careful about going beyond that statement. Here the Hausman test is probably picking up something that differs within and between person, which would cast doubt on our robust regression model in terms of interpreting $[.036, .042]$ to contain the rate of return for keeping a job, economywide, for all women, without exception.

The formula for the robust estimator of variance is

$$\widehat{\mathcal{V}} = \widehat{\mathbf{V}}\Big(\sum_{j=1}^{N} \mathbf{u}_j'\mathbf{u}_j\Big)\widehat{\mathbf{V}}$$

where $\widehat{\mathbf{V}} = (-\partial^2 \ln L/\partial\boldsymbol{\beta}^2)^{-1}$ (the conventional estimator of variance) and $\mathbf{u}_j$ (a row vector) is the contribution from the jth observation to $\partial \ln L/\partial\boldsymbol{\beta}$.

In the example above, observations are assumed to be independent. Assume for a moment that the observations denoted by j are not independent but that they can be divided into M groups G_1, G_2, ..., G_M that are independent. The robust estimator of variance is

$$\widehat{\mathcal{V}} = \widehat{\mathbf{V}}\Big(\sum_{k=1}^{M} \mathbf{u}_k^{(G)\prime}\mathbf{u}_k^{(G)}\Big)\widehat{\mathbf{V}}$$

where $\mathbf{u}_k^{(G)}$ is the contribution of the kth group to $\partial \ln L/\partial\boldsymbol{\beta}$. That is, application of the robust variance formula merely involves using a different decomposition of $\partial \ln L/\partial\boldsymbol{\beta}$, namely, $\mathbf{u}_k^{(G)}$, $k = 1, \dots, M$ rather than $\mathbf{u}_j$, $j = 1, \dots, N$. Moreover, if the log-likelihood function is additive in the observations denoted by j

$$\ln L = \sum_{j=1}^{N} \ln L_j$$

then $\mathbf{u}_j = \partial \ln L_j/\partial\boldsymbol{\beta}$, so

$$\mathbf{u}_k^{(G)} = \sum_{j \in G_k} \mathbf{u}_j$$

That is what the `vce(cluster` *clustvar*`)` option does. (This point was first made in writing by Rogers [1993], although he considered the point an obvious generalization of Huber [1967] and the calculation—implemented by Rogers—had appeared in Stata a year earlier.)

❑ Technical note

What is written above is asymptotically correct but ignores a finite-sample adjustment to $\widehat{\mathcal{V}}$. For maximum likelihood estimators, when you specify `vce(robust)` but not `vce(cluster` *clustvar*`)`, a better estimate of variance is $\widehat{\mathcal{V}}^* = \{N/(N-1)\}\widehat{\mathcal{V}}$. When you also specify the `vce(cluster` *clustvar*`)` option, this becomes $\widehat{\mathcal{V}}^* = \{M/(M-1)\}\widehat{\mathcal{V}}$.

For linear regression, the finite-sample adjustment is $N/(N-k)$ without `vce(cluster` *clustvar*`)`—where k is the number of regressors—and $\{M/(M-1)\}\{(N-1)/(N-k)\}$ with `vce(cluster` *clustvar*`)`. Also, two data-dependent modifications to the calculation for $\widehat{\mathcal{V}}^*$, suggested by MacKinnon and White (1985), are also provided by `regress`; see [R] **regress**. Angrist and Pischke (2009, chap. 8) is devoted to robust covariance matrix estimation and offers practical guidance on the use of `vce(robust)` and `vce(cluster` *clustvar*`)` in both cross-sectional and panel-data applications.

❑

20.21 Obtaining scores

Many of the estimation commands that provide the `vce(robust)` option also provide the ability to generate equation-level score variables via the `predict` command. With the `score` option, `predict` returns an important ingredient into the robust variance calculation that is sometimes useful in its own right. As explained in [U] **20.20 Obtaining robust variance estimates** above, ignoring the finite-sample corrections, the robust estimate of variance is

$$\widehat{\mathcal{V}} = \widehat{\mathbf{V}}\Big(\sum_{j=1}^{N} \mathbf{u}_j'\mathbf{u}_j\Big)\widehat{\mathbf{V}}$$

where $\widehat{\mathbf{V}} = (-\partial^2 \ln L/\partial\boldsymbol{\beta}^2)^{-1}$ is the conventional estimator of variance. Let's consider likelihood functions that are additive in the observations

$$\ln L = \sum_{j=1}^{N} \ln L_j$$

then $\mathbf{u}_j = \partial \ln L_j/\partial\boldsymbol{\beta}$. In general, function L_j is a function of $\mathbf{x}_j$ and $\boldsymbol{\beta}$, $L_j(\boldsymbol{\beta};\mathbf{x}_j)$. For many likelihood functions, however, it is only the linear form $\mathbf{x}_j\boldsymbol{\beta}$ that enters the function. In those cases,

$$\frac{\partial \ln L_j(\mathbf{x}_j\boldsymbol{\beta})}{\partial\boldsymbol{\beta}} = \frac{\partial \ln L_j(\mathbf{x}_j\boldsymbol{\beta})}{\partial(\mathbf{x}_j\boldsymbol{\beta})}\frac{\partial(\mathbf{x}_j\boldsymbol{\beta})}{\partial\boldsymbol{\beta}} = \frac{\partial \ln L_j(\mathbf{x}_j\boldsymbol{\beta})}{\partial(\mathbf{x}_j\boldsymbol{\beta})}\mathbf{x}_j$$

By writing $u_j = \partial \ln L_j(\mathbf{x}_j\boldsymbol{\beta})/\partial(\mathbf{x}_j\boldsymbol{\beta})$, this becomes simply $u_j\mathbf{x}_j$. Thus the formula for the robust estimate of variance can be rewritten as

$$\widehat{\mathcal{V}} = \widehat{\mathbf{V}}\Big(\sum_{j=1}^{N} u_j^2\mathbf{x}_j'\mathbf{x}_j\Big)\widehat{\mathbf{V}}$$

We refer to u_j as the equation-level score (in the singular), and it is u_j that is returned when you use `predict` with the `score` option. u_j is like a residual in that

1. $\sum_j u_j = 0$ and
2. correlation of u_j and $\mathbf{x}_j$, calculated over $j = 1, \ldots, N$, is 0.

In fact, for linear regression, u_j is the residual, normalized,

$$\begin{aligned} \frac{\partial \ln L_j}{\partial(\mathbf{x}_j\boldsymbol{\beta})} &= \frac{\partial}{\partial(\mathbf{x}_j\boldsymbol{\beta})} \ln f\left\{(y_j - \mathbf{x}_j\boldsymbol{\beta})/\sigma\right\} \\ &= (y_j - \mathbf{x}_j\boldsymbol{\beta})/\sigma \end{aligned}$$

where $f()$ is the standard normal density.

▷ Example 19

`probit` provides both the `vce(robust)` option and `predict, score`. Equation-level scores play an important role in calculating the robust estimate of variance, but we can use `predict, score` regardless of whether we specify `vce(robust)`:

```
. use http://www.stata-press.com/data/r12/auto
. probit foreign mpg weight
Iteration 0:   log likelihood =  -45.03321
Iteration 1:   log likelihood = -29.244141
Iteration 2:   log likelihood = -27.041557
Iteration 3:   log likelihood =  -26.84658
Iteration 4:   log likelihood = -26.844189
Iteration 5:   log likelihood = -26.844189
Probit regression                                 Number of obs   =         74
                                                  LR chi2(2)      =      36.38
                                                  Prob > chi2     =     0.0000
Log likelihood = -26.844189                       Pseudo R2       =     0.4039

------------------------------------------------------------------------------
     foreign |      Coef.   Std. Err.      z    P>|z|     [95% Conf. Interval]
-------------+----------------------------------------------------------------
         mpg |  -.1039503   .0515689    -2.02   0.044    -.2050235   -.0028772
      weight |  -.0023355   .0005661    -4.13   0.000     -.003445   -.0012261
       _cons |   8.275464   2.554142     3.24   0.001     3.269438    13.28149
------------------------------------------------------------------------------

. predict double u, scores
. summarize u
    Variable |       Obs        Mean    Std. Dev.       Min        Max
-------------+--------------------------------------------------------
           u |        74   -3.87e-16    .5988325  -1.655439   1.660787
. correlate u mpg weight
(obs=74)

             |        u      mpg   weight
-------------+---------------------------
           u |   1.0000
         mpg |  -0.0000   1.0000
      weight |  -0.0000  -0.8072   1.0000

. list make foreign mpg weight u if abs(u)>1.65

     +--------------------------------------------------------+
     | make            foreign   mpg   weight               u |
     |--------------------------------------------------------|
 24. | Ford Fiesta    Domestic    28    1,800      -1.6554395 |
 64. | Peugeot 604     Foreign    14    3,420       1.6607871 |
     +--------------------------------------------------------+
```

The light, high-mileage Ford Fiesta is surprisingly domestic, whereas the heavy, low-mileage Peugeot 604 is surprisingly foreign.

◁

❑ Technical note

For some estimation commands, one score is not enough. Consider a likelihood that can be written as $L_j(\mathbf{x}_j\boldsymbol{\beta}_1, \mathbf{z}_j\boldsymbol{\beta}_2)$, a function of two linear forms (or linear equations). Then $\partial\ln L_j/\partial\boldsymbol{\beta}$ can be written $(\partial\ln L_j/\partial\boldsymbol{\beta}_1, \partial\ln L_j/\partial\boldsymbol{\beta}_2)$. Each of the components can in turn be written as $[\partial\ln L_j/\partial(\boldsymbol{\beta}_1\mathbf{x})]\mathbf{x} = u_1\mathbf{x}$ and $[\partial\ln L_j/\partial(\boldsymbol{\beta}_2\mathbf{z})]\mathbf{z} = u_2\mathbf{z}$. There are then two equation-level scores, u_1 and u_2, and, in general, there could be more.

Stata's `streg, distribution(weibull)` command is an example of this: it estimates $\boldsymbol{\beta}$ and a shape parameter, $\ln p$, the latter of which can be thought of as a degenerate linear form $(\ln p)\mathbf{z}$ with $\mathbf{z} = \mathbf{1}$. `predict, scores` after this command requires that you specify two new variable names, or you can specify *stub*`*`, which will generate two variables, *stub*`1` and *stub*`2`; the first will be defined containing u_1—the score associated with $\boldsymbol{\beta}$—and the second will be defined containing u_2—the score associated with $\ln p$.

❑

❑ Technical note

Using Stata's matrix commands—see [P] **matrix**—we can make the robust variance calculation for ourselves and then compare it with that made by Stata.

```
. use http://www.stata-press.com/data/r12/auto, clear
(1978 Automobile Data)

. quietly probit foreign mpg weight, score(u)

. matrix accum S = mpg weight [iweight=u^2*74/73]
(obs=26.53642547)

. matrix rV = e(V)*S*e(V)

. matrix list rV

symmetric rV[3,3]
                   foreign:     foreign:     foreign:
                       mpg       weight        _cons
   foreign:mpg   .00352299
foreign:weight   .00002216    2.434e-07
 foreign:_cons  -.14090346   -.00117031    6.4474174

. quietly probit foreign mpg weight, vce(robust)

. matrix list e(V)

symmetric e(V)[3,3]
                   foreign:     foreign:     foreign:
                       mpg       weight        _cons
   foreign:mpg   .00352299
foreign:weight   .00002216    2.434e-07
 foreign:_cons  -.14090346   -.00117031    6.4474174
```

The results are the same.

There is an important lesson here for programmers. Given the scores, conventional variance estimates can be easily transformed to robust estimates. If we were writing a new estimation command, it would not be difficult to include a `vce(robust)` option.

It is, in fact, easy if we ignore clustering. With clustering, it is more work because the calculation involves forming sums within clusters. For programmers interested in implementing robust variance calculations, Stata provides an `_robust` command to ease the task. This is documented in [P] **_robust**.

To use _robust, you first produce conventional results (a vector of coefficients and covariance matrix) along with a variable containing the scores u_j (or variables if the likelihood function has more than one stub). You then call _robust, and it will transform your conventional variance estimate into the robust estimate. _robust will handle the work associated with clustering and the details of the finite-sample adjustment, and it will even label your output so that the word *Robust* appears above the standard error when the results are displayed.

Of course, this is all even easier if you write your commands with Stata's ml maximum likelihood optimization, in which case you merely pass the vce(robust) option on to ml. ml will then call _robust itself and do all the work for you.

❑

❑ Technical note

For some estimation commands, predict, score computes parameter-level scores $\partial L_j/\partial\beta$ instead of equation-level scores $\partial L_j/\partial \mathbf{x}_j\beta$. Those commands are

asclogit

asmprobit

asroprobit

nlogit

stcox

stcrreg

xtmixed

These models share the characteristic that there are multiple observations per independent event.

In making the robust variance calculation, what are really needed are parameter-level scores $\partial L_j/\partial\beta$, and so you may be asking yourself why predict, score does not always produce parameter-level scores. In the usual case we can obtain them from equation-level scores via the chain rule, and fewer variables are required if we adopt this approach. In the cases above, however, the likelihood is calculated at the group level and is not split into contributions from the individual observations. Thus the chain rule cannot be used and we must use the parameter level scores directly.

_robust can be tricked into using them if each parameter appears to be in its own equation as a constant. This requires resetting the row and column stripes on the covariance matrix before _robust is called. The equation names for each row and column must each be unique and the variable names must all be _cons.

❑

20.22 Weighted estimation

[U] **11.1.6 weight** introduced the syntax for weights. Stata provides four kinds of weights: fweights, or frequency weights; pweights, or sampling weights; aweights, or analytic weights; and iweights, or importance weights. The syntax for using each is the same. Type

```
. regress y x1 x2
```

and you obtain unweighted estimates; type

```
. regress y x1 x2 [pweight=pop]
```

and you obtain (in this example) pweighted estimation.

The sections below explain how each type of weight is used in estimation.

20.22.1 Frequency weights

Frequency weights—`fweights`—are integers and are nothing more than replication counts. The weight is statistically uninteresting, but from a data processing perspective it is important. Consider the following data

```
 y     x1     x2
22      1      0
22      1      0
22      1      1
23      0      1
23      0      1
23      0      1
```

and the estimation command

```
. regress y x1 x2
```

Equivalent is the following, more compressed data

```
 y     x1     x2     pop
22      1      0       2
22      1      1       1
23      0      1       3
```

and the corresponding estimation command

```
. regress y x1 x2 [fweight=pop]
```

When you specify frequency weights, you are treating each observation as one or more real observations.

❑ Technical note

You might occasionally run across a command that does not allow weights at all, especially among user-written commands. `expand` (see [D] **expand**) can be used with such commands to obtain frequency-weighted results. The `expand` command duplicates observations so that the data become self-weighting. Suppose that you want to run the command `usercmd`, which does something or other, and you would like to type `usercmd y x1 x2 [fw=pop]`. Unfortunately, `usercmd` does not allow weights. Instead, you type

```
. expand pop
. usercmd y x1 x2
```

to obtain your result. Moreover, there is an important principle here: the results of running any command with frequency weights should be the same as running the command on the unweighted, expanded data. Unweighted, duplicated data and frequency-weighted data are merely two ways of recording identical information.

❑

20.22.2 Analytic weights

Analytic weights—*analytic* is a term we made up—statistically arise in one particular problem: linear regression on data that are themselves observed means. That is, think of the model

$$y_i = \mathbf{x}_i\boldsymbol{\beta} + \epsilon_i, \qquad \epsilon_i \sim N(0, \sigma^2)$$

and now think about fitting this model on data $(\overline{y}_j, \overline{\mathbf{x}}_j)$ that are themselves observed averages. For instance, a piece of the underlying data for $(y_i, \mathbf{x}_i)$ might be $(3, 1)$, $(4, 2)$, and $(2, 2)$, but you do not know that. Instead, you have one observation $\{(3+4+2)/3, (1+2+2)/3\} = (3, 1.67)$ and know only that the $(3, 1.67)$ arose as the average of 3 underlying observations. All your data are like that.

`regress` with `aweights` is the solution to that problem:

```
. regress y x [aweight=pop]
```

There is a history of misusing such weights. A researcher does not have cell-mean data but instead has a probability-weighted random sample. Long before Stata existed, some researchers were using `aweights` to produce estimates from such samples. We will come back to this point in [U] **20.22.3 Sampling weights** below.

Anyway, the statistical problem that `aweights` resolve can be written as

$$y_i = \mathbf{x}_i\boldsymbol{\beta} + \epsilon_i, \qquad \epsilon_i \sim N(0, \sigma^2/w_i)$$

where the w_i are the analytic weights. The details of the solution are to make linear regression calculations using the weights as if they were `fweights` but to normalize them to sum to N before doing that.

Most commands that allow `aweights` handle them in this manner. That is, if you specify `aweights`, they are

1. normalized to sum to N and then
2. inserted in the calculation formulas in the same way as `fweights`.

20.22.3 Sampling weights

Sampling weights—probability weights or `pweights`—refer to probability-weighted random samples. Actually, what you specify in `[pweight=...]` is a variable recording the number of subjects in the full population that the sampled observation in your data represents. That is, an observation that had probability $1/3$ of being included in your sample has pweight 3.

Some researchers have used `aweights` with this kind of data. If they do, they are probably making a mistake. Consider the regression model

$$y_i = \mathbf{x}_i\boldsymbol{\beta} + \epsilon_i, \qquad \epsilon_i \sim N(0, \sigma^2)$$

Begin by considering the exact nature of the problem of fitting this model on cell-mean data—for which `aweights` are the solution: heteroskedasticity arising from the grouping. The error term ϵ_i is homoskedastic (meaning that it has constant variance σ^2). Say that the first observation in the data is the mean of 3 underlying observations. Then

$$\begin{aligned} y_1 &= \mathbf{x}_1\boldsymbol{\beta} + \epsilon_1, \qquad & \epsilon_i &\sim N(0, \sigma^2) \\ y_2 &= \mathbf{x}_2\boldsymbol{\beta} + \epsilon_2, \qquad & \epsilon_i &\sim N(0, \sigma^2) \\ y_3 &= \mathbf{x}_3\boldsymbol{\beta} + \epsilon_3, \qquad & \epsilon_i &\sim N(0, \sigma^2) \end{aligned}$$

and taking the mean

$$(y_1 + y_2 + y_3)/3 = \{(\mathbf{x}_1 + \mathbf{x}_2 + \mathbf{x}_3)/3\}\boldsymbol{\beta} + (\epsilon_1 + \epsilon_2 + \epsilon_3)/3$$

For another observation in the data—which may be the result of summing of a different number of observations—the variance will be different. Hence, the model for the data is

$$\overline{y}_j = \overline{x}_j\boldsymbol{\beta} + \overline{\epsilon}_j, \qquad \overline{\epsilon}_j \sim N(0, \sigma^2/N_j)$$

This makes intuitive sense. Consider 2 observations, one recording means over two subjects and the other means over 100,000 subjects. You would expect the variance of the residual to be less in the 100,000-subject observation; that is, there is more information in the 100,000-subject observation than in the two-subject observation.

Now instead say that you are fitting the same model, $y_i = \mathbf{x}_i\boldsymbol{\beta} + \epsilon_i$, $\epsilon_i \sim N(0, \sigma^2)$, on probability-weighted data. Each observation in your data is one subject, but the different subjects have different chances of being included in your sample. Therefore, for each subject in your data

$$y_i = \mathbf{x}_i\boldsymbol{\beta} + \epsilon_i, \qquad \epsilon_i \sim N(0, \sigma^2)$$

That is, there is no heteroskedasticity problem. The use of the `aweighted` estimator cannot be justified on these grounds.

As a matter of fact, from the argument just given, you do not need to adjust for the weights at all, although the argument does not justify not making an adjustment. If you do not adjust, you are holding tightly to the assumed truth of your model. Two issues arise when considering adjustment for sampling weights:

1. the efficiency of the point estimate $\widehat{\boldsymbol{\beta}}$ of $\boldsymbol{\beta}$ and

2. the reported standard errors (and, more generally, the variance matrix of $\widehat{\boldsymbol{\beta}}$).

Efficiency argues in favor of adjustment, and that, by the way, is why many researchers have used `aweights` with `pweighted` data. The adjustment implied by `pweights` to the point estimates is the same as the adjustment implied by `aweights`.

With regard to the second issue, the use of `aweights` produces incorrect results because it interprets larger weights as designating more accurately measured points. For `pweights`, however, the point is no more accurately measured—it is still just one observation with one residual ϵ_j and variance σ^2. In [U] **20.20 Obtaining robust variance estimates** above, we introduced another estimator of variance that measures the variation that would be observed if the data collection followed by the estimation were repeated. Those same formulas provide the solution to `pweights`, and they have the added advantage that they are not conditioned on the model's being true. If we have any hopes of measuring the variation that would be observed were the data collection followed by estimation repeated, we must include the probability of the observations being sampled in the calculation.

In Stata, when you type

```
. regress y x1 x2 [pw=pop]
```

the results are the same as if you had typed

```
. regress y x1 x2 [pw=pop], vce(robust)
```

That is, specifying `pweights` implies the `vce(robust)` option and, hence, the robust variance calculation (but weighted). In this example, we use `regress` simply for illustration. The same is true of `probit` and all of Stata's estimation commands. Estimation commands that do not have a `vce(robust)` option (there are a few) do not allow `pweights`.

`pweights` are adequate for handling random samples where the probability of being sampled varies. `pweights` may be all you need. If, however, the observations are not sampled independently but are sampled in groups—called clusters in the jargon—you should specify the estimator's `vce(cluster` *clustvar*`)` option as well:

```
. regress y x1 x2 [pw=pop], vce(cluster block)
```

There are two ways of thinking about this:

1. The robust estimator answers the question of which variation would be observed were the data collection followed by the estimation repeated; if that question is to be answered, the estimator must account for the clustered nature of how observations are selected. If observations 1 and 2 are in the same cluster, then you cannot select observation 1 without selecting observation 2 (and, by extension, you cannot select observations like 1 without selecting observations like 2).
2. If you prefer, you can think about potential correlations. Observations in the same cluster may not really be independent—that is an empirical question to be answered by the data. For instance, if the clusters are neighborhoods, it would not be surprising that the individual neighbors are similar in their income, their tastes, and their attitudes, and even more similar than two randomly drawn persons from the area at large with similar characteristics, such as age and sex.

Either way of thinking leads to the same (robust) estimator of variance.

Sampling weights usually arise from complex sampling designs, which often involve not only unequal probability sampling and cluster sampling but also stratified sampling. There is a family of commands in Stata designed to work with the features of complex survey data, and those are the commands that begin with `svy`. To fit a linear regression model with stratification, for example, you would use the `svy:regress` command.

Non-`svy` commands that allow `pweights` and clustering give essentially identical results to the `svy` commands. If the sampling design is simple enough that it can be accommodated by the non-`svy` command, that is a fine way to perform the analysis. The `svy` commands differ in that they have more features, and they do all the little details correctly for real survey data. See [SVY] **survey** for a brief discussion of some of the issues involved in the analysis of survey data and a list of all the differences between the `svy` and non-`svy` commands.

Not all model estimation commands in Stata allow `pweights`. This is often because they are computationally or statistically difficult to implement.

20.22.4 Importance weights

Stata's `iweights`—importance weights—are the emergency exit. These weights are for those who want to take control and create special effects. For example, programmers have used `regress` with `iweights` to compute iteratively reweighted least-squares solutions for various problems.

`iweights` are treated much like `aweights`, except that they are not normalized. Stata's iweight rule is that

1. the weights are not normalized and
2. they are generally inserted into calculation formulas in the same way as `fweights`. There are exceptions; see the *Methods and formulas* for the particular command.

`iweights` are used mostly by programmers who are often on the way to implementing one of the other kinds of weights.

20.23 A list of postestimation commands

The following commands can be used after estimation:

[R] **contrast**	contrasts and ANOVA-style joint tests of estimates
[R] **estat**	AIC, BIC, VCE, and estimation sample summary
[R] **estimates**	cataloging estimation results
[R] **hausman**	Hausman specification test
[R] **lincom**	point estimates, standard errors, testing, and inference for linear combinations of coefficients
[R] **linktest**	specification link test for single-equation models
[R] **lrtest**	likelihood-ratio test
[R] **margins**	marginal means, predictive margins, and marginal effects
[R] **marginsplot**	graph the results from margins (profile plots, interaction plots, etc.)
[R] **nlcom**	point estimates, standard errors, testing, and inference for generalized predictions
[R] **predict**	predictions, residuals, influence statistics, and other diagnostic measures
[R] **predictnl**	point estimates, standard errors, testing, and inference for generalized predictions
[R] **pwcompare**	pairwise comparisons of estimates
[R] **suest**	seemingly unrelated estimation
[R] **test**	Wald tests of simple and composite linear hypotheses
[R] **testnl**	Wald tests of nonlinear hypotheses

Also see [U] **13.5 Accessing coefficients and standard errors** for accessing coefficients and standard errors.

The commands above are general-purpose postestimation commands that can be used after almost all estimation commands. Many estimation commands provide other estimator-specific postestimation commands. The full list of postestimation commands available for an estimator can be found in an entry titled *estimator* `postestimation` that immediately follows each estimator's entry in the [R], [MI], [MV], [ST], [SVY], [TS], and [XT] reference manuals.

Halbert Lynn White Jr. (1950–) was born in Kansas City. After receiving economics degrees at Princeton and MIT, he taught and researched economics at the University of Rochester and, since 1980, has taught at the University of California in San Diego. He is also active as a consultant. White's research interests include artificial neural networks, finance, econometrics, and forecasting, and his hobbies include music and antiques.

Peter Jost Huber (1934–) was born in Wohlen (Aargau, Switzerland). He gained mathematics degrees from ETH Zürich, including a PhD thesis on homotopy theory, and then studied statistics at Berkeley on postdoctoral fellowships. This visit yielded a celebrated 1964 paper on robust estimation, and Huber's later monographs on robust statistics were crucial in directing that field. Thereafter his career took him back and forth across the Atlantic, with periods at Cornell, ETH Zürich, Harvard, MIT, and Bayreuth. His work has touched several other major parts of statistics, theoretical and applied, including regression, exploratory multivariate analysis, large datasets, and statistical computing. Huber also has a major long-standing interest in Babylonian astronomy.

20.24 References

Afifi, A. A., and S. P. Azen. 1979. *Statistical Analysis: A Computer Oriented Approach.* 2nd ed. New York: Academic Press.

Angrist, J. D., and J.-S. Pischke. 2009. *Mostly Harmless Econometrics: An Empiricist's Companion.* Princeton, NJ: Princeton University Press.

Baum, C. F. 2009. *An Introduction to Stata Programming.* College Station, TX: Stata Press.

Binder, D. A. 1983. On the variances of asymptotically normal estimators from complex surveys. *International Statistical Review* 51: 279–292.

Buja, A., and H. R. Künsch. 2008. A conversation with Peter Huber. *Statistical Science* 23: 120–135.

Deaton, A. 1997. *The Analysis of Household Surveys: A Microeconometric Approach to Development Policy.* Baltimore, MD: Johns Hopkins University Press.

Fuller, W. A. 1975. Regression analysis for sample survey. *Sankhyā, Series C* 37: 117–132.

Gail, M. H., W. Y. Tan, and S. Piantadosi. 1988. Tests for no treatment effect in randomized clinical trials. *Biometrika* 75: 57–64.

Hampel, F. R. 1992. Introduction to Huber (1964) "Robust estimation of a location parameter". In *Breakthroughs in Statistics. Volume II: Methodology and Distribution*, ed. S. Kotz and N. L. Johnson, 479–491. New York: Springer.

Huber, P. J. 1967. The behavior of maximum likelihood estimates under nonstandard conditions. In Vol. 1 of *Proceedings of the Fifth Berkeley Symposium on Mathematical Statistics and Probability*, 221–233. Berkeley: University of California Press.

Kent, J. T. 1982. Robust properties of likelihood ratio tests. *Biometrika* 69: 19–27.

Kish, L., and M. R. Frankel. 1974. Inference from complex samples. *Journal of the Royal Statistical Society, Series B* 36: 1–37.

Lin, D. Y., and L. J. Wei. 1989. The robust inference for the Cox proportional hazards model. *Journal of the American Statistical Association* 84: 1074–1078.

Long, J. S., and J. Freese. 2000a. sg145: Scalar measures of fit for regression models. *Stata Technical Bulletin* 56: 34–40. Reprinted in *Stata Technical Bulletin Reprints*, vol. 10, pp. 197–205. College Station, TX: Stata Press.

——. 2000b. sg152: Listing and interpreting transformed coefficients from certain regression models. *Stata Technical Bulletin* 57: 27–34. Reprinted in *Stata Technical Bulletin Reprints*, vol. 10, pp. 231–240. College Station, TX: Stata Press.

MacKinnon, J. G., and H. White. 1985. Some heteroskedasticity-consistent covariance matrix estimators with improved finite sample properties. *Journal of Econometrics* 29: 305–325.

Rogers, W. H. 1993. sg17: Regression standard errors in clustered samples. *Stata Technical Bulletin* 13: 19–23. Reprinted in *Stata Technical Bulletin Reprints*, vol. 3, pp. 88–94. College Station, TX: Stata Press.

Royall, R. M. 1986. Model robust confidence intervals using maximum likelihood estimators. *International Statistical Review* 54: 221–226.

Weesie, J. 2000. sg127: Summary statistics for estimation sample. *Stata Technical Bulletin* 53: 32–35. Reprinted in *Stata Technical Bulletin Reprints*, vol. 9, pp. 275–277. College Station, TX: Stata Press.

White, H. 1980. A heteroskedasticity-consistent covariance matrix estimator and a direct test for heteroskedasticity. *Econometrica* 48: 817–838.

——. 1982. Maximum likelihood estimation of misspecified models. *Econometrica* 50: 1–25.

Advice

21 Inputting and importing data

Contents

21.1 Overview

To input data into Stata, you can use

[D] **edit** and [D] **input**	to enter data from the keyboard
[D] **insheet**	to read tab- or comma-separated data
[D] **import excel**	to read Excel files
[D] **import sasxport**	to read datasets in SAS XPORT format
[D] **infile (free format)**	to read unformatted text data
[D] **infile (fixed format)** or [D] **infix (fixed format)**	to read formatted text data
[D] **infile (fixed format)**	to read EBCDIC data
[D] **odbc**	to read from an ODBC source
[TS] **haver**	to read data in Haver Analytics' format
[D] **xmlsave** (where `xmluse` is documented)	to use datasets in XML format
[U] **21.4 Transfer programs**	to transfer data

Because dataset formats differ, you should familiarize yourself with each method.

[D] **infile (fixed format)** and [D] **infix (fixed format)** are two different commands that do the same thing. Read about both, and then use whichever appeals to you.

Alternatively, `edit` and `input` both allow you to enter data from the keyboard. `edit` opens a Viewer, and `input` allows you to type as the command line.

After you have read this chapter, also see [D] **import** for more examples of the different commands to input data.

21.2 Determining which input method to use

Below are several rules that, when applied sequentially, will direct you to the appropriate method for entering your data. After the rules is a description of each command, as well as a reference to the corresponding entry in the *Reference* manuals.

1. If you have a few data and simply wish to type the data directly into Stata at the keyboard, see [D] **edit**—doing so should be easy. Also see [D] **input**.
2. If your dataset is in binary format or the internal format of some software package, you have several options:
   a. If the data are in a spreadsheet, copy and paste the data into Stata's Data Editor; see [D] **edit** for details.
   b. If the data are in an Excel spreadsheet, use `import excel` to read them; see [D] **import excel**.
   c. If the data are in SAS XPORT format, use `import sasxport` to read the data; see [D] **import sasxport**.
   d. If the data in Haver Analytics' `.dat` format (Haver Analytics provides economics and financial databases), and you are using Stata for Windows, use `haver` to read the data; see [TS] **haver**.
   e. Translate the data into text (also known as character) format by using the other software. For instance, in most software, you can save data as tab-delimited or comma-separated text. Then, see [D] **insheet**.
   f. If the data are located in an ODBC source, which typically includes databases and spreadsheets, you can use the `odbc load` command to import the data; see [D] **odbc**. Currently `odbc` is available for Windows, Mac, and Linux versions of Stata.
   g. Other software packages are available that will convert non–Stata format data files into Stata-format files; see [U] **21.4 Transfer programs**.
3. If the dataset has one observation per line and the data are tab- or comma separated, use `insheet`; see [D] **insheet**. This is the easiest way to read data.
4. If the dataset is formatted and that formatting information is required to interpret the data, you can use `infile` with a dictionary or `infix`; see [D] **infile (fixed format)** or [D] **infix (fixed format)**.
5. If there are no string variables, you can use `infile` without a dictionary: see [D] **infile (free format)**.
6. If all the string variables in the data are enclosed in (single or double) quotes, you can use `infile` without a dictionary; see [D] **infile (free format)**.
7. If the undelimited string variables have no blanks, you can use `infile` without a dictionary; see [D] **infile (free format)**.
8. If the data are in EBCDIC format, see [D] **infile (fixed format)**.
9. If you make it to here, see [D] **infile (fixed format)** or [D] **infix (fixed format)**.

21.2.1 Entering data interactively

If you have a few data, you can type the data directly into Stata; see [D] **edit** or [D] **input**. Otherwise, we assume that your data are stored on disk.

21.2.2 Copying and pasting data

If your data are in another program and you wish to analyze them with Stata, first see if the program you are using allows you to copy the data to the clipboard. If it does, do so, and then open the Data Editor in Stata and select **Edit** > **Paste** to paste the data into Stata.

21.2.3 If the dataset is in binary format

Stata can read text datasets, which is technical jargon for datasets composed of characters—datasets that can be typed on your screen or printed on your printer. The alternative, binary datasets, can only sometimes be read by Stata. Binary datasets are popular, and almost every software package has its own binary format. Stata `.dta` datasets are an example of a binary format that Stata can read. The Excel `.xls` and `.xlsx` formats are other binary formats that Stata can read. The OpenOffice `.ods` format is a binary format that Stata cannot read.

If your dataset is in binary format or in the internal format of another software package that Stata cannot import, you must translate it into plain text or use some other program for conversion to Stata format. If this dataset is an Excel `.xls` or `.xlsx` file, you can read it by using Stata's `import excel` command; see [D] **import excel**. If this dataset is located in a database or an ODBC source, see [U] **21.5 ODBC sources**. If the dataset is in SAS XPORT format, you can read it by using Stata's `import sasxport` command; see [D] **import sasxport**. If the dataset is in Haver Analytics' `.dat` format, you can read it by using Stata's `haver` command; see [TS] **haver**. If the dataset is in EBCDIC format, you can read it by using Stata's `infile` command; see [D] **infile (fixed format)**.

Detecting whether data are stored in binary format can be tricky. For instance, many Windows users wish to read data that have been entered into a word processor—let's assume Word. Unwittingly, they have stored the dataset as a Word document. The dataset looks like text to them: When they look at it in Word, they see readable characters. The dataset seems to even pass the printing test in that Word can print it. Nevertheless, the dataset is not text; it is stored in an internal Word format, and the data cannot really pass the printing test because only Word can print it. To read the dataset, Windows users must use it in Word and then store it as a plain text (`.txt`) file.

So, how do you know whether your dataset is binary? Here's a simple test: regardless of the operating system you use, start Stata and type `type` followed by the name of the file:

```
. type myfile.raw
```
output will appear

You do not have to list the entire file; press *Break* when you have seen enough.

Do you see things that look like hieroglyphics? If so, the dataset is binary. See [U] **21.4 Transfer programs** below.

If it looks like data, however, the file is (probably) plain text.

Let's assume that you have a text dataset that you wish to read. The data's format will determine the command you need to use. The different formats are discussed in the following sections.

21.2.4 If the data are simple

The easiest way to read text data is with `insheet`; see [D] **insheet**.

`insheet` is smart: it looks at the dataset, determines what it contains, and then reads it. That is, `insheet` is smart given certain restrictions, such as that the dataset has one observation per line and that the values are tab- or comma separated. `insheet` can read this

begin data1.raw
```
M,Joe Smith,288,14
M,K Marx,238,12
F,Farber,211,7
```
end data1.raw

or this (which has variable names on the first line)

begin data2.raw
```
sex, name, dept, division
M,Joe Smith,288,14
M,K Marx,238,12
F,Farber,211,7
```
end data2.raw

or this (which has one tab character separating the values):

begin data3.raw
```
M       Joe Smith       288     14
M       K Marx  238     12
F       Farber  211     7
```
end data3.raw

This looks odd because of how tabs work; `data3.raw` could similarly have a variable header, but `insheet` cannot read

begin data4.raw
```
M       Joe Smith       288     14
M       K Marx          238     12
F       Farber          211     7
```
end data4.raw

which has spaces rather than tabs.

There is a way to tell `data3.raw` from `data4.raw`: Ask Stata to type the data and show the tabs by typing

```
. type data3.raw, showtabs
M<T>Joe Smith<T>288<T>14
M<T>K Marx<T>238<T>12
F<T>Farber<T>211<T>7

. type data4.raw, showtabs
M       Joe Smith       288     14
M       K Marx          238     12
F       Farber          211     7
```

21.2.5 If the dataset is formatted and the formatting is significant

If the dataset is formatted and formatting information is required to interpret the data, see [D] **infile (fixed format)** or [D] **infix (fixed format)**.

Using `infix` or `infile` with a data dictionary is something new users want to avoid if at all possible.

The purpose of this section is only to take you to the most complicated of all cases if there is no alternative. Otherwise, you should wait and see if it is necessary. Do not misinterpret this section and say, "Ah, my dataset is formatted, so at last I have a solution."

Just because a dataset is formatted does not mean that you have to exploit the formatting information. The following dataset is formatted

begin data5.raw
```
  1   27.39    12
  2    1.00     4
  3  100.10   100
```
end data5.raw

in that the numbers line up in neat columns, but you do not need to know the information to read it. Alternatively, consider the same data run together:

begin data6.raw
```
1 27.39 12
2  1.00  4
3100.10100
```
end data6.raw

This dataset is formatted, too, and you must know the formatting information to make sense of "`3100.10100`". You must know that variable 2 starts in column 4 and is six characters long to extract the 100.10. It is datasets like `data6.raw` that you should be looking for at this stage—datasets that make sense only if you know the starting and ending columns of data elements. To read data such as `data6.raw`, you must use either `infix` or `infile` with a data dictionary.

Reading unformatted data is easier. If you need the formatting information to interpret the data, then you must communicate that information to Stata, which means that you will have to type it. This is the hardest kind of data to read, but Stata can do it. See [D] **infile (fixed format)** or [D] **infix (fixed format)**.

Looking back at `data4.raw`,

begin data4.raw
```
M       Joe Smith       288     14
M       K Marx          238     12
F       Farber          211     7
```
end data4.raw

you may be uncertain whether you have to read it with a data dictionary. If you are uncertain, do not jump yet.

Finally, here is an obvious example of unformatted data:

begin data7.raw
```
1 27.39           12
2 1 4
3 100.1 100
```
end data7.raw

Here blanks separate one data element from the next and, in one case, many blanks, although there is no special meaning attached to more than one blank.

The following sections discuss datasets that are unformatted or formatted in a way that do not require a data dictionary.

21.2.6 If there are no string variables

If there are no string variables, see [D] **infile (free format)**.

Although the dataset `data7.raw` is unformatted, it can still be read using `infile` without a dictionary. This is not the case with `data4.raw` because this dataset contains undelimited string variables with embedded blanks.

❑ Technical note

Some Stata users prefer to read data with a data dictionary, even when we suggest differently, as above. They like the convenience of the data dictionary—they can sit in front of an editor and carefully compose the list of variables and attach variable labels rather than having to type the variable list (correctly) on the Stata command line. However, they can create a do-file containing the `infile` statement and thus have all the advantages of a data dictionary without some of the (extremely technical) disadvantages of data dictionaries.

Nevertheless, we do tend to agree with such users—we, too, prefer data dictionaries. Our recommendations, however, are designed to work in all cases. If the dataset is unformatted and contains no string variables, it can always be read without a data dictionary, whereas only sometimes can it be read with a data dictionary.

The distinction is that `infile` without a data dictionary performs stream I/O, whereas with a data dictionary it performs record I/O. The difference is intentional—it guarantees that you will be able to read your data into Stata somehow. Some datasets require stream I/O, others require record I/O, and still others can be read either way. Recommendations 1–5 identify datasets that either require stream I/O or can be read either way.

❑

We are now left with datasets that contain at least one string variable.

21.2.7 If all the string variables are enclosed in quotes

If all the string variables in the data are enclosed in (single or double) quotes, see [D] **infile (free format)**.

See [U] **23 Working with strings** for a formal definition of strings, but as a quick guide, a string variable is a variable that takes on values like "`bob`" or "`joe`", as opposed to numeric variables that take on values like 1, 27.5, and –17.393. Undelimited strings—strings not enclosed in quotes—can be difficult to read.

Here is an example including delimited string variables:

begin data8.raw
```
"M" "Joe Smith" 288 14
"M" "K Marx" 238 12
"F" "Farber" 211 7
```
end data8.raw

or

begin data8.raw, alternative format

```
"M" "Joe Smith" 288  14
"M" "K Marx"    238  12
"F" "Farber"    211   7
```

end data8.raw, alternative format

Both of these are merely variations on `data4.raw` except that the strings are enclosed in quotes. Here `infile` without a dictionary can be used to read the data.

Here is another version of `data4.raw` without delimiters or even formatting:

begin data9.raw

```
M Joe Smith 288 14
M K Marx 238 12
F Farber 211 7
```

end data9.raw

What makes these data difficult? Blanks sometimes separate values and sometimes are nothing more than a blank within a string. For instance, you cannot tell whether Farber has first initial F with missing sex or is instead female with a missing first initial.

Fortunately, such data rarely happen. Either the strings are delimited, as we showed in `data8.raw`, or the data are in columns, as in `data4.raw`.

21.2.8 If the undelimited strings have no blanks

There is a case in which uncolumnized, undelimited strings cause no confusion—when they contain no blanks. For instance, if our data contained only last names,

begin data10.raw

```
Smith 288 14
Marx 238 12
Farber 211 7
```

end data10.raw

Stata could read it without a data dictionary. Caution: the last names must contain no blanks—no Van Owen's or von Beethoven's.

If the undelimited string variables have no blanks, see [D] **infile (free format)**.

21.2.9 If you have EBCDIC data

You may rarely encounter data from a mainframe that is encoded in extended binary coded decimal interchange code (EBCDIC). EBCDIC is used on some IBM mainframe operating systems.

If you have EBCDIC data, you should have information on that data specifying where each field begins and ends and what type of data is in that field. You can read EBCDIC data in the same way that you read fixed-format ASCII data, using `infile` (see [D] **infile (fixed format)**. You create a data dictionary that tells Stata which columns to read for each field, and you merely specify the `ebcdic` option with the `infile` command to read the data.

Alternatively, you can convert an EBCDIC file to an ASCII file with the `filefilter` command. See [D] **filefilter**.

21.2.10 If you make it to here

If you make it to here, see [D] **infile (fixed format)** or [D] **infix (fixed format)**.

Remember `data4.raw`?

begin data4.raw
```
M       Joe Smith       288     14
M       K Marx          238     12
F       Farber          211     7
```
end data4.raw

It can be read using either `infile` with a dictionary or `infix`.

21.3 If you run out of memory

You may need to tweak a setting; see [U] **6 Managing memory** and [D] **memory**.

You can also try to conserve memory.

When you read the data, did you specify variable types? Stata can store integers more compactly than floats and small integers more compactly than large integers; see [U] **12 Data**.

If that is not sufficient, you will have to resort to reading the data in pieces. Both `infile` and `infix` allow you to specify an `in` *range* modifier, and, here the range is interpreted as the observation range to read. Thus, `infile` ... `in 1/100` would read observations 1–100 of your data and stop.

`infile` ... `in 101/200` would read observations 101–200. The end of the range may be specified as larger than the actual number of observations in the data. If the dataset contained only 150 observations, `infile` ... `in 101/200` would read observations 101–150.

Another way of reading the data in pieces is to specify the `if` *exp* modifier. Say that your data contained an equal number of males and females, coded as the variable `sex` (which you will read) being 0 or 1, respectively. You could type `infile` ... `if sex==0` to read the males. `infile` will read an observation, determine if `sex` is zero, and if not, throw the observation away. You could read just the females by typing `infile` ... `if sex==1`.

If the dataset is really big, perhaps you need only a random sample of the data—you never intended to analyze the entire dataset. Because `infile` and `infix` allow `if` *exp*, you could type `infile` ... `if runiform()<.1`. `runiform()` is the uniformly distributed random-number generator; see [D] **functions**. This method would read an approximate 10% sample of the data. If you are serious about using random samples, do not forget to set the seed before using `runiform()`; see [R] **set seed**.

The final approach is to read all the observations but only some of the variables. When reading data without a data dictionary, you can specify `_skip` for variables, indicating that the variable is to be skipped. When reading with a data dictionary or using `infix`, you can specify the actual columns to read, skipping any columns you wish to ignore.

If you are using `import excel`, you can read a subset of an Excel worksheet by using the `cellrange()` option. See [D] **import excel**.

21.4 Transfer programs

To import data from, say, Access, you can save the data as a text file and then read it into Stata according to the rules above, read it via an ODBC source, or purchase a program to translate the dataset from the Access format to Stata's format.

One such program is Stat/Transfer, which is available for Windows, Mac OS X, and Unix. It reads and writes data in a variety of formats, including Microsoft Access, dBASE, Epi Info, Excel, FoxPro, GAUSS, JMP, LIMDEP, Lotus 1-2-3, MATLAB, MineSet, ODBC, Paradox, Quattro Pro, S-Plus, SAS, SPSS, Statistica, SYSTAT, and, of course, Stata.

Stat/Transfer, available from...	is manufactured by...
StataCorp	Circle Systems
4905 Lakeway Drive	1001 Fourth Avenue, Suite 3200
College Station, Texas 77845	Seattle, Washington 98154
Telephone: 979-696-4600	Telephone: 206-682-3783
Fax: 979-696-4601	Fax: 206-508-9030
Email: stata@stata.com	sales@circlesys.com

21.5 ODBC sources

If your dataset is located in a network database or shared spreadsheet, you may be able to import your data via ODBC. Open Database Connectivity (ODBC) is a standard for exchanging data between programs. Stata supports the ODBC standard for importing data via the `odbc` command and can read from any ODBC source on your computer.

This process requires a data source, such as a database located on a network. To use the `odbc` command to import data from a database requires that the database first be set up as an ODBC source on the same machine that is running Stata. The database itself does not have to be on the same machine, just the definition of that database as the ODBC source. On a Windows machine, an ODBC source is added via a Control Panel called "Data Sources". Also, typing `odbc list` from Stata displays all the ODBC sources that are provided by the computer.

If the database is functioning and the appropriate data source has been set up on the same machine as Stata, one call using `odbc load` is all that is needed to import data. For a more thorough description of this process, see [D] **odbc**.

21.6 Reference

Swagel, P. 1994. os14: A program to format raw data files. *Stata Technical Bulletin* 20: 10–12. Reprinted in *Stata Technical Bulletin Reprints*, vol. 4, pp. 80–82. College Station, TX: Stata Press.

22 Combining datasets

You have two datasets that you wish to combine. Below, we will draw a dataset as a box where, in the box, the variables go across and the observations go down.

See [D] **append** if you want to combine datasets vertically:

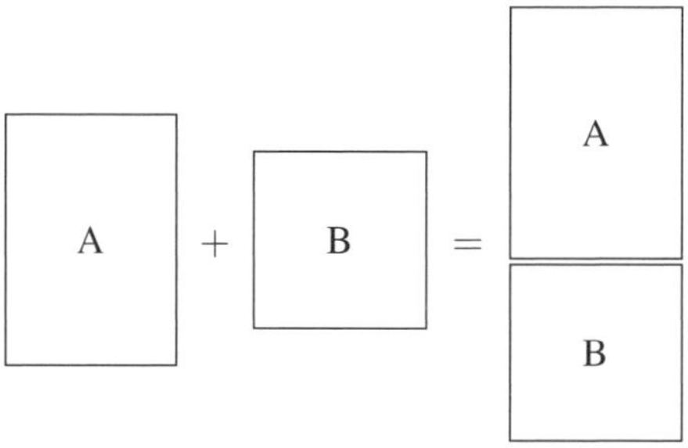

`append` adds observations to the existing variables. That is an oversimplification because `append` does not require that the datasets have the same variables. `append` is appropriate, for instance, when you have data on hospital patients and then receive data on more patients.

See [D] **merge** if you want to combine datasets horizontally:

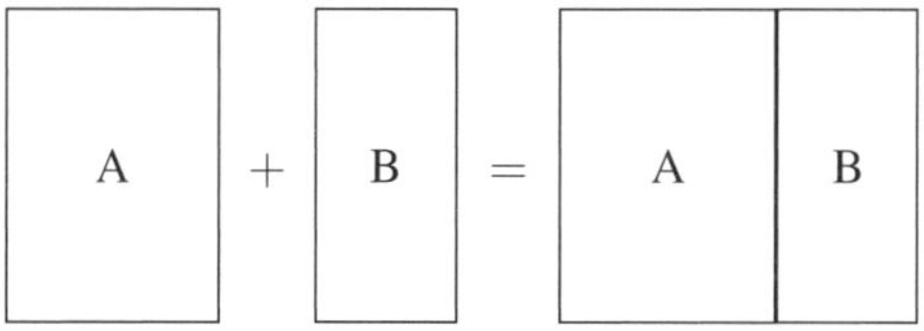

`merge` adds variables to the existing observations. That is an oversimplification because `merge` does not require that the datasets have the same observations. `merge` is appropriate, for instance, when you have data on survey respondents and then receive data on part 2 of the questionnaire.

See [D] **joinby** when you want to combine datasets horizontally but form all pairwise combinations within group:

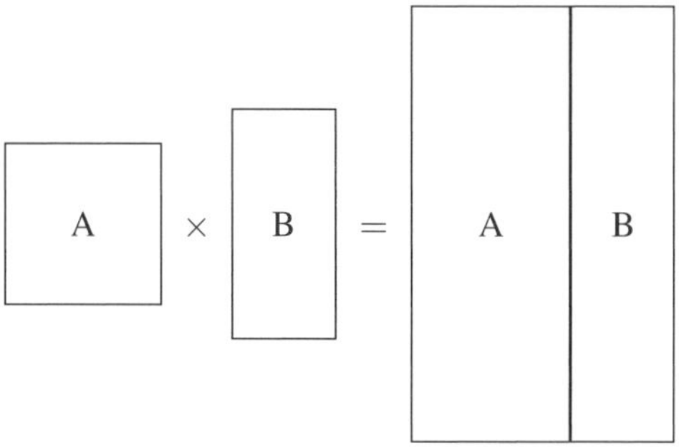

`joinby` is similar to `merge` but forms all combinations of the observations where it makes sense. `joinby` would be appropriate, for instance, where A contained data on parents and B contained data on their children. `joinby` *familyid* would form a dataset of each parent joined with each of his or her children.

Also see [D] **cross** for a less frequently used command that forms every pairwise combination of two datasets.

See Mitchell (2010, chap. 6) for more information on combining datasets in Stata.

22.1 References

Golbe, D. L. 2010. Stata tip 83: Merging multilingual datasets. *Stata Journal* 10: 152–156.

Gould, W. W. 2011a. Merging data, part 1: Merges gone bad. The Stata Blog: Not Elsewhere Classified. http://blog.stata.com/2011/04/18/merging-data-part-1-merges-gone-bad/

——. 2011b. Merging data, part 2: Multiple-key merges. The Stata Blog: Not Elsewhere Classified. http://blog.stata.com/2011/05/27/merging-data-part-2-multiple-key-merges/

Mitchell, M. N. 2010. *Data Management Using Stata: A Practical Handbook*. College Station, TX: Stata Press.

23 Working with strings

Contents

Please read [U] **12 Data** before reading this entry.

23.1 Description

The word *string* is shorthand for a string of characters. "Male" and "Female", "yes" and "no", and "R. Smith" and "P. Jones" are examples of strings. The alternative to strings is numbers—0, 1, 2, 5.7, and so on. Variables containing strings—called *string variables*—occur in data for a variety of reasons. Four of these reasons are listed below.

A variable might contain strings because it is an identifying variable. Employee names in a payroll file, patient names in a hospital file, and city names in a city data file are all examples of this. This is a proper use of string variables.

A variable might contain strings because it records categorical information. "Male" and "Female" and "yes" and "no" are examples of such use, but this is not an appropriate use of string variables. It is not appropriate because the same information could be coded numerically, and, if it were, it would take less memory to store the data and the data would be more useful. We will explain how to convert categorical strings to categorical numbers below.

Also, a variable might contain strings because of a mistake. For example, the variable contains things like 1, 5, 8.2, but because of an error in reading the data, the data were mistakenly put into a string variable. We will explain how to fix such mistakes.

Finally, a variable might contain strings because the data simply could not be coerced into being stored numerically. "15 Jan 1992", "1/15/92", and "1A73" are examples of such use. We will explain how to deal with such complexities.

23.2 Categorical string variables

A variable might contain strings because it records categorical information.

Suppose that you have read in a dataset that contains a variable called `sex`, recorded as "male" and "female", yet when you attempt to run a linear regression, the following message is displayed:

```
. use http://www.stata-press.com/data/r12/hbp2
. regress hbp sex
no observations
r(2000);
```

There are no observations because `regress`, along with most of Stata's "analytic" commands, cannot deal with string variables. Commands want to see numbers, and when they do not, they treat the variable as if it contained numeric missing values. Despite this limitation, it is possible to obtain tables:

```
. encode sex, gen(gender)
. regress hbp gender
      Source |       SS       df       MS              Number of obs =    1128
-------------+------------------------------           F(  1,  1126) =   14.04
       Model |  .644485682     1  .644485682           Prob > F      =  0.0002
    Residual |  51.6737767  1126  .045891454           R-squared     =  0.0123
-------------+------------------------------           Adj R-squared =  0.0114
       Total |  52.3182624  1127  .046422593           Root MSE      =  .21422

------------------------------------------------------------------------------
         hbp |      Coef.   Std. Err.      t    P>|t|     [95% Conf. Interval]
-------------+----------------------------------------------------------------
      gender |   .0491501   .0131155     3.75   0.000     .0234166    .0748837
       _cons |  -.0306744   .0221353    -1.39   0.166    -.0741054    .0127566
------------------------------------------------------------------------------
```

The magic here is to convert the string variable `sex` into a numeric variable called `gender` with an associated value label, a trick accomplished by `encode`; see [U] **12.6.3 Value labels** and [D] **encode**.

23.3 Mistaken string variables

A variable might contain strings because of a mistake.

Suppose that you have numeric data in a variable called `x`, but because of a mistake, `x` was made a string variable when you read the data. When you `list` the variable, it looks fine:

```
. list x

     +-----+
     |   x |
     |-----|
  1. |   2 |
  2. | 2.5 |
  3. |  17 |
 (output omitted)
     +-----+
```

Yet, when you attempt to obtain summary statistics on `x`,

```
. summarize x
    Variable |       Obs        Mean    Std. Dev.       Min        Max
-------------+--------------------------------------------------------
           x |         0
```

If this happens to you, type `describe` to confirm that `x` is stored as a string:

```
. describe
Contains data
  obs:            10
 vars:             3
 size:           160
-------------------------------------------------------------------------------
              storage  display     value
variable name   type   format      label      variable label
-------------------------------------------------------------------------------
x               str4   %9s
y               float  %9.0g
z               float  %9.0g
-------------------------------------------------------------------------------
Sorted by:
```

`x` is stored as a `str4`.

The problem is that `summarize` does not know how to calculate the mean of string variables—how to calculate the mean of "Joe" plus "Bill" plus "Roger"—even when the string variable contains what could be numbers. By using the `destring` command ([D] **destring**), the variable mistakenly stored as a `str4` can be converted to a numeric variable.

```
. destring x, replace
. summarize x
    Variable |       Obs        Mean    Std. Dev.       Min        Max
-------------+--------------------------------------------------------
        newx |        10        1.76    .8071899         .7          3
```

An alternative to using the `destring` command is to use `generate` with the `real()` function; see [D] **functions**.

23.4 Complex strings

A variable might contain strings because the data simply could not be coerced into being stored numerically.

A complex string is a string that contains more than one piece of information. The most common example of a complex string is a date: "15 Jan 1992" contains three pieces of information—a day, a month, and a year. If your complex strings are dates or times, see [U] **24 Working with dates and times**.

Although Stata has functions for dealing with dates, you will have to deal with other complex strings yourself. Assume that you have data that include part numbers:

```
. list partno
```

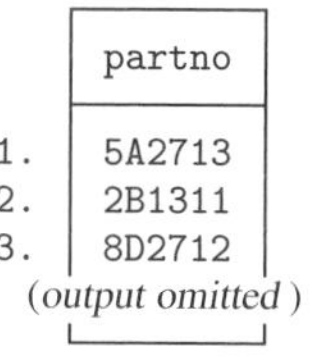

```
     +---------+
     | partno  |
     |---------|
  1. | 5A2713  |
  2. | 2B1311  |
  3. | 8D2712  |
```
(output omitted)

The first digit of the part number is a division number, and the character that follows identifies the plant at which the part was manufactured. The next three digits represent the major part number and the last digit is a modifier indicating the color. This complex variable can be decomposed using the `substr()` and `real()` functions described in [D] **functions**:

```
. gen byte div = real(substr(partno,1,1))
. gen str1 plant = substr(partno,2,1)
. gen int part = real(substr(partno,3,3))
. gen byte color = real(substr(partno,6,1))
```

We use the `substr()` function to extract pieces of the string and use the `real()` function, when appropriate, to translate the piece into a number.

For an extended discussion of numeric and string data types and how to convert from one kind to another, see Cox (2002).

23.5 Reference

Cox, N. J. 2002. Speaking Stata: On numbers and strings. *Stata Journal* 2: 314–329.

24 Working with dates and times

Contents

24.1 Overview

Full documentation on Stata's date and time capabilities—including documentation on relevant functions and display formats—can be found in [D] **datetime**.

Stata can work with dates such as 21nov2006, with times such as 13:42:02.213, and with dates and times such as 21nov2006 13:42:02.213. You can write these dates and times however you wish, such as 11/21/2006, November 21, 2006, and 1:42 p.m.

Stata stores dates, times, and dates and times as integers such as −4,102, 0, 82, 4,227, and 1,479,735,745,213. It works like this:

1. You begin with the datetime variables in your data however they are recorded, such as 21nov2006 or 11/21/2006 or November 21, 2006, or 13:42:02.213 or 1:42 p.m. The original values are usually best stored in string variables.
2. Using functions we will describe below, you translate the original into the integers that Stata understands and store those values in a new variable.
3. You specify the appropriate display format for the new variable so that, rather than displaying as the integer values that they are, they display in a way you can read them such as 21nov2006 or 11/21/2006 or November 21, 2006, or 13:42:02.213 or 1:42 p.m.

The numeric encoding that Stata uses is centered around the first millisecond of 01jan1960, that is, 01jan1960 00:00:00.000. That datetime is assigned integer value 0.

Integer value 1 is the millisecond after that: 01jan1960 00:00:00.001.

Integer value −1 is the millisecond before that: 31dec1959 23:59:59.999.

By that logic, 21nov2006 13:42:02.213 is integer value 1,479,735,722,213 or, at least, it is if we ignore the leap seconds that have been inserted to keep clocks in alignment with astronomical observation. If we account for leap seconds, 21nov2006 13:42:02.213 would be 23 seconds later, namely, 1,479,735,745,213. Stata can work either way.

Obtaining the number of milliseconds associated with a datetime is easy because Stata provides functions that translate things like 21nov2006 13:42:02.213 (written however you wish) to 1,479,735,722,213 or 1,479,735,745,213.

Just remember, Stata records datetime values as the number of milliseconds since the first millisecond of 01jan1960.

Stata records pure time values (clock times independent of date) the same way. Rather than thinking of the numeric value as the number of milliseconds since 01jan1960, however, think of it as the number of milliseconds since the beginning of the day. For instance, at 2 p.m. every day, the airplane takes off from Houston for London. The numeric value associated with 2 p.m. is 50,400,000 because there are that many milliseconds between the beginning of the day (00:00:00.000) and 2 p.m.

The advantage of thinking this way is that you can add dates and times. What is the datetime value for when the plane takes off on 21nov2006? Well, 21nov2006 00:00:00.000 is 1,479,686,400,000 (ignoring leap seconds), and 1,479,686,400,000 + 50,400,000 is 1,479,736,800,000.

Subtracting datetime values is useful, too. How many hours are there between 21jan1952 7:23 a.m. and 21nov2006 3:14 p.m.? Answer: $(1{,}479{,}741{,}240{,}000 - (-250{,}706{,}220{,}000))/3{,}600{,}000 =$ 480,679.85 hours.

Variables that record the number of milliseconds since 01jan1960 and ignore leap seconds are called `%tc` variables.

Variables that record the number of milliseconds since 01jan1960 and account for leap seconds are called `%tC` variables.

Stata has seven other kinds of `%t` variables.

In many applications, calendar dates by themselves are sufficient. The applicant was hired on 15jan2006, for instance. You could use a `%tc` variable to record that value, assigning some arbitrary time that you would ignore, but it is better and easier to use a `%td` variable. In `%td` variables, 0 still corresponds to 01jan1960, but a unit change now represents an entire day rather than a millisecond. The value 1 represents 02jan1960. The value −1 represents 31dec1959. When you subtract `%td` variables, you obtain the number of days between dates.

In a financial application, you might use `%tq` variables. In `%tq`, 0 represents the first quarter of 1960, 1 represents the second quarter, and −1 represents the last quarter of 1959. When you subtract `%tq` variables, you obtain the number of quarters between dates.

Stata understands nine `%t` formats:

Format	Base	Units	Comment
`%tc`	01jan1960	milliseconds	ignores leap seconds
`%tC`	01jan1960	milliseconds	accounts for leap seconds
`%td`	01jan1960	days	calendar date format
`%tw`	1960-w1	weeks	52nd week may have more than 7 days
`%tm`	jan1960	months	calendar month format
`%tq`	1960-q1	quarters	financial quarter
`%th`	1960-h1	half-years	1 half-year = 2 quarters
`%ty`	0 A.D	year	1960 means year 1960
`%tb`	–	days	user defined

All formats except `%ty` and `%tb` are based on the beginning of January 1960. The value 0 means the first millisecond, day, week, month, quarter, or half-year of 1960, depending on format. The value 1 is the millisecond, day, week, month, quarter, or half-year after that. The value −1 is the millisecond, day, week, month, quarter, or half-year before that.

Stata's `%ty` format records years as numeric values and it codes them the natural way: rather than 0 meaning 1960, 1960 means 1960, and so 2006 also means 2006.

24.2 Inputting dates and times

Dates and time variables are best read as strings. You then use one of the string-to-numeric conversion functions to convert the string to an appropriate `%t` value:

Format	String-to-numeric conversion function
`%tc`	`clock(`*string*`,` *mask*`)`
`%tC`	`Clock(`*string*`,` *mask*`)`
`%td`	`date(`*string*`,` *mask*`)`
`%tw`	`weekly(`*string*`,` *mask*`)`
`%tm`	`monthly(`*string*`,` *mask*`)`
`%tq`	`quarterly(`*string*`,` *mask*`)`
`%th`	`halfyearly(`*string*`,` *mask*`)`
`%ty`	`yearly(`*string*`,` *mask*`)`

The full documentation of these functions can be found in [D] **datetime translation**.

In the above table, *string* is the string variable to be translated, and *mask* specifies the order in which the components of the date and/or time appear in *string*. For instance, the *mask* in `%td` function `date()` is made up of the letters `M`, `D`, and `Y`.

`date(`*string*`, "DMY")` specifies *string* contains dates in the order of day, month, year. With that specification, `date()` can translate 21nov2006, 21 November 2006, 21-11-2006, 21112006, and other strings that contain dates in the order day, month, year.

`date(`*string*`, "MDY")` specifies *string* contains dates in the order of month, day, year. With that specification, `date()` can translate November 21, 2006, 11/21/2006, 11212006, and other strings that contain dates in the order month, day, year.

You can specify a two-digit prefix in front of `Y` to handle two-digit years. `date(`*string*`, "MD19Y")` specifies *string* contains dates in the order of month, day, and year, and that if the year contains only two digits, it is to be prefixed with 19. With that specification, `date()` could not only translate November 21, 2006, 11/21/2006, and 11212006, but also Feb. 15 '98, 2/15/98, and 21598. (There is another way to deal with two-digit years so that 98 becomes 1998 while 06 becomes 2006; it involves specifying an optional third argument. See *Working with two-digit years* in [D] **datetime translation**.)

Let's consider some `%td` data. We have the following raw-data file:

begin bdays.raw

```
Bill  21 Jan 1952  22
May   11 Jul 1948  18
Sam   12 Nov 1960  25
Kay    9 Aug 1975  16
```

end bdays.raw

We could read these data by typing

```
. infix str name 1-5  str bday 7-17  x 20-21 using bdays
(4 observations read)
```

We read the date not as three separate variables but as one variable. Variable `bday` contains the entire date:

```
. list

     +-----------------------------+
     | name          bday       x  |
     |-----------------------------|
  1. | Bill   21 Jan 1952      22  |
  2. |  May   11 Jul 1948      18  |
  3. |  Sam   12 Nov 1960      25  |
  4. |  Kay    9 Aug 1975      16  |
     +-----------------------------+
```

The data look fine, but if we set about using them, we would quickly discover there is not much we could do with variable `bday`. Variable `bday` looks like a date, but it is just a string. We need to turn `bday` into a `%t` variable that Stata understands:

```
. gen birthday = date(bday, "DMY")

. list

     +-------------------------------------------+
     | name          bday     x       birthday   |
     |-------------------------------------------|
  1. | Bill   21 Jan 1952    22          -2902   |
  2. |  May   11 Jul 1948    18          -4191   |
  3. |  Sam   12 Nov 1960    25            316   |
  4. |  Kay    9 Aug 1975    16           5699   |
     +-------------------------------------------+
```

New variable `birthday` is a `%td` variable. The problem now is that, whereas the new variable is perfectly understandable to Stata, it is not understandable to us. Naturally enough, a `%td` variable needs a `%td` format:

```
. format birthday %td

. list

     +--------------------------------------------+
     | name          bday     x        birthday   |
     |--------------------------------------------|
  1. | Bill   21 Jan 1952    22       21jan1952   |
  2. |  May   11 Jul 1948    18       11jul1948   |
  3. |  Sam   12 Nov 1960    25       12nov1960   |
  4. |  Kay    9 Aug 1975    16       09aug1975   |
     +--------------------------------------------+
```

Using our new `%td` variable, we can create a variable recording how old each of these subjects was on 01jan2000:

```
. gen age2000 = (td(1jan2000)-birthday)/365.25

. list

     +-------------------------------------------------------+
     | name          bday     x      birthday      age2000   |
     |-------------------------------------------------------|
  1. | Bill   21 Jan 1952    22     21jan1952     47.94524   |
  2. |  May   11 Jul 1948    18     11jul1948     51.47433   |
  3. |  Sam   12 Nov 1960    25     12nov1960     39.13484   |
  4. |  Kay    9 Aug 1975    16     09aug1975     24.39699   |
     +-------------------------------------------------------+
```

`td()` is a function that makes it easy to type `%td` dates. There are also functions `tc()`, `tC()`, `tw()`, `tm()`, `tq()`, and `th()` for the other `%t` formats; see [D] **datetime**.

Let's consider one more example. We have the following data:

```
. use http://www.stata-press.com/data/r12/datexmpl2
. list
```

	id	timestamp	action
1.	1001	Tue Nov 14 08:59:43 CST 2006	15
2.	1002	Wed Nov 15 07:36:49 CST 2006	15
3.	1003	Wed Nov 15 09:21:07 CST 2006	11
4.	1002	Wed Nov 15 14:57:36 CST 2006	16
5.	1005	Thu Nov 16 08:22:53 CST 2006	12
6.	1001	Thu Nov 16 08:36:44 CST 2006	16

Variable `timestamp` is a string which we want to convert to a `%tc` variable. From the table above, we know we will use function `clock()`. The *mask* in `clock()` uses the letters `D`, `M`, `Y`, and `h`, `m`, `s`, which specify the order of the day, month, year and hours, minutes, seconds. `timestamp` contains more than that and so cannot directly be converted using `clock()`. First, we must create a variable that `clock()` understands:

```
. gen str ts = substr(timestamp, 5, 15) + " " + substr(timestamp, 25, 4)
. list ts
```

	ts
1.	Nov 14 08:59:43 2006
2.	Nov 15 07:36:49 2006
3.	Nov 15 09:21:07 2006
4.	Nov 15 14:57:36 2006
5.	Nov 16 08:22:53 2006
6.	Nov 16 08:36:44 2006

New variable `ts` can be translated using `clock(ts, "MD hms Y")`. `"MD hms Y"` specifies that the order of the components in `ts` is month, day, hours, minutes, seconds, and year. There is no meaning to the spaces; we could just as well have specified `clock(ts, "MDhmsY")`. You can specify spaces when they help to make what you type more readable.

Because `%tc` values can be so large, whenever you use the function `clock()`, you must store the results in a `double`, as we do below:

```
. gen double dt = clock(ts, "MD hms Y")
. list id dt action
```

	id	dt	action
1.	1001	1.479e+12	15
2.	1002	1.479e+12	15
3.	1003	1.479e+12	11
4.	1002	1.479e+12	16
5.	1005	1.479e+12	12
6.	1001	1.479e+12	16

Don't panic. New variable `dt` contains numeric values, and large ones, which is why it was so important that we stored the values as `doubles`. That output above just shows us what a `%tc` variable looks like with default formatting. If we wanted to see the numeric values better, we could change dt to have a %20.0gc format. We would then see that the first value is 1,479,113,983,000, the second 1,479,195,409,000, and so on. We will not do that. Instead, we will put a %tc format on our %tc variable:

```
. format dt %tc

. list id dt action

     ┌─────────────────────────────────────┐
     │   id                   dt   action  │
     ├─────────────────────────────────────┤
  1. │ 1001   14nov2006 08:59:43       15  │
  2. │ 1002   15nov2006 07:36:49       15  │
  3. │ 1003   15nov2006 09:21:07       11  │
  4. │ 1002   15nov2006 14:57:36       16  │
  5. │ 1005   16nov2006 08:22:53       12  │
     ├─────────────────────────────────────┤
  6. │ 1001   16nov2006 08:36:44       16  │
     └─────────────────────────────────────┘
```

Variable dt is a variable we can use. Say we wanted to know how many hours it had been since the previous action:

```
. sort dt

. gen hours = hours(dt - dt[_n-1])
(1 missing value generated)

. format hours %9.2f

. list id dt action hours

     ┌───────────────────────────────────────────────┐
     │   id                   dt   action     hours  │
     ├───────────────────────────────────────────────┤
  1. │ 1001   14nov2006 08:59:43       15         .  │
  2. │ 1002   15nov2006 07:36:49       15     22.62  │
  3. │ 1003   15nov2006 09:21:07       11      1.74  │
  4. │ 1002   15nov2006 14:57:36       16      5.61  │
  5. │ 1005   16nov2006 08:22:53       12     17.42  │
     ├───────────────────────────────────────────────┤
  6. │ 1001   16nov2006 08:36:44       16      0.23  │
     └───────────────────────────────────────────────┘
```

We subtracted the previous value of dt from dt, which results in the number of milliseconds. Converting milliseconds to hours is easy enough: we just have to divide by $60 \times 60 \times 1{,}000 = 3{,}600{,}000$. It is easy to forget or mistype that constant, so we used Stata's hours() function, which converts milliseconds to hours. hours(), and other useful functions, are documented in [D] **datetime**.

24.3 Displaying dates and times

A %td variable should have a %td format, a %tc variable should have a %tc format, and so on for all the other %t variable–format pairs. You do that by typing format *varname* %td, format *varname* %tc, etc.

Formats %tc, %tC, %td, %tw, %tm, %tq, %th, and %ty are called the default %t formats. By specifying codes following them, you can control how the variable is to be displayed.

In the previous example, we started with a string variable that contained a time stamp and looked like "Tue Nov 14 08:59:43 CST 2006". After creating a %tc variable from it, and putting the default %tc format on it, our datetimes looked like "14nov2006 08:59:43". Below we specify a %tc format that makes our new variable look just like the original:

```
. format dt %tcDay_Mon_DD_HH:MM:SS_!C!S!T_CCYY
. list id dt action hours

     +-----------------------------------------------------------+
     |   id                               dt   action    hours |
     |-----------------------------------------------------------|
  1. | 1001    Tue Nov 14 08:59:43 CST 2006       15        .  |
  2. | 1002    Wed Nov 15 07:36:49 CST 2006       15    22.62  |
  3. | 1003    Wed Nov 15 09:21:07 CST 2006       11     1.74  |
  4. | 1002    Wed Nov 15 14:57:36 CST 2006       16     5.61  |
  5. | 1005    Thu Nov 16 08:22:53 CST 2006       12    17.42  |
     |-----------------------------------------------------------|
  6. | 1001    Thu Nov 16 08:36:44 CST 2006       16     0.23  |
     +-----------------------------------------------------------+
```

`%t` display formats are documented in [D] **datetime display formats**.

24.4 Typing dates and times (datetime literals)

You will sometimes need to type dates and times in expressions. When we needed to calculate the age of subjects as of 01jan2000 in a previous example, for instance, we typed,

```
. gen age2000 = (td(1jan2000)-birthday)/365.25
```

although we could just as well have typed,

```
. gen age2000 = (14610-birthday)/365.25
```

because 14,610 is the `%td` value of 01jan2000. Typing `td(1jan2000)` is easier and less error-prone.

Similarly, if we needed 10:55 a.m. on 01jan1960 as a `%tc` value, rather than typing 39,300,000, we could type `tc(01jan1960 10:55)`. See *Conveniently typing SIF values* in [D] **datetime** for details.

24.5 Extracting components of dates and times

Once you have a `%t` variable, you can use the extraction functions to obtain components of the variable. For instance, the following functions are appropriate for use with `%td` variables:

`year(`*date*`)`	returns four-digit year; e.g., 1980, 2002
`month(`*date*`)`	returns month; 1, 2, ..., 12
`day(`*date*`)`	returns day within month; 1, 2, ..., 31
`halfyear(`*date*`)`	returns the half of year; 1 or 2
`quarter(`*date*`)`	returns quarter of year; 1, 2, 3, or 4
`week(`*date*`)`	returns week of year; 1, 2, ..., 52
`dow(`*date*`)`	returns day of week; 0, 1, ..., 6; 0 = Sunday
`doy(`*date*`)`	returns day of year; 1, 2, ..., 366

There are other functions useful with other `%t` variables. See *Extracting time-of-day components from SIFs* and *Extracting date components from SIFs* in [D] **datetime**.

24.6 Converting between date and time values

You can convert between `%t` values. For instance, the `cofd()` function converts a `%td` value to a `%tc` value. `cofd()` of 17,126 (21nov2006) returns 1,479,686,400,000 (21nov2006 00:00:00). Function `dofc()` of 1,479,736,920,000 (21nov2006 14:02) returns 17,126 (21nov2006).

There are other functions for converting between other `%t` values; see *SIF-to-SIF conversion* in [D] **datetime**.

24.7 Business dates and calendars

In addition to the built-in date types above, such as `%tc` and `%td`, Stata provides a type you can define, denoted as `%tb` and called business dates and calendars.

A business calendar is like an ordinary calendar with some dates crossed out. The crossed-out dates correspond to the dates on which the business is closed:

November 2011

Su	Mo	Tu	We	Th	Fr	Sa
		1	2	3	4	X
X	7	8	9	10	11	X
X	14	15	16	17	18	19
X	21	22	23	X	25	X
X	28	29	30			

With respect to a business date, *yesterday* is the last day the business was open, and *tomorrow* is the next day the business will be open.

Consider *date* = 25nov2011. If *date* is a regular (`%td`) date variable,

$$\textit{yesterday} = \textit{date} - 1 = \text{24nov2011}$$

$$\textit{tomorrow} = \textit{date} + 1 = \text{26nov2011}$$

If *date* is a business (`%tb`) date variable,

$$\textit{yesterday} = \textit{date} - 1 = \text{23nov2011}$$

$$\textit{tomorrow} = \textit{date} + 1 = \text{28nov2011}$$

That is important because variables containing dates are often used with Stata's lag and lead operators; see [U] **13.9 Time-series operators**. If variable `trading_date` is an ordinary `%td` variable, then `L.trading_date` really is yesterday and `F.trading_date` really is tomorrow. But if `trading_date` has an appropriately defined `%tb` format, `L.trading_date` is the previous trading date and `F.trading_date` is the next trading date.

To make a business calendar, you create a file named *calname*`.stbcal`, such as `nyse.stbcal`. After that, Stata deeply understands the new format `%tbnyse`. For more information, see [D] **datetime business calendars**.

Business dates work just like regular dates, it is just that some dates are crossed out.

24.8 References

Cox, N. J. 2006. Speaking Stata: Time of day. *Stata Journal* 6: 124–137.

——. 2010. Stata tip 68: Week assumptions. *Stata Journal* 10: 682–685.

25 Working with categorical data and factor variables

Contents

25.1 Continuous, categorical, and indicator variables

Although to Stata a variable is a variable, it is helpful to distinguish among three conceptual types:

- A *continuous variable* measures something. Such a variable might measure a person's age, height, or weight; a city's population or land area; or a company's revenues or costs.
- A *categorical variable* identifies a group to which the thing belongs. You could categorize persons according to their race or ethnicity, cities according to their geographic location, or companies according to their industry. Sometimes, categorical variables are stored as strings.
- An *indicator variable* denotes whether something is true. For example, is a person a veteran, does a city have a mass transit system, or is a company profitable?

Indicator variables are a special case of categorical variables. Consider a variable that records a person's sex. Examined one way, it is a categorical variable. A categorical variable identifies the group to which a thing belongs, and here the thing is a person and the basis for categorization is anatomy. Looked at another way, however, it is an indicator variable. It indicates whether the person is female.

We can use the same logic on any categorical variable that divides the data into two groups. It is a categorical variable because it identifies whether an observation is a member of this or that group; it is an indicator variable because it denotes the truth value of the statement "the observation is in this group".

All indicator variables are categorical variables, but the opposite is not true. A categorical variable might divide the data into more than two groups. For clarity, let's reserve the term *categorical variable* for variables that divide the data into more than two groups, and let's use the term *indicator variable* for categorical variables that divide the data into exactly two groups.

Stata can convert continuous variables to categorical and indicator variables and categorical variables to indicator variables.

25.1.1 Converting continuous variables to indicator variables

Stata treats logical expressions as taking on the values *true* or *false*, which it identifies with the numbers 1 and 0; see [U] **13 Functions and expressions**. For instance, if you have a continuous variable measuring a person's age and you wish to create an indicator variable denoting persons aged 21 and over, you could type

```
. generate age21p = age>=21
```

The variable `age21p` takes on the value 1 for persons aged 21 and over and 0 for persons under 21.

Because `age21p` can take on only 0 or 1, it would be more economical to store the variable as a `byte`. Thus it would be better to type

```
. generate byte age21p = age>=21
```

This solution has a problem. The value of `age21` is set to 1 for all persons whose `age` is missing because Stata defines missing to be larger than all other numbers. In our data, we might have no such missing ages, but it still would be safer to type

```
. generate byte age21p = age>=21 if age<.
```

That way, persons whose age is missing would also have a missing `age21p`.

❑ Technical note

Put aside missing values and consider the following alternative to `generate age21p = age>=21` that may have occurred to you:

```
. generate age21p = 1 if age>=21
```

That does not produce the desired result. This statement makes `age21p` 1 (*true*) for all persons aged 21 and above but makes `age21p` missing for everyone else.

If you followed this second approach, you would have to combine it with

```
. replace age21p = 0 if age<21
```

❑

25.1.2 Converting continuous variables to categorical variables

Suppose that you wish to categorize persons into four groups on the basis of their age. You want a variable to denote whether a person is 21 or under, between 22 and 38, between 39 and 64, or 65 and above. Although most people would label these categories 1, 2, 3, and 4, there is really no reason to restrict ourselves to such a meaningless numbering scheme. Let's call this new variable `agecat` and make it so that it takes on the topmost value for each group. Thus persons in the first group will be identified with an `agecat` of 21, persons in the second with 38, persons in the third with 64, and persons in the last (drawing a number out of the air) with 75. Here is a way to create the variable that will work, but it is not the best method for doing so:

```
. use http://www.stata-press.com/data/r12/agexmpl
. generate byte agecat=21 if age<=21
(176 missing values generated)
. replace agecat=38 if age>21 & age<=38
(148 real changes made)
. replace agecat=64 if age>38 & age<=64
(24 real changes made)
. replace agecat=75 if age>64 & age<.
(4 real changes made)
```

We created the categorical variable according to the definition by using the `generate` and `replace` commands. The only thing that deserves comment is the opening `generate`. We (wisely) told Stata to `generate` the new variable `agecat` as a `byte`, thus conserving memory.

We can create the same result with one command using the `recode()` function:

```
. use http://www.stata-press.com/data/r12/agexmpl, clear
. generate byte agecat=recode(age,21,38,64,75)
```

`recode()` takes three or more arguments. It examines the first argument (here `age`) against the remaining arguments in the list. It returns the first element in the list that is greater than or equal to the first argument or, failing that, the last argument in the list. Thus, for each observation, `recode()` asked if `age` was less than or equal to 21. If so, the value is 21. If not, is it less than or equal to 38? If so, the value is 38. If not, is it less than or equal to 64? If so, the value is 64. If not, the value is 75.

Most researchers typically make tables of categorical variables, so we will `tabulate` the result:

```
. tabulate agecat

     agecat |      Freq.     Percent        Cum.
------------+-----------------------------------
         21 |         28       13.73       13.73
         38 |        148       72.55       86.27
         64 |         24       11.76       98.04
         75 |          4        1.96      100.00
------------+-----------------------------------
      Total |        204      100.00
```

There is another way to convert continuous variables into categorical variables, and it is even more automated: `autocode()` works like `recode()`, except that all you tell the function is the range and the total number of cells that you want that range broken into:

```
. use http://www.stata-press.com/data/r12/agexmpl, clear
. generate agecat=autocode(age,4,18,65)
. tabulate agecat

     agecat |      Freq.     Percent        Cum.
------------+-----------------------------------
      29.75 |         82       40.20       40.20
       41.5 |         96       47.06       87.25
      53.25 |         16        7.84       95.10
         65 |         10        4.90      100.00
------------+-----------------------------------
      Total |        204      100.00
```

In one instruction, we told Stata to break `age` into four evenly spaced categories from 18 to 65. When we `tabulate agecat`, we see the result. In particular, we see that the breakpoints of the four categories are 29.75, 41.5, 53.25, and 65. The first category contains everyone aged 29.75 years or less; the second category contains persons over 29.75 who are 41.5 years old or less; the third category contains persons over 41.5 who are 53.25 years old or less; and the last category contains all persons over 53.25.

❑ Technical note

We chose the range 18–65 arbitrarily. Although you cannot tell from the table above, there are persons in this dataset who are under 18, and there are persons over 65. Those persons are counted in the first and last cells, but we have not divided the age range in the data evenly. We could split the full age range into four categories by obtaining the overall minimum and maximum ages (by typing `summarize`) and substituting the overall minimum and maximum for the 18 and 65 in the `autocode()` function:

```
. use http://www.stata-press.com/data/r12/agexmpl, clear
. summarize age
    Variable |       Obs        Mean    Std. Dev.       Min        Max
-------------+--------------------------------------------------------
         age |       204    31.57353    10.28986          2         66
. generate agecat2=autocode(age,4,2,66)
```

We could also `sort` the data into ascending order of `age` and tell Stata to construct four categories over the range `age[1]` (the minimum) to `age[_N]` (the maximum):

```
. use http://www.stata-press.com/data/r12/agexmpl, clear
. sort age
. generate agecat2=autocode(age,4,age[1],age[_N])
. tabulate agecat2
    agecat2 |      Freq.     Percent        Cum.
------------+-----------------------------------
         18 |         10        4.90        4.90
         34 |        138       67.65       72.55
         50 |         41       20.10       92.65
         66 |         15        7.35      100.00
------------+-----------------------------------
      Total |        204      100.00
```

❑

25.2 Estimation with factor variables

Stata handles categorical variables as factor variables; see [U] **11.4.3 Factor variables**. Categorical variables refer to the variables in your data that take on categorical values, variables such as `sex`, `group`, and `region`. Factor variables refer to Stata's treatment of categorical variables. Factor variables create indicator variables for the levels (categories) of categorical variables and, optionally, for their interactions.

In what follows, the word level means the value that a categorical variable takes on. The variable `sex` might take on levels 0 and 1, with 0 representing male and 1 representing female. We could say that `sex` is a two-level factor variable.

The regressors created by factor variables are called indicators or, more explicitly, virtual indicator variables. They are called virtual because the machinery for factor variables seldom creates new variables in your dataset, even though the indicators will appear just as if they were variables in your estimation results.

To be used as a factor variable, a categorical variable must take on nonnegative integer values. If you have variables with negative values, recode them; see [D] **recode**. If you have string variables, you can use `egen`'s `group()` function to recode them,

```
. egen newcatvar= group(mystringcatvar)
```

If you also specify the `label` option, `egen` will create a value label for the numeric code it produces so that your output will be subsequently more readable:

```
. egen newcatvar= group(mystringcatvar), label
```

Alternatively, you can use `encode` to convert string categorical variables to numeric ones:

```
. encode mystringcatvar, generate(newcatvar)
```

`egen group(), label` and `encode` do the same thing. We tend to use `egen group(), label`. See [D] **egen** and [D] **encode**.

In the unlikely event that you have a noninteger categorical variable, use the `egen` solution. More likely, however, is that you need to read [U] **25.1.2 Converting continuous variables to categorical variables**.

❑ Technical note

If you should ever need to create your own indicator variables from a string or numeric variable—and it is difficult to imagine why you would—type

```
. tabulate var, gen(newstub)
```

Typing that will create indicator variables named *newstub*1, *newstub*2, ...; see [R] **tabulate oneway**. ❑

We will be using linear regression in the examples that follow just because it is so easy to explain and to interpret. We could, however, just as well have used logistic regression, Heckman selectivity, or even Cox proportional-hazards regression with shared frailties. Stata's factor-variable features work with nearly every estimation command.

25.2.1 Including factor variables

The fundamental building block of factor variables is the treatment of each factor variable as if it represented a collection of indicators, with one indicator for each level of the variable. To treat a variable as a factor variable, you add `i.` in front of the variable's name:

```
. use http://www.stata-press.com/data/r12/fvex, clear
(Artificial factor variables' data)
. regress y  i.group age
      Source |       SS       df       MS              Number of obs =    3000
-------------+------------------------------           F(  3,  2996) =   31.67
       Model |  42767.8126     3  14255.9375           Prob > F      =  0.0000
    Residual |  1348665.19  2996  450.155272           R-squared     =  0.0307
-------------+------------------------------           Adj R-squared =  0.0298
       Total |  1391433.01  2999  463.965657           Root MSE      =  21.217

------------------------------------------------------------------------------
           y |      Coef.   Std. Err.      t    P>|t|     [95% Conf. Interval]
-------------+----------------------------------------------------------------
       group |
          2  |  -2.395169   .9497756    -2.52   0.012    -4.257447   -.5328905
          3  |   .2966833   1.200423     0.25   0.805    -2.057054     2.65042
             |
         age |   -.318005    .039939    -7.96   0.000    -.3963157   -.2396943
       _cons |    83.2149   1.963939    42.37   0.000      79.3641    87.06571
------------------------------------------------------------------------------
```

In these data, variable group takes on the values 1, 2, and 3.

Because we typed

```
. regress y  i.group age
```

rather than

```
. regress y   group age
```

instead of fitting the regression as a continuous function of group's values, regress fit the regression on indicators for each level of group included as a separate covariate. In the left column of the coefficient table in the output, the numbers 2 and 3 identify the coefficients that correspond to the values of 2 and 3 of the group variable. Using the more precise terminology of [U] **11.4.3 Factor variables**, the coefficients reported for 2 and 3 are the coefficients for virtual variables 2.group and 3.group, the indicator variables for group = 2 and group = 3.

If group took on the values 2, 10, 11, and 125 rather than 1, 2, and 3, then we would see 2, 10, 11, and 125 below group in the table, corresponding to virtual variables 2.group, 10.group, 11.group, and 125.group.

We can use as many sets of indicators as we need in a varlist. Thus we can type

```
. regress y  i.group i.sex i.arm ...
```

25.2.2 Specifying base levels

In the above results, group = 1 was used as the base level and regress omitted reporting that fact in the output. Somehow, you are just supposed to know that, and usually you do. We can see base levels identified explicitly, however, if we specify the baselevels option, either at the time we estimate the model or, as we do now, when we replay results:

```
. regress, baselevels

      Source |       SS       df       MS              Number of obs =    3000
-------------+------------------------------           F(  3,  2996) =   31.67
       Model |  42767.8126     3  14255.9375           Prob > F      =  0.0000
    Residual |  1348665.19  2996  450.155272           R-squared     =  0.0307
-------------+------------------------------           Adj R-squared =  0.0298
       Total |  1391433.01  2999  463.965657           Root MSE      =  21.217

------------------------------------------------------------------------------
           y |      Coef.   Std. Err.      t    P>|t|     [95% Conf. Interval]
-------------+----------------------------------------------------------------
       group |
          1  |          0  (base)
          2  |  -2.395169   .9497756    -2.52   0.012    -4.257447   -.5328905
          3  |   .2966833   1.200423     0.25   0.805    -2.057054     2.65042
             |
         age |   -.318005    .039939    -7.96   0.000    -.3963157   -.2396943
       _cons |    83.2149   1.963939    42.37   0.000      79.3641    87.06571
------------------------------------------------------------------------------
```

The smallest value of the factor variable is used as the base by default. Using the notation explained in [U] **11.4.3.2 Base levels**, we can request another base level, such as `group` = 2, by typing

```
. regress y  ib2.group age
```

or, such as the largest value of `group`,

```
. regress y  ib(last).group age
```

Changing the base does not fundamentally alter the estimates in the sense that predictions from the model would be identical no matter which base levels we use. Changing the base does change the interpretation of coefficients. In the regression output above, the reported coefficients measure the differences from `group` = 1. Group 2 differs from group 1 by −2.4, and that difference is significant at the 5% level. Group 3 is not significantly different from group 1.

If we fit the above using `group` = 3 as the base,

```
. regress y  ib3.group age
  (output omitted )
```

the coefficients on `group` = 1 and `group` = 2 would be −0.297 and −2.692. Note that the difference between group 2 and group 1 would still be −2.692 − (−0.296) = −2.4. Results may look different, but when looked at correctly, they are the same. Similarly, the significance of `group` = 2 would now be 0.805 rather than 0.012, but that is because what is being tested is different. In the output above, the test against 0 is a test of whether group 2 differs from group 1. In the output that we omit, the test is whether group 2 differs from group 3. If, after running the `ib3.group` specification, we were to type

```
. test 2.group = 1.group
```

we would obtain the same 0.012 result. Similarly, after running the shown result, if we typed `test 3.group = 1.group`, we would obtain 0.805.

25.2.3 Setting base levels permanently

As explained directly above, you can temporarily change the base level by using the `ib.` operator; also see [U] **11.4.3.2 Base levels**. You can change the base level permanently by using the `fvset` command; see [U] **11.4.3.3 Setting base levels permanently**.

25.2.4 Testing significance of a main effect

In the example we have been using,

```
. use http://www.stata-press.com/data/r12/fvex
. regress y i.group age
```

many disciplines refer to the coefficients on the set of indicators for `i.group` as a main effect. Because we have no interactions, the main effect of `i.group` refers to the effect of the levels of `group` taken as a whole. We can test the significance of the indicators by using `contrast` (see [R] **contrast**):

```
. contrast group
Contrasts of marginal linear predictions
Margins      : asbalanced
```

	df	F	P>F
group	2	4.89	0.0076
Residual	2996		

When we specify the name of a factor variable used in the previous estimation command in the `contrast` command, it will perform a joint test on the effects of that variable. Here we are testing whether the coefficients for the group indicators are jointly zero. We reject the hypothesis.

25.2.5 Specifying indicator (dummy) variables as factor variables

We are using the model

```
. use http://www.stata-press.com/data/r12/fvex
. regress y  i.group age
```

We are going to add `sex` to our model. Variable `sex` is a 0/1 variable in our data, a type of variable we call an indicator variable and which many people call a dummy variable. We could type

```
. regress y    sex i.group age
```

but we are going to type

```
. regress y  i.sex i.group age
```

It is better to include indicator variables as factor variables, which is to say, to include indicator variables with the `i.` prefix.

You will obtain the same estimation results either way, but by specifying `i.sex` rather than `sex`, you will communicate to postestimation commands that care that `sex` is not a continuous variable, and that will save you typing later should you use one of those postestimation commands. `margins` (see [R] **margins**) is an example of a postestimation command that cares.

Below we type `regress y i.sex i.group age`, and we will specify the `baselevels` option just to make explicit how `regress` is interpreting our request. Ordinarily, we would not specify the `baselevels` option.

```
. regress y  i.sex i.group age, baselevels
      Source |       SS       df       MS              Number of obs =    3000
-------------+------------------------------           F(  4,  2995) =  136.51
       Model |  214569.509     4  53642.3772           Prob > F      =  0.0000
    Residual |   1176863.5  2995  392.942737           R-squared     =  0.1542
-------------+------------------------------           Adj R-squared =  0.1531
       Total |  1391433.01  2999  463.965657           Root MSE      =  19.823

------------------------------------------------------------------------------
           y |      Coef.   Std. Err.      t    P>|t|     [95% Conf. Interval]
-------------+----------------------------------------------------------------
         sex |
          0  |          0  (base)
          1  |   18.44069   .8819175    20.91   0.000     16.71146    20.16991
             |
       group |
          1  |          0  (base)
          2  |   5.178636   .9584485     5.40   0.000     3.299352    7.057919
          3  |   13.45907   1.286127    10.46   0.000     10.93729    15.98085
             |
         age |  -.3298831   .0373191    -8.84   0.000    -.4030567   -.2567094
       _cons |   68.63586   1.962901    34.97   0.000     64.78709    72.48463
------------------------------------------------------------------------------
```

As with all factor variables, by default the first level of `sex` serves as its base, so the coefficient 18.4 measures the increase in y for `sex` = 1 as compared with `sex` = 0. In these data, `sex` = 1 represents females and `sex` = 0 represents males.

25.2.6 Including interactions

We are using the model

```
. use http://www.stata-press.com/data/r12/fvex
. regress y  i.sex i.group age
```

If we are not certain that the levels of `group` have the same effect for females as they do for males, we should add to our model interactions for each combination of the levels in `sex` and `group`. We would need to add indicators for

`sex` = 0	and	`group` = 1
`sex` = 0	and	`group` = 2
`sex` = 0	and	`group` = 3
`sex` = 1	and	`group` = 1
`sex` = 1	and	`group` = 2
`sex` = 1	and	`group` = 3

Doing this would allow each combination of `sex` and `group` to have a different effect on y.

Interactions like those listed above are produced using the `#` operator. We could type

```
. regress y  i.sex i.group i.sex#i.group age
```

The `#` operator assumes that the variables on either side of it are factor variables, so we can omit the `i.` prefixes and obtain the same result by typing

```
. regress y  i.sex i.group sex#group age
```

We must continue to specify the prefix on the main effects `i.sex` and `i.group`, however.

In the output below, we add the `allbaselevels` option to that. The `allbaselevels` option is much like `baselevels`, except `allbaselevels` lists base levels in interactions as well as in main effects. Specifying `allbaselevels` will make the output easier to understand the first time, and after that, you will probably never specify it again.

```
. regress y  i.sex i.group sex#group age, allbaselevels

      Source |       SS       df       MS              Number of obs =    3000
-------------+------------------------------           F(  6,  2993) =   92.52
       Model |  217691.706     6  36281.9511           Prob > F      =  0.0000
    Residual |   1173741.3  2993  392.162145           R-squared     =  0.1565
-------------+------------------------------           Adj R-squared =  0.1548
       Total |  1391433.01  2999  463.965657           Root MSE      =  19.803

------------------------------------------------------------------------------
           y |      Coef.   Std. Err.      t    P>|t|     [95% Conf. Interval]
-------------+----------------------------------------------------------------
         sex |
           0 |          0  (base)
           1 |   21.71794   1.490858    14.57   0.000     18.79473    24.64115
             |
       group |
           1 |          0  (base)
           2 |   8.420661   1.588696     5.30   0.000     5.305615    11.53571
           3 |   16.47226     1.6724     9.85   0.000     13.19309    19.75143
             |
   sex#group |
         0 1 |          0  (base)
         0 2 |          0  (base)
         0 3 |          0  (base)
         1 1 |          0  (base)
         1 2 |  -4.658322   1.918195    -2.43   0.015    -8.419436   -.8972081
         1 3 |  -6.736936   2.967391    -2.27   0.023    -12.55527   -.9186038
             |
         age |  -.3305546    .0373032    -8.86   0.000    -.4036972   -.2574121
       _cons |   65.97765   2.198032    30.02   0.000     61.66784    70.28745
------------------------------------------------------------------------------
```

Look at the `sex#group` term in the output. There are six combinations of `sex` and `group`, just as we expected. That four of the cells are labeled base and that only two extra coefficients were estimated should not surprise us, at least after we think about. There are 3×2 `sex#age` groups, and thus $3 \times 2 = 6$ means to be estimated, and we indeed estimated six coefficients, including a constant, plus a seventh for continuous variable `age`. Now look at which combinations were treated as base. Treated as base were all combinations that were the base of `sex`, plus all combinations that were the base of `group`. The `sex` = 0-and-`group` = 1 combination was omitted for both reasons, and the other combinations were omitted for one or the other reason.

We entered a two-way interaction between `sex` and `group`. If we believed that the effects of `sex#group` were themselves dependent on the treatment arm of an experiment, we would want the three-way interaction, which we could obtain by typing `sex#group#arm`. Stata allows up to eight-way interactions among factor variables and another eight-ways of interaction among continuous covariates.

❑ Technical note

The virtual variables associated with the interaction terms have the names `1.sex#2.group` and `1.sex#3.group`.

❑

25.2.7 Testing significance of interactions

We are using the model

```
. use http://www.stata-press.com/data/r12/fvex
. regress y  i.sex i.group sex#group age
```

We can test the overall significance of the `sex#group` interaction by typing

```
. contrast sex#group
Contrasts of marginal linear predictions
Margins      : asbalanced
```

	df	F	P>F
sex#group	2	3.98	0.0188
Residual	2993		

We can type the interaction term to be tested—`sex#group`—in the same way as we typed it to include it in the regression. The interaction is significant beyond the 5% level. That is not surprising because both interaction indicators were significant in the regression.

25.2.8 Including factorial specifications

We have the model

```
. use http://www.stata-press.com/data/r12/fvex
. regress y  i.sex i.group sex#group age
```

The above model is called a factorial specification with respect to `sex` and `group` because `sex` and `group` appear by themselves and an interaction. Were it not for `age` being included in the model, we could call this model a full-factorial specification. In any case, Stata provides a shorthand for factorial specifications. We could fit the model above by typing

```
. regress y  sex##group age
```

When you type *A*##*B*, Stata takes that to mean *A B A*#*B*.

When you type *A*##*B*##*C*, Stata takes that to mean *A B C A*#*B A*#*C B*#*C A*#*B*#*C*.

And so on. Up to eight-way interactions are allowed.

The `##` notation is just a shorthand. Estimation results are unchanged. This time we will not specify the `allbaselevels` option:

```
. regress y  sex##group age

      Source |       SS       df       MS              Number of obs =    3000
-------------+------------------------------           F(  6,  2993) =   92.52
       Model |  217691.706     6  36281.9511           Prob > F      =  0.0000
    Residual |   1173741.3  2993  392.162145           R-squared     =  0.1565
-------------+------------------------------           Adj R-squared =  0.1548
       Total |  1391433.01  2999  463.965657           Root MSE      =  19.803

------------------------------------------------------------------------------
           y |      Coef.   Std. Err.      t    P>|t|     [95% Conf. Interval]
-------------+----------------------------------------------------------------
       1.sex |   21.71794   1.490858    14.57   0.000     18.79473    24.64115
             |
       group |
          2  |   8.420661   1.588696     5.30   0.000     5.305615    11.53571
          3  |   16.47226     1.6724     9.85   0.000     13.19309    19.75143
             |
   sex#group |
        1 2  |  -4.658322   1.918195    -2.43   0.015    -8.419436   -.8972081
        1 3  |  -6.736936   2.967391    -2.27   0.023    -12.55527   -.9186038
             |
         age |  -.3305546   .0373032    -8.86   0.000    -.4036972   -.2574121
       _cons |   65.97765   2.198032    30.02   0.000     61.66784    70.28745
------------------------------------------------------------------------------
```

25.2.9 Including squared terms and polynomials

may be used to interact continuous variables if you specify the `c.` indicator in front of them. The command

```
. regress y  age c.age#c.age
```

fits y as a quadratic function of age. Similarly,

```
. regress y  age c.age#c.age c.age#c.age#c.age
```

fits a third-order polynomial.

Using the # operator is preferable to generating squared and cubed variables of `age` because when # is used, Stata understands the relationship between `age` and `c.age#c.age` and `c.age#c.age#c.age`. Postestimation commands can take advantage of this to produce smarter answers; see, for example, *Requirements for model specification* in [R] **margins**.

25.2.10 Including interactions with continuous variables

and ## may be used to create interactions of categorical variables with continuous variables if the continuous variables are prefixed with `c.`, such as `sex#c.age` in

```
. regress y  i.sex age sex#c.age
. regress y  sex##c.age
. regress y  i.sex sex#c.age
```

The result of fitting the first of these models (equivalent to the second) is shown below. We include `allbaselevels` to make results more understandable the first time you see them.

```
. regress y  i.sex age sex#c.age, allbaselevels
      Source |       SS       df       MS              Number of obs =    3000
-------------+------------------------------           F(  3,  2996) =  139.91
       Model |  170983.675     3  56994.5583           Prob > F      =  0.0000
    Residual |  1220449.33  2996   407.35959           R-squared     =  0.1229
-------------+------------------------------           Adj R-squared =  0.1220
       Total |  1391433.01  2999  463.965657           Root MSE      =  20.183

------------------------------------------------------------------------------
           y |      Coef.   Std. Err.      t    P>|t|     [95% Conf. Interval]
-------------+----------------------------------------------------------------
         sex |
           0 |          0  (base)
           1 |   14.92308   2.789012     5.35   0.000     9.454508    20.39165
             |
         age |  -.4929608   .0480944   -10.25   0.000    -.5872622   -.3986595
             |
   sex#c.age |
           0 |          0  (base)
           1 |  -.0224116   .0674167    -0.33   0.740    -.1545994    .1097762
             |
       _cons |   82.36936   1.812958    45.43   0.000      78.8146    85.92413
------------------------------------------------------------------------------
```

The coefficient on the interaction (-0.022) is the difference in the slope of `age` for females (`sex` = 1) as compared with the slope for males. It is far from significant at any reasonable level, so we cannot distinguish the two slopes.

A different but equivalent parameterization of this model would be to omit the main effect of age, the result of which would be that we would estimate the separate slope coefficients of age for males and females:

```
. regress y  i.sex sex#c.age
      Source |       SS       df       MS              Number of obs =    3000
-------------+------------------------------           F(  3,  2996) =  139.91
       Model |  170983.675     3  56994.5583           Prob > F      =  0.0000
    Residual |  1220449.33  2996   407.35959           R-squared     =  0.1229
-------------+------------------------------           Adj R-squared =  0.1220
       Total |  1391433.01  2999  463.965657           Root MSE      =  20.183

------------------------------------------------------------------------------
           y |      Coef.   Std. Err.      t    P>|t|     [95% Conf. Interval]
-------------+----------------------------------------------------------------
       1.sex |   14.92308   2.789012     5.35   0.000     9.454508    20.39165
             |
   sex#c.age |
           0 |  -.4929608   .0480944   -10.25   0.000    -.5872622   -.3986595
           1 |  -.5153724   .0472435   -10.91   0.000    -.6080054   -.4227395
             |
       _cons |   82.36936   1.812958    45.43   0.000      78.8146    85.92413
------------------------------------------------------------------------------
```

It is now easier to see the slopes themselves, although the test of the equality of the slopes no longer appears in the output. We can obtain the comparison of slopes by using the `lincom` postestimation command:

```
. lincom 1.sex#c.age - 0.sex#c.age
 ( 1)  - 0b.sex#c.age + 1.sex#c.age = 0
```

y	Coef.	Std. Err.	t	P>|t|	[95% Conf.	Interval]
(1)	-.0224116	.0674167	-0.33	0.740	-.1545994	.1097762

As noted earlier, it can be difficult at first to know how to refer to individual parameters when you need to type them on postestimation commands. The solution is to replay your estimation results specifying the `coeflegend` option:

```
. regress, coeflegend
```

Source	SS	df	MS
Model	170983.675	3	56994.5583
Residual	1220449.33	2996	407.35959
Total	1391433.01	2999	463.965657

Number of obs	=	3000
F(3, 2996)	=	139.91
Prob > F	=	0.0000
R-squared	=	0.1229
Adj R-squared	=	0.1220
Root MSE	=	20.183

y	Coef.	Legend
1.sex	14.92308	_b[1.sex]
sex#c.age		
0	-.4929608	_b[0b.sex#c.age]
1	-.5153724	_b[1.sex#c.age]
_cons	82.36936	_b[_cons]

The legend suggests that we type

```
. lincom _b[1.sex#c.age] - _b[0b.sex#c.age]
```

instead of `lincom 1.sex#c.age - 0.sex#c.age`. That is, the legend suggests that we bracket terms in `_b[]` and explicitly recognize base levels. The latter does not matter. Concerning bracketing, some commands allow you to omit brackets, and others do not. All commands will allow bracketing, which is why the legend suggests it.

25.2.11 Parentheses binding

Factor-variable operators can be applied to groups of variables if those variables are bound in parentheses. For instance, you can type

```
. regress y  sex##(group c.age c.age#c.age)
```

rather than

```
. regress y  i.sex i.group sex#group age sex#c.age c.age#c.age sex#c.age#c.age
```

Parentheses may be nested. The parenthetically bound notation does not let you specify anything you could not specify without it, but it can save typing and, as importantly, make what you type more understandable. Consider

```
. regress y  i.sex i.group sex#group age sex#c.age c.age#c.age sex#c.age#c.age
. regress y  sex##(group c.age c.age#c.age)
```

The second specification is shorter and easier to read. We can see that all the covariates have different parameters for males and females.

25.2.12 Including indicators for single levels

Consider the following regression of statewide marriage rates (marriages per 100,000) on the median age in the state of the United States:

```
. use http://www.stata-press.com/data/r12/censusfv
(1980 Census data by state)

. regress marriagert  medage

      Source |       SS       df       MS              Number of obs =      50
-------------+------------------------------           F(  1,    48) =    0.00
       Model |  148.944706     1  148.944706           Prob > F      =  0.9949
    Residual |   173402855    48  3612559.48           R-squared     =  0.0000
-------------+------------------------------           Adj R-squared = -0.0208
       Total |   173403004    49  3538836.82           Root MSE      =  1900.7

------------------------------------------------------------------------------
  marriagert |      Coef.   Std. Err.      t    P>|t|     [95% Conf. Interval]
-------------+----------------------------------------------------------------
      medage |   1.029541   160.3387     0.01   0.995    -321.3531    323.4122
       _cons |   1301.307   4744.027     0.27   0.785    -8237.199    10839.81
------------------------------------------------------------------------------
```

There appears to be no effect of median age. We know, however, that couples from around the country flock to Nevada to be married in Las Vegas, which skews our results. Nevada corresponds to `state` = 30 in these data. We can add a single indicator for `state` = 30 by typing

```
. regress marriagert  medage i30.state

      Source |       SS       df       MS              Number of obs =      50
-------------+------------------------------           F(  2,    47) = 2311.15
       Model |   171657575     2  85828787.6           Prob > F      =  0.0000
    Residual |  1745428.85    47   37136.784           R-squared     =  0.9899
-------------+------------------------------           Adj R-squared =  0.9895
       Total |   173403004    49  3538836.82           Root MSE      =  192.71

------------------------------------------------------------------------------
  marriagert |      Coef.   Std. Err.      t    P>|t|     [95% Conf. Interval]
-------------+----------------------------------------------------------------
      medage |  -61.23095    16.2825    -3.76   0.000    -93.98711   -28.47479
    30.state |   13255.81   194.9742    67.99   0.000     12863.57    13648.05
       _cons |   2875.366   481.5533     5.97   0.000     1906.606    3844.126
------------------------------------------------------------------------------
```

These results are more reasonable. By the way, we learned that Nevada corresponds to `state` = 30 by 1) describing our data, from which we learned that variable `state` had a value label named `st`, and 2) using `label list st` to see the list of value labels.

There is a subtlety to specifying individual levels. Let's add another indicator, this time for California. The following will not produce the desired results, and we specify the `baselevels` option to help you understand the issue. First, however, here is the result:

```
. regress marriagert  medage i6.state i30.state, baselevels

      Source |       SS       df       MS              Number of obs =      50
-------------+------------------------------           F(  2,    47) = 2311.15
       Model |   171657575     2  85828787.6           Prob > F      =  0.0000
    Residual |  1745428.85    47   37136.784           R-squared     =  0.9899
-------------+------------------------------           Adj R-squared =  0.9895
       Total |   173403004    49  3538836.82           Root MSE      =  192.71

------------------------------------------------------------------------------
  marriagert |      Coef.   Std. Err.      t    P>|t|     [95% Conf. Interval]
-------------+----------------------------------------------------------------
      medage |  -61.23095    16.2825    -3.76   0.000    -93.98711   -28.47479
             |
       state |
          6  |          0  (base)
         30  |   13255.81   194.9742    67.99   0.000     12863.57    13648.05
             |
       _cons |   2875.366   481.5533     5.97   0.000     1906.606    3844.126
------------------------------------------------------------------------------
```

Look at the result for `state`. Rather than obtaining a coefficient for `6.state` as we expected, Stata instead chose to omit it as the base category.

Stata considers all the individual specifiers for a factor variable together as being related. In our command, we specified that we wanted `i6.state` and `i30.state` by typing

```
. regress marriagert  medage i6.state i30.state
```

and Stata put that together as "include `state`, levels 6 and 30". Then Stata applied its standard logic for dealing with factor variables: treat the smallest level as the base category.

To achieve the desired result, we need to tell Stata that we want no base, which we do by typing the "base none" (bn) modifier:

```
. regress marriagert  medage i6bn.state i30.state
```

We need to specify `bn` only once, and it does not matter where we specify it. We could type

```
. regress marriagert  medage i6.state i30bn.state
```

and we would obtain the same result. We can specify `bn` more than once:

```
. regress marriagert  medage i6bn.state i30bn.state
```

The result of typing any one of these commands is

```
. regress marriagert  medage i6bn.state i30.state, baselevels

      Source |       SS       df       MS              Number of obs =      50
-------------+------------------------------           F(  3,    46) = 1514.31
       Model |   171664796     3  57221598.6           Prob > F      =  0.0000
    Residual |  1738208.24    46  37787.1357           R-squared     =  0.9900
-------------+------------------------------           Adj R-squared =  0.9893
       Total |   173403004    49  3538836.82           Root MSE      =  194.39

------------------------------------------------------------------------------
  marriagert |      Coef.   Std. Err.      t    P>|t|     [95% Conf. Interval]
-------------+----------------------------------------------------------------
      medage |  -60.66128   16.47607    -3.68   0.001    -93.82589   -27.49667
             |
       state |
          6  |   86.12456   197.0209     0.44   0.664    -310.4578     482.707
         30  |   13257.18   196.6991    67.40   0.000     12861.25    13653.12
             |
       _cons |   2856.788   487.6072     5.86   0.000     1875.286     3838.29
------------------------------------------------------------------------------
```

25.2.13 Including subgroups of levels

We just typed

```
. regress marriagert  medage i6bn.state i30.state
```

You can specify specific levels by using numlists. We could have typed

```
. regress marriagert  medage i(6 30)bn.state
```

There may be a tendency to think of `i(6 30)bn.state` as specifying something extra to be added to the regression. In the example above, it is just that. But consider

```
. regress y  i.arm i.agegroup arm#i(3/4)bn.agegroup
```

The goal may be to restrict the interaction term just to levels 3 and 4 of `agegroup`, but the effect will be to restrict `agegroup` to levels 3 and 4 throughout, which includes the term `i.agegroup`.

Try the example for yourself:

```
. use http://www.stata-press.com/data/r12/fvex
. regress y  i.arm i.agegroup arm#i(3/4)bn.agegroup
```

If you really wanted to restrict the interaction term `arm#agegroup` to levels 3 and 4 of `agegroup`, while leaving `i.agegroup` to include all the levels, you need to fool Stata:

```
. generate agegrp = agegroup
. regress y i.arm i.agegroup arm#i(3/4)bn.agegrp
```

In the above, we use `agegroup` for the main effect, but `agegrp` in the interaction.

25.2.14 Combining factor variables and time-series operators

You can combine factor-variable operators with the time-series operators `L.` and `F.` to lag and lead factor variables. Terms like `iL.group` (or `Li.group`), `cL.age#cL.age` (or `Lc.age#Lc.age`), and `F.arm#L.group` are all legal as long as the data are `tsset` or `xtset`. See [U] **11.4.3.6 Using factor variables with time-series operators**.

25.2.15 Treatment of empty cells

Consider the following data:

```
. use http://www.stata-press.com/data/r12/estimability, clear
(margins estimability)
. table sex group
```

sex	group 1	2	3	4	5
0	2	9	27	8	2
1	9	9	3		

In these data, there are no observations for `sex` = 1 and `group` = 4, and for `sex` = 1 and `group` = 5. Here is what happens when you use these data to fit an interacted model:

```
. regress y  sex##group
note: 1.sex#4.group identifies no observations in the sample
note: 1.sex#5.group identifies no observations in the sample

      Source |       SS       df       MS              Number of obs =      69
-------------+------------------------------           F(  7,    61) =    4.88
       Model |  839.550121     7  119.935732           Prob > F      =  0.0002
    Residual |  1500.65278    61  24.6008652           R-squared     =  0.3588
-------------+------------------------------           Adj R-squared =  0.2852
       Total |   2340.2029    68  34.4147485           Root MSE      =  4.9599

------------------------------------------------------------------------------
           y |      Coef.   Std. Err.      t    P>|t|     [95% Conf. Interval]
-------------+----------------------------------------------------------------
       1.sex |  -5.666667   3.877352    -1.46   0.149    -13.41991    2.086579
             |
       group |
          2  |  -13.55556   3.877352    -3.50   0.001     -21.3088    -5.80231
          3  |        -13   3.634773    -3.58   0.001    -20.26818   -5.731822
          4  |    -12.875   3.921166    -3.28   0.002    -20.71586   -5.034145
          5  |        -11   4.959926    -2.22   0.030    -20.91798   -1.082015
             |
   sex#group |
        1 2  |   12.11111   4.527772     2.67   0.010     3.057271    21.16495
        1 3  |         10   4.913786     2.04   0.046     .1742775    19.82572
        1 4  |          0  (empty)
        1 5  |          0  (empty)
             |
       _cons |         32   3.507197     9.12   0.000     24.98693    39.01307
------------------------------------------------------------------------------
```

Stata reports that the results for sex = 1 and group = 4 and for sex = 1 and group = 5 are empty; no coefficients can be estimated.

Empty cells are of no concern when fitting models and interpreting results. If, however, you subsequently perform tests or form linear or nonlinear combinations involving any of the coefficients in the interaction, you should be aware that those tests or combinations may depend on how you parameterized your model. See *Estimability of margins* in [R] **margins**.

26 Overview of Stata estimation commands

Contents

26.1 Introduction

Estimation commands fit models such as linear regression and probit. Stata has many such commands, so it is easy to overlook a few. Some of these commands differ greatly from each other, others are gentle variations on a theme, and still others are equivalent to each other.

Estimation commands share features that this chapter will not discuss; see [U] **20 Estimation and postestimation commands**. Especially see [U] **20.20 Obtaining robust variance estimates**, which discusses an alternative calculation for the estimated variance matrix (and hence standard errors) that many of Stata's estimation commands provide, and [U] **20.12 Performing hypothesis tests on the coefficients**.

Here, however, this chapter will put aside all of that—and all issues of syntax—and deal solely with matching commands to their statistical concepts. This chapter will not cross-reference specific commands. To find the details on a particular command, look up its name in the index.

26.2 Linear regression with simple error structures

Consider models of the form

$$y_j = \mathbf{x}_j\boldsymbol{\beta} + \epsilon_j$$

for a continuous y variable. In this category, estimation is restricted to when σ_ϵ^2 is constant across observations j. The model is called the linear regression model, and the estimator is often called the (ordinary) least-squares (OLS) estimator.

`regress` is Stata's linear regression command. (`regress` produces the robust estimate of variance as well as the conventional estimate, and `regress` has a collection of commands that can be run after it to explore the nature of the fit.)

Also, the following commands will do linear regressions, as does `regress`, but offer special features:

1. `ivregress` fits models in which some of the regressors are endogenous, using either instrumental variables or generalized method of moments (GMM) estimators.

2. `areg` fits models $y_j = \mathbf{x}_j\boldsymbol{\beta}+\mathbf{d}_j\boldsymbol{\gamma}+\epsilon_j$, where $\mathbf{d}_j$ is a mutually exclusive and exhaustive dummy variable set. `areg` obtains estimates of $\boldsymbol{\beta}$ (and associated statistics) without ever forming $\mathbf{d}_j$, meaning that it also does not report the estimated $\boldsymbol{\gamma}$. If your interest is in fitting fixed-effects models, Stata has a better command—`xtreg`—discussed in [U] **26.17.1 Linear regression with panel data**. Most users who find `areg` appealing will probably want to use `xtreg` because it provides more useful summary and test statistics. `areg` duplicates the output that `regress` would produce if you were to generate all the dummy variables. This means, for instance, that the reported R^2 includes the effect of $\boldsymbol{\gamma}$.

3. `boxcox` obtains maximum likelihood estimates of the coefficients and the Box–Cox transform parameters in a model of the form

$$y_i^{(\theta)} = \beta_0 + \beta_1 x_{i1}^{(\lambda)} + \beta_2 x_{i2}^{(\lambda)} + \cdots + \beta_k x_{ik}^{(\lambda)} + \gamma_1 z_{i1} + \gamma_2 z_{i2} + \cdots + \gamma_l z_{il} + \epsilon_i$$

   where $\epsilon \sim N(0, \sigma^2)$. Here the y is subject to a Box–Cox transform with parameter θ. Each of the $x_1, x_2, \ldots, x_k$ is transformed by a Box–Cox transform with parameter λ. The $z_1, z_2, \ldots, z_l$ are independent variables that are not transformed. In addition to the general form specified above, `boxcox` can fit three other versions of this model defined by the restrictions $\lambda = \theta$, $\lambda = 1$, and $\theta = 1$.

4. `tobit` allows estimation of linear regression models when y_i has been subject to left-censoring, right-censoring, or both. Say that y_i is not observed if $y_i < 1{,}000$, but for those observations, it is known that $y_i < 1{,}000$. `tobit` fits such models.

   `ivtobit` does the same but allows for endogenous regressors.

5. `intreg` (interval regression) is a generalization of `tobit`. In addition to allowing open-ended intervals, `intreg` allows closed intervals. Rather than observing y_j, it is assumed that y_{0j} and y_{1j} are observed, where $y_{0j} \le y_j \le y_{1j}$. Survey data might report that a subject's monthly income was in the range \$1,500–\$2,500. `intreg` allows such data to be used to fit a regression model. `intreg` allows $y_{0j} = y_{1j}$ and so can reproduce results reported by `regress`. `intreg` allows y_{0j} to be $-\infty$ and y_{1j} to be $+\infty$ and so can reproduce results reported by `tobit`.

6. `truncreg` fits the regression model when the sample is drawn from a restricted part of the population and so is similar to `tobit`, except that here the independent variables are not observed. Under the normality assumption for the whole population, the error terms in the truncated regression model have a truncated-normal distribution.
7. `cnsreg` allows you to place linear constraints on the coefficients.
8. `eivreg` adjusts estimates for errors in variables.
9. `nl` provides the nonlinear least-squares estimator of $y_j = f(\mathbf{x}_j, \boldsymbol{\beta}) + \epsilon_j$.
10. `rreg` fits robust regression models, which are not to be confused with regression with robust standard errors. Robust standard errors are discussed in [U] **20.20 Obtaining robust variance estimates**. Robust regression concerns point estimates more than standard errors, and it implements a data-dependent method for downweighting outliers.
11. `qreg` produces quantile-regression estimates, a variation that is not linear regression at all but is an estimator of $y_j = \mathbf{x}_j\boldsymbol{\beta} + \epsilon_j$. In the basic form of this model, sometimes called median regression, $\mathbf{x}_j\boldsymbol{\beta}$ measures not the predicted mean of y_j conditional on $\mathbf{x}_j$, but its median. As such, `qreg` is of most interest when ϵ_j does not have constant variance. `qreg` allows you to specify the quantile, so you can produce linear estimates for the predicted 1st, 2nd, ..., 99th percentile.

    Another command, `bsqreg`, is identical to `qreg` but presents bootstrap standard errors.

    The `sqreg` command estimates multiple quantiles simultaneously; standard errors are obtained via the bootstrap.

    The `iqreg` command estimates the difference between two quantiles; standard errors are obtained via the bootstrap.
12. `vwls` (variance-weighted least squares) produces estimates of $y_j = \mathbf{x}_j\boldsymbol{\beta} + \epsilon_j$, where the variance of ϵ_j is calculated from group data or is known a priori. `vwls` is therefore of most interest to categorical-data analysts and physical scientists.

26.3 Structural equation modeling (SEM)

SEM stands for structural equation modeling. The `sem` command fits SEM; see the [SEM] *Stata Structural Equation Modeling Reference Manual*. For those of you unfamiliar with SEM, skip to the next paragraph. For those of you familiar with SEM, let us just say that `sem` is a complete implementation. `sem` can fit models ranging from linear regression to measurement models to simultaneous equations, including along the way confirmatory factor analysis (CFA), correlated uniqueness models, latent growth models, and multiple indicators and multiple causes (MIMIC). You can use the GUI or command interface to specify your model. You can obtain standardized or unstandardized results, direct and indirect effects, goodness-of-fit statistics, modification indices, scores tests, Wald tests, linear and nonlinear tests of estimated parameters, and linear and nonlinear combinations of estimated parameters with confidence intervals. You can perform estimation across groups with easy model specification and easy-to-use tests for group invariance. All this may be done using raw or summary statistics data. In addition, `sem` optionally can use full information maximum-likelihood (FIML) estimation to handle observations containing missing values.

For those of you unfamiliar with SEM, it is worth your time to learn about it if you ever fit linear regressions, multivariate linear regressions, seemingly unrelated regressions, simultaneous systems, measurement error models, or latent-factor models, or if you are interested in generalized method of moments (GMM).

As you have already figured out, SEM is based on the linear model. What SEM brings to the table are unobserved (latent) variables that can be treated (almost) as if they were observed and flexible specification—nearly anything can be allowed to be correlated or can be constrained to be uncorrelated.

SEM fits the first and second moments of the distribution of observed variables—means, variances, and covariances—-rather than fitting the observed values themselves. Both maximum likelihood and GMM methods are available; `sem` uses a weighting matrix corresponding to asymptotic distribution free estimation in the SEM literature.

You still think of the model in the same way as usual, but in a model such as

$$y_j = \beta_0 + \beta_1 x_{1j} + \cdots + \beta_k x_{kj} + \epsilon_j$$

lets now call ϵ_j the error. Reserve the word residual for the true residual of the SEM, which is the difference between the observed and predicted moments.

When SEM is used to fit models that can be fit by the other linear estimators, results are either the same, asymptotically the same (meaning that the finite-sample differences go away as the sample size increases), or different. In the rare cases in which the results differ, the SEM results are better for theoretical reasons.

See the [SEM] *Stata Structural Equation Modeling Reference Manual.*

26.4 ANOVA, ANCOVA, MANOVA, and MANCOVA

ANOVA and ANCOVA are related to linear regression, but we classify them separately. The related Stata commands are `anova`, `oneway`, and `loneway`. The `manova` command provides MANOVA and MANCOVA (multivariate ANOVA and ANCOVA).

`anova` fits ANOVA and ANCOVA models, one-way and up—including two-way factorial, three-way factorial, etc.—and fits nested and mixed-design models and repeated-measures models.

`oneway` fits one-way ANOVA models. It is quicker at producing estimates than `anova`, although `anova` is so fast that this probably does not matter. The important difference is that `oneway` can report multiple-comparison tests.

`loneway` is an alternative to `oneway`. The results are numerically the same, but `loneway` can deal with more levels (limited only by dataset size; `oneway` is limited to 376 levels and `anova` to 798, but for `anova` to reach 798 requires a lot of memory), and `loneway` reports some additional statistics, such as the intraclass correlation.

`manova` fits MANOVA and MANCOVA models, one-way and up—including two-way factorial, three-way factorial, etc.—and it fits nested and mixed-design models.

26.5 Generalized linear models

The generalized linear model is

$$g\{E(y_j)\} = \mathbf{x}_j\boldsymbol{\beta}, \qquad y_j \sim F$$

where $g()$ is called the link function and F is a member of the exponential family, both of which you specify before estimation. `glm` fits this model.

The GLM framework encompasses a surprising array of models known by other names, including linear regression, Poisson regression, exponential regression, and others. Stata provides dedicated estimation commands for many of these. Stata has, for instance, `regress` for linear regression, `poisson` for Poisson regression, and `streg` for exponential regression, and that is not all of the overlap.

`glm` by default uses maximum likelihood estimation and alternatively estimates via iterated reweighted least squares (IRLS) when the `irls` option is specified. For each family, F, there is a corresponding link function, $g()$, called the canonical link, for which IRLS estimation produces results identical to maximum likelihood estimation. You can, however, match families and link functions as you wish, and, when you match a family to a link function other than the canonical link, you obtain a different but valid estimator of the standard errors of the regression coefficients. The estimator you obtain is asymptotically equivalent to the maximum likelihood estimator, which, in small samples, produces slightly different results.

For example, the canonical link for the binomial family is logit. `glm, irls` with that combination produces results identical to the maximum-likelihood `logit` (and `logistic`) command. The binomial family with the probit link produces the probit model, but probit is not the canonical link here. Hence, `glm, irls` produces standard error estimates that differ slightly from those produced by Stata's maximum-likelihood `probit` command.

Many researchers feel that the maximum-likelihood standard errors are preferable to IRLS estimates (when they are not identical), but they would have a difficult time justifying that feeling. Maximum likelihood probit is an estimator with (solely) asymptotic properties; `glm, irls` with the binomial family and probit link is an estimator with (solely) asymptotic properties, and in finite samples, the standard errors differ a little.

Still, we recommend that you use Stata's dedicated estimators whenever possible. IRLS—the theory—and `glm, irls`—the command—are all encompassing in their generality, meaning that they rarely use the right jargon or provide things in the way you wish they would. The narrower commands, such as `logit`, `probit`, and `poisson`, focus on the issue at hand and are invariably more convenient.

`glm` is useful when you want to match a family to a link function that is not provided elsewhere.

`glm` also offers several estimators of the variance–covariance matrix that are consistent, even when the errors are heteroskedastic or autocorrelated. Another advantage of a `glm` version of a model over a model-specific version is that many of these VCE estimators are available only for the `glm` implementation. You can also obtain the ML–based estimates of the VCE from `glm`.

26.6 Binary-outcome qualitative dependent-variable models

There are many ways to write these models, one of which is

$$\Pr(y_j \neq 0) = F(\mathbf{x}_j\boldsymbol{\beta})$$

where F is some cumulative distribution. Two popular choices for $F()$ are the normal and logistic, and the models are called the probit and logit (or logistic regression) models. A third is the complementary log–log function; maximum likelihood estimates are obtained by Stata's `cloglog` command.

The two parent commands for the maximum likelihood estimator of probit and logit are `probit` and `logit`, although `logit` has a sibling, `logistic`, that provides the same estimates but displays results in a slightly different way. There is also an exact logistic estimator; see [U] **26.11 Exact estimators**.

Do not read anything into the names logit and logistic. Logit and logistic have two interchanged definitions in two scientific camps. In the medical sciences, logit means the minimum χ^2 estimator, and logistic means maximum likelihood. In the social sciences, it is the other way around. From our experience, it appears that neither reads the other's literature, because both talk (and write books) asserting that logit means one thing and logistic the other. Our solution is to provide both `logit` and `logistic`, which do the same thing, so that each camp can latch on to the maximum likelihood command under the name each expects.

There are two slight differences between `logit` and `logistic`. `logit` reports estimates in the coefficient metric, whereas `logistic` reports exponentiated coefficients—odds ratios. This is in accordance with the expectations of each camp and makes no substantial difference. The other difference is that `logistic` has a family of post`logistic` commands that you can run to explore the nature of the fit. Actually, that is not exactly true because all the commands for use after `logistic` can also be used after `logit`.

If you have not already selected `logit` or `logistic` as your favorite, we recommend that you try `logistic`. Logistic regression (logit) models are more easily interpreted in the odds-ratio metric.

In addition to `logit` and `logistic`, Stata provides `glogit`, `blogit`, and `binreg` commands.

`blogit` is the maximum likelihood estimator (the same as `logit` or `logistic`) but applied on data organized in a different way. Rather than having individual observations, your data are organized so that each observation records the number of observed successes and failures.

`glogit` is the weighted-regression, grouped-data estimator.

`binreg` can be used to model either individual-level or grouped data in an application of the generalized linear model. The family is assumed to be binomial, and each link provides a distinct parameter interpretation. Also, `binreg` offers several options for setting the link function according to the desired biostatistical interpretation. The available links and interpretation options are

Option	Implied link	Parameter
`or`	`logit`	Odds ratios = exp(β)
`rr`	`log`	Risk ratios = exp(β)
`hr`	`log complement`	Health ratios = exp(β)
`rd`	`identity`	Risk differences = β

Related to logit, the skewed logit estimator scobit adds a power to the logit link function and is estimated by Stata's `scobit` command.

Turning to probit, you have two choices: `probit` and `ivprobit`. `probit` fits a maximum-likelihood probit model. `ivprobit` fits a probit model where one or more of the regressors are endogenously determined.

Stata also provides `bprobit` and `gprobit`. The `bprobit` command is a maximum likelihood estimator—equivalent to `probit`—but works with data organized in the different way outlined above. `gprobit` is the weighted-regression, grouped-data estimator.

Continuing with probit: `hetprob` fits heteroskedastic probit models. In these models, the variance of the error term is parameterized.

`heckprob` fits probit models with sample selection.

Also, Stata's `biprobit` command fits bivariate probit models, meaning two correlated outcomes. `biprobit` also fits partial-observability models in which only the outcomes $(0,0)$ and $(1,1)$ are observed.

26.7 ROC analysis

ROC stands for receiver operating characteristics. ROC deals with specificity and sensitivity, the number of false positives and undetected true positives of a diagnostic test. The term ROC dates back to the early days of radar when there was a knob on the radar receiver labeled "ROC". If you turned the knob one way, the receiver became more sensitive, which meant it was more likely to show airplanes that really were there and, simultaneously, more likely to show returns where there were no airplanes (false positives). If you turned the knob the other way, you suppressed many of the false positives, but unfortunately, also suppressed the weak returns from real airplanes (undetected positives). These days, in the statistical applications we imagine, one does not turn a knob but instead chooses a value of the diagnostic test, above which is declared to be a positive, and below which, a negative.

ROC analysis is applied to binary outcomes such as those appropriate for probit or logistic regression. After fitting a model, one can obtain predicted probabilities of a positive outcome. One chooses a value, above which the predicted probability is declared a positive and below which, a negative.

ROC analysis is about modeling the tradeoff of sensitivity and specificity as the threshold value is chosen.

Stata's suite for ROC analysis consists of six commands: `roctab`, `roccomp`, `rocfit`, `rocgold`, `rocreg`, and `rocregplot`.

`roctab` provides nonparametric estimation of the ROC curve and produces Bamber and Hanley confidence intervals for the area under the curve.

`roccomp` provides tests of equality of ROC areas. It can estimate nonparametric and parametric binormal ROC curves.

`rocfit` fits maximum likelihood models for a single classifier, an indicator of the latent binormal variable for the true status.

`rocgold` performs tests of equality of ROC area against a "gold standard" ROC curve and can adjust significance levels for multiple tests across classifiers via Šidák's method.

`rocreg` performs ROC regression; it can adjust both sensitivity and specificity for prognostic factors such as age and gender; it is by far the most general of all the ROC commands.

`rocregplot` graphs ROC curves as modeled by `rocreg`. ROC curves may be drawn across covariate values, across classifiers, and across both.

See [R] **roc**.

26.8 Conditional logistic regression

`clogit` is Stata's conditional logistic regression estimator. In this model, observations are assumed to be partitioned into groups, and a predetermined number of events occur in each group. The model measures the risk of the event according to the observation's covariates, $\mathbf{x}_j$. The model is used in matched case–control studies (`clogit` allows $1:1$, $1:k$, and $m:k$ matching) and is used in natural experiments whenever observations can be grouped into pools in which a fixed number of events occur. `clogit` is also used to fix logistic regression with fixed group effects.

26.9 Multiple-outcome qualitative dependent-variable models

For more than two outcomes, Stata provides ordered logit, ordered probit, rank-ordered logit, multinomial logistic regression, multinomial probit regression, McFadden's choice model (conditional fixed-effects logistic regression), and nested logistic regression.

oprobit and ologit provide maximum-likelihood ordered probit and logit. These are generalizations of probit and logit models known as the proportional odds model and are used when the outcomes have a natural ordering from low to high. The idea is that there is an unmeasured $z_j = \mathbf{x}_j\boldsymbol{\beta}$, and the probability that the kth outcome is observed is $\Pr(c_{k-1} < z_j < c_k)$, where $c_0 = -\infty$, $c_k = +\infty$, and $c_1, \ldots, c_{k-1}$ along with $\boldsymbol{\beta}$ are estimated from the data.

rologit fits the rank-ordered logit model for rankings. This model is also known as the Plackett–Luce model, the exploded logit model, and choice-based conjoint analysis.

asroprobit fits the probit model for rankings, a more flexible estimator than rologit because asroprobit allows covariances among the rankings. asroprobit is also similar to asmprobit (below), which is used for outcomes that have no natural ordering. The as in the name signifies that asroprobit also allows alternative-specific regressors—variables that have different coefficients for each alternative.

slogit fits the stereotype logit model for data that are not truly ordered, as data are for ologit, but for which you are not sure that it is unordered, in which case mlogit would be appropriate.

mlogit fits maximum-likelihood multinomial logistic models, also known as polytomous logistic regression. It is intended for use when the outcomes have no natural ordering and you know only the characteristics of the outcome chosen (and, perhaps, the chooser).

asclogit fits McFadden's choice model, also known as conditional logistic regression. In the context denoted by the name *McFadden's choice model*, the model is used when the outcomes have no natural ordering, just as multinomial logistic regression, but the characteristics of the outcomes chosen and not chosen are known (along with, perhaps, the characteristics of the chooser).

In the context denoted by the name *conditional logistic regression*—mentioned above—subjects are members of pools, and one or more are chosen, typically to be infected by some disease or to have some other unfortunate event befall them. Thus the characteristics of the chosen and not chosen are known, and the issue of the characteristics of the chooser never arises. Either way, it is the same model.

In their choice-model interpretations, mlogit and clogit assume that the odds ratios are independent of other alternatives, known as the independence of irrelevant alternatives (IIA) assumption. This assumption is often rejected by the data and the nested logit model relaxes this assumption. nlogit is also popular for fitting the random utility choice model.

asmprobit is for use with outcomes that have no natural ordering and with regressors that are alternative specific. It is weakly related to mlogit. Unlike mlogit, asmprobit does not assume the IIA.

mprobit is also for use with outcomes that have no natural ordering but with models that do not have alternative-specific regressors.

26.10 Count dependent-variable models

These models concern dependent variables that count the number of occurrences of an event. In this category, we include Poisson and negative binomial regression. For the Poisson model,

$$E(\text{count}) = E_j \exp(\mathbf{x}_j\boldsymbol{\beta})$$

where E_j is the exposure time. `poisson` fits this model; see [R] **poisson**. There is also an exact Poisson estimator; see [U] **26.11 Exact estimators**.

Negative-binomial regression refers to estimating with data that are a mixture of Poisson counts. One derivation of the negative binomial model is that individual units follow a Poisson regression model but there is an omitted variable that follows a gamma distribution with parameter α. Negative-binomial regression estimates $\boldsymbol{\beta}$ and α. `nbreg` fits such models. A variation on this, unique to Stata, allows you to model α. `gnbreg` fits those models. See [R] **nbreg**.

Truncation refers to count models in which the outcome count variable is observed only above a certain threshold. In truncated data, the threshold is typically zero. Commands `tpoisson` and `tnbreg` fit such models; see [R] **tpoisson** and [R] **tnbreg**.

Zero inflation refers to count models in which the number of zero counts is more than would be expected in the regular model. The excess zeros are explained by a preliminary probit or logit process. If the preliminary process produces a positive outcome, the usual counting process occurs, and otherwise the count is zero. Thus whenever the preliminary process produces a negative outcome, excess zeros are produced. The `zip` and `zinb` commands fit such models; see [R] **zip** and [R] **zinb**.

26.11 Exact estimators

Exact estimators refer to models which, rather than being estimated by asymptotic formulas, are estimated by enumerating the conditional distribution of the sufficient statistics and then computing the maximum likelihood estimate using that distribution. Standard errors cannot be estimated, but confidence intervals can be and are obtained from the enumerations.

`exlogistic` fits logistic models of binary data in this way.

`expoisson` fits Poisson models of count data in this way.

In small samples, exact estimates have better coverage than the asymptotic estimates, and exact estimates are the only way to obtain estimates, tests, and confidence intervals of covariates that perfectly predict the observed outcome.

26.12 Linear regression with heteroskedastic errors

We now consider the model $y_j = \mathbf{x}_j\boldsymbol{\beta} + \epsilon_j$, where the variance of ϵ_j is nonconstant.

First, `regress` can fit such models if you specify the `vce(robust)` option. What Stata calls robust is also known as the White correction for heteroskedasticity.

For scientists who have data where the variance of ϵ_j is known a priori, `vwls` is the command. `vwls` produces estimates for the model given each observation's variance, which is recorded in a variable in the data.

If you wish to model the heteroskedasticity on covariates, use the `het()` option of the `arch` command. Although `arch` is written primarily to analyze time-series data, it can be used with cross-sectional data. Before using `arch` with cross-sectional data, set the data as time series, by typing `gen faketime = _n` and then typing `tsset faketime`.

Finally, `qreg` performs quantile regression, which in the presence of heteroskedasticity is most of interest. Median regression (one of `qreg`'s capabilities) is an estimator of $y_j = \mathbf{x}_j\boldsymbol{\beta} + \epsilon_j$ when ϵ_j is heteroskedastic. Even more useful, you can fit models of other quantiles and so model the heteroskedasticity. Also see the `sqreg` and `iqreg` commands; `sqreg` estimates multiple quantiles simultaneously. `iqreg` estimates differences in quantiles.

26.13 Stochastic frontier models

frontier fits stochastic production or cost frontier models on cross-sectional data. The model can be expressed as

$$y_i = \mathbf{x}_i\boldsymbol{\beta} + v_i - su_i$$

where

$$s = \begin{cases} 1 & \text{for production functions} \\ -1 & \text{for cost functions} \end{cases}$$

u_i is a nonnegative disturbance standing for technical inefficiency in the production function or cost inefficiency in the cost function. Although the idiosyncratic error term v_i is assumed to have a normal distribution, the inefficiency term is assumed to be one of the three distributions: half-normal, exponential, or truncated-normal. Also, when the nonnegative component of the disturbance is assumed to be either half-normal or exponential, frontier can fit models in which the error components are heteroskedastic conditional on a set of covariates. When the nonnegative component of the disturbance is assumed to be from a truncated-normal distribution, frontier can also fit a conditional mean model, where the mean of the truncated-normal distribution is modeled as a linear function of a set of covariates.

For panel-data stochastic frontier models, see [U] **26.17.1 Linear regression with panel data**.

26.14 Regression with systems of equations

For systems of equations with endogenous covariates, use the three-stage least-squares (3SLS) estimator reg3. The reg3 command can produce constrained and unconstrained estimates.

When we have correlated errors across equations but no endogenous right-hand-side variables,

$$\begin{aligned} y_{1j} &= \mathbf{x}_{1j}\boldsymbol{\beta} + \epsilon_{1j} \\ y_{2j} &= \mathbf{x}_{2j}\boldsymbol{\beta} + \epsilon_{2j} \\ &\vdots \\ y_{mj} &= \mathbf{x}_{mj}\boldsymbol{\beta} + \epsilon_{mj} \end{aligned}$$

where $\epsilon_{k\cdot}$ and $\epsilon_{l\cdot}$ are correlated with correlation ρ_{kl}, a quantity to be estimated from the data. This is called Zellner's seemingly unrelated regressions, and sureg fits such models. When $\mathbf{x}_{1j} = \mathbf{x}_{2j} = \cdots = \mathbf{x}_{mj}$, the model is known as multivariate regression, and the corresponding command is mvreg.

The equations need not be linear; if they are not linear, use nlsur.

26.15 Models with endogenous sample selection

What has become known as the Heckman model refers to linear regression in the presence of sample selection: $y_j = \mathbf{x}_j\boldsymbol{\beta} + \epsilon_j$ is not observed unless some event occurs that itself has probability $p_j = F(\mathbf{z}_j\gamma + \nu_j)$, where ϵ and ν might be correlated and $\mathbf{z}_j$ and $\mathbf{x}_j$ may contain variables in common.

heckman fits such models by maximum likelihood or Heckman's original two-step procedure.

This model has recently been generalized to replace the linear regression equation with another probit equation, and that model is fit by heckprob.

Another important case of endogenous sample selection is the treatment-effects model, which considers the effect of an endogenously chosen binary treatment on another endogenous, continuous variable, conditional on two sets of independent variables. treatreg fits a treatment-effects model by using either a two-step consistent estimator or full maximum likelihood.

26.16 Models with time-series data

ARIMA refers to models with autoregressive integrated moving-average processes, and Stata's arima command fits models with ARIMA disturbances via the Kalman filter and maximum likelihood. These models may be fit with or without covariates. arima also fits ARMA models. See [TS] **arima**.

ARFIMA stands for autoregressive fractionally integrated moving average and handles long-memory processes. ARFIMA generalizes the ARMA and ARIMA models. ARMA models assume short memory; after a shock, the process reverts back to its trend relatively quickly. ARIMA models assume shocks are permanent and memory never fades. ARFIMA provides a middle ground in the length of the process's memory. The arfima command fits ARFIMA models. In addition to one-step and dynamic forecasts, arfima can predict fractionally integrated series. See [TS] **arfima**.

UCM stands for unobserved components model and decomposes a time series into trend, seasonal, cyclic, and idiosyncratic components after controlling for optional exogenous variables. UCM provides a flexible and formal approach to smoothing and decomposition problems. The ucm command fits UCM models. See [TS] **ucm**.

Relatedly, band-pass and high-pass filters are also used to decompose a time series into trend and cyclic components, even though the tsfilter commands are not estimation commands; see [TS] **tsfilter**. Provided are Baxter–King, Butterworth, Christiano–Fitzgerald, and Hodrick–Prescott filters.

Concerning ARIMA, ARFIMA, and UCM, the estimated parameters are sometimes more easily interpreted in terms of the implied spectral density. psdensity transforms results; see [TS] **psdensity**.

Stata's prais command performs regression with AR(1) disturbances using the Prais–Winsten or Cochrane–Orcutt transformation. Both two-step and iterative solutions are available, as well as a version of the Hildreth–Lu search procedure. See [TS] **prais**.

newey produces linear regression estimates with the Newey–West variance estimates that are robust to heteroskedasticity and autocorrelation of specified order. See [TS] **newey**.

Stata provides estimators for ARCH, GARCH, univariate, and multivariate models. These models are for time-varying volatility. ARCH models allow for conditional heteroskedasticity by including lagged variances. GARCH models also include lagged second moments of the innovations (errors). ARCH stands for autoregressive conditional heteroskedasticity. GARCH stands for generalized ARCH.

arch fits univariate ARCH and GARCH models, and the command provides many popular extensions, including multiplicative conditional heteroskedasticity. Errors may be normal or Student's t or may follow a generalized error distribution. Robust standard errors are optionally provided. See [TS] **arch**.

mgarch fits multivariate ARCH and GARCH models, including the diagonal vech model and the constant, dynamic, and varying conditional correlation models. Errors may be multivariate normal or multivariate Student's t. Robust standard errors are optionally provided. See [TS] **mgarch**.

Stata provides VAR, SVAR, and VEC estimators for modeling multivariate time series. VAR and SVAR deal with stationary series, and SVAR places additional constraints on the VAR model that identifies the impulse–response functions. VEC is for cointegrating VAR models. VAR stands for vector autoregression. SVAR stands for structural VAR. VEC stands for vector error-correction model.

var fits VAR models, svar fits SVAR models, and vec fits VEC models. These commands share many of the same features for specification testing, forecasting, and parameter interpretation; see [TS] **var intro** for both var and svar, [TS] **vec intro** for vec, and [TS] **irf** for all three impulse–response functions and forecast-error variance decomposition. For lag-order selection, residual analysis, and Granger causality tests, see [TS] **var intro** (for var and svar) and [TS] **vec intro**.

sspace estimates the parameters of multivariate state-space models using the Kalman filter. The state-space representation of time-series models is extremely flexible and can be used to estimate the parameters of many different models, including vector autoregressive moving-average (VARMA) models, dynamic-factor (DF) models, and structural time-series (STS) models. It can also solve some stochastic dynamic-programming problems. See [TS] **sspace**.

dfactor estimates the parameters of dynamic-factor models. These flexible models for multivariate time series provide for a vector-autoregressive structure in both observed outcomes and in unobserved factors. They also allow exogenous covariates for observed outcomes or unobserved factors. See [TS] **dfactor**.

26.17 Panel-data models

26.17.1 Linear regression with panel data

This section could just as well be called "linear regression with complex error structures". Commands in this class begin with the letters xt.

xtreg fits models of the form

$$y_{it} = \mathbf{x}_{it}\boldsymbol{\beta} + \nu_i + \epsilon_{it}$$

xtreg can produce the between-regression estimator, the within-regression (fixed effects) estimator, or the GLS random-effects (matrix-weighted average of between and within results) estimator. It can also produce the maximum-likelihood random-effects estimator.

xtregar can produce the within estimator and a GLS random-effects estimator when the ϵ_{it} are assumed to follow an AR(1) process.

xtivreg contains the between-2SLS estimator, the within-2SLS estimator, the first-differenced-2SLS estimator, and two GLS random-effects-2SLS estimators to handle cases in which some of the covariates are endogenous.

xtmixed is a generalization of xtreg that allows for multiple levels of panels, random coefficients, and variance-component estimation in general. In the xtmixed framework, residuals (random effects) can occur anywhere and have any level of subscript.

xtabond is for use with dynamic panel-data models (models in which there are lagged dependent variables) and can produce the one-step, one-step robust, and two-step Arellano–Bond estimators. xtabond can handle predetermined covariates, and it reports both the Sargan and autocorrelation tests derived by Arellano and Bond.

xtdpdsys is an extension of xtabond and produces estimates with smaller bias when the coefficients of the AR process are large. xtpdsys is also more efficient than xtabond. Whereas xtabond uses moment conditions based on the differenced errors, xtpdsys uses moment conditions based on both the differenced errors and their levels.

xtdpd is an extension of xtdpdsys and can be used to estimate the parameters of a broader class of dynamic panel-data models. xtdpd can be used to fit models with serially correlated idiosyncratic errors, whereas xtdpdsys and xtabond assume no serial correlation. Or xtdpd can be used with models where the structure of the predetermined variables is more complicated than that assumed by xtdpdsys or xtabond.

xtgls produces generalized least-squares estimates for models of the form

$$y_{it} = \mathbf{x}_{it}\boldsymbol{\beta} + \epsilon_{it}$$

where you may specify the variance structure of ϵ_{it}. If you specify that ϵ_{it} is independent for all i and t, xtgls produces the same results as regress up to a small-sample degrees-of-freedom correction applied by regress but not by xtgls.

You may choose among three variance structures concerning i and three concerning t, producing a total of nine different models. Assumptions concerning i deal with heteroskedasticity and cross-sectional correlation. Assumptions concerning t deal with autocorrelation and, more specifically, AR(1) serial correlation.

Alternative methods report the OLS coefficients and a version of the GLS variance–covariance estimator. xtpcse produces panel-corrected standard error (PCSE) estimates for linear cross-sectional time-series models, where the parameters are estimated by OLS or Prais–Winsten regression. When you are computing the standard errors and the variance–covariance estimates, the disturbances are, by default, assumed to be heteroskedastic and contemporaneously correlated across panels.

In the jargon of GLS, the random-effects model fit by xtreg has exchangeable correlation within i—xtgls does not model this particular correlation structure. xtgee, however, does.

xtgee fits population-averaged models, and it optionally provides robust estimates of variance. Moreover, xtgee allows other correlation structures. One that is of particular interest to those with many data goes by the name *unstructured*. The within-panel correlations are simply estimated in an unconstrained way. [U] **26.17.3 Generalized linear models with panel data** will discuss this estimator further because it is not restricted to linear regression models.

xthtaylor uses instrumental variables estimators to estimate the parameters of panel-data random-effects models of the form

$$y_{it} = \mathbf{X}_{1it}\boldsymbol{\beta}_1 + \mathbf{X}_{2it}\boldsymbol{\beta}_2 + \mathbf{Z}_{1i}\boldsymbol{\delta}_1 + \mathbf{Z}_{2i}\boldsymbol{\delta}_2 + u_i + e_{it}$$

The individual effects u_i are correlated with the explanatory variables $\mathbf{X}_{2it}$ and $\mathbf{Z}_{2i}$ but are uncorrelated with $\mathbf{X}_{1it}$ and $\mathbf{Z}_{1i}$, where $\mathbf{Z}_1$ and $\mathbf{Z}_2$ are constant within panel.

xtfrontier fits stochastic production or cost frontier models for panel data. You may choose from a time-invariant model or a time-varying decay model. In both models, the nonnegative inefficiency term is assumed to have a truncated-normal distribution. In the time-invariant model, the inefficiency term is constant within panels. In the time-varying decay model, the inefficiency term is modeled as a truncated-normal random variable multiplied by a specific function of time. In both models, the idiosyncratic error term is assumed to have a normal distribution. The only panel-specific effect is the random inefficiency term.

26.17.2 Censored linear regression with panel data

xttobit fits random-effects tobit models and generalizes that to observation-specific censoring.

xtintreg performs random-effects interval regression and generalizes that to observation-specific censoring. Interval regression, in addition to allowing open-ended intervals, also allows closed intervals.

26.17.3 Generalized linear models with panel data

[U] **26.5 Generalized linear models** above discussed the model

$$g\{E(y_j)\} = \mathbf{x}_j\boldsymbol{\beta}, \qquad y_j \sim F \tag{1}$$

where $g()$ is the link function and F is a member of the exponential family, both of which you specify before estimation.

There are two ways to extend the generalized linear model to panel data. They are the generalized linear mixed model (GLMM) and generalized estimation equations (GEE).

A GLMM uses random effects to model within-panel correlation. GLMMs may be fit with the commands `xtmixed`, `xtmelogit`, and `xtmepoisson`; see [XT] **xtmixed**, [XT] **xtmelogit**, and [XT] **xtmepoisson**.

GEE uses a working correlation structure to model within-panel correlation. GEEs may be fit with the `xtgee` command; see [XT] **xtgee**.

26.17.4 Qualitative dependent-variable models with panel data

`xtprobit` fits random-effects probit regression via maximum likelihood. It also fits population-averaged models via GEE. This last is nothing more than `xtgee` with the binomial family, probit link, and exchangeable error structure.

`xtlogit` fits random-effects logistic regression models via maximum likelihood. It also fits conditional fixed-effects models via maximum likelihood. Finally, as with `xtprobit`, it fits population-averaged models via GEE.

`xtcloglog` estimates random-effects complementary log-log regression via maximum likelihood. It also fits population-averaged models via GEE.

`xtmelogit` fits multilevel mixed-effects models for binary/binomial responses.

`xtmepoisson` fits multilevel mixed-effects models for count responses.

26.17.5 Count dependent-variable models with panel data

`xtpoisson` fits two different random-effects Poisson regression models via maximum likelihood. The two distributions for the random effect are gamma and normal. It also fits conditional fixed-effects models, and it fits population-averaged models via GEE. This last is nothing more than `xtgee` with the Poisson family, log link, and exchangeable error structure.

`xtnbreg` fits random-effects negative binomial regression models via maximum likelihood (the distribution of the random effects is assumed to be beta). It also fits conditional fixed-effects models, and it fits population-averaged models via GEE.

26.17.6 Random-coefficients model with panel data

`xtrc` fits Swamy's random-coefficients linear regression model. In this model, rather than only the intercept varying across groups, all the coefficients are allowed to vary. `xtrc` is a special case of `xtmixed`.

26.18 Survival-time (failure-time) models

Commands are provided to fit Cox proportional-hazards models, competing-risks regression, and several parametric survival models including exponential, Weibull, Gompertz, lognormal, loglogistic, and generalized gamma; see [ST] **stcox**, [ST] **stcrreg**, and [ST] **streg**. The commands for Cox and parametric regressions, `stcox` and `streg`, are appropriate for single- or multiple-failure-per-subject data. The command for competing-risks regression, `stcrreg`, is appropriate only for single-failure data. Conventional, robust, bootstrap, and jackknife standard errors are available with all three commands, with the exception that for `stcrreg`, robust standard errors are the conventional standard errors.

Both the Cox model and the parametric models (as fit using Stata) allow for two additional generalizations. First, the models may be modified to allow for latent random effects, or *frailties*. Second, the models may be stratified in that the baseline hazard function may vary completely over a set of strata. The parametric models also allow for the modeling of ancillary parameters.

Competing-risks regression, as fit using Stata, is a useful alternative to Cox regression for datasets where more than one type of failure occurs, in other words, for data where failure events compete with one another. In such situations, competing-risks regression allows you to easily assess covariate effects on the incidence of the failure type of interest without having to make strong assumptions concerning the independence of failure types.

`stcox`, `stcrreg`, and `streg` require that the data be `stset` so that the proper response variables may be established. After you `stset` the data, the time/censoring response is taken as understood, and you need only supply the regressors (and other options) to `stcox`, `stcrreg`, and `streg`. With `stcrreg`, one required option deals with specifying which events compete with the failure event of interest that was previously `stset`.

26.19 Generalized method of moments (GMM)

`gmm` fits models using generalized method of moments (GMM). With the interactive version of the command, you enter your moment equations directly into the dialog box or command line using substitutable expressions just like with `nl` or `nlsur`. The moment-evaluator program version gives you greater flexibility in exchange for increased complexity; with this version, you write a program that calculates the moments based on a vector of parameters passed to it.

`gmm` can fit both single- and multiple-equation models, and you can combine moment conditions of the form $E\{\mathbf{z}_i u_i(\beta)\} = \mathbf{0}$, where $\mathbf{z}_i$ is a vector of instruments and $u_i(\beta)$ is often an additive regression error term, as well as more general moment conditions of the form $E\{\mathbf{h}_i(\mathbf{z}_i;\beta)\} = \mathbf{0}$. In the former case, you specify the expression for $u_i(\beta)$ and use the `instruments()` and `xtinstruments()` options to specify $\mathbf{z}_i$. In the latter case, you specify the expression for $\mathbf{h}_i(\mathbf{z}_i;\beta)$; because that expression incorporates your instruments, you do not use the `instruments()` or `xtinstruments()` option.

`gmm` supports cross-sectional, time-series, and panel data. You can request weight matrices and VCEs that are suitable for independent and identically distributed errors, that are suitable for heteroskedastic errors, that are appropriate for clustered observations, or that are heteroskedasticity- and autocorrelation-consistent (HAC). For HAC weight matrices and VCEs, `gmm` lets you specify the bandwidth or request an automatic bandwidth selection algorithm.

26.20 Estimation with correlated errors

By correlated errors, we mean that observations are grouped, and that within group, the observations might be correlated but, across groups, they are uncorrelated. `regress` with the `vce(cluster` *clustvar*`)` option can produce "correct" estimates, that is, inefficient estimates with correct standard errors and lots of robustness; see [U] **20.20 Obtaining robust variance estimates**. Obviously, if you know the correlation structure (and are not mistaken), you can do better, so `xtreg` and `xtgls` are also of interest here; we discuss them in [U] **26.17.1 Linear regression with panel data**.

Estimation in the presence of autocorrelated errors is discussed in [U] **26.16 Models with time-series data**.

26.21 Survey data

Stata's `svy` command fits statistical models for complex survey data. `svy` is a prefix command, so to obtain linear regression, you type

```
. svy: regress ...
```

or to obtain probit regression, you type

```
. svy: probit ...
```

but first you must type a `svyset` command to define the survey design characteristics. Prefix `svy` works with many estimation commands, and everything is documented together in the *Survey Data Reference Manual.*

`svy` supports the following variance-estimation methods:

- Taylor-series linearization
- Bootstrap
- Balanced repeated replication (BRR)
- Jackknife
- Successive difference replication (SDR)

See [SVY] **variance estimation** for details.

`svy` supports the following survey design characteristics:

- With- and without- replacement sampling
- Sampling weights
- Stratification
- Poststratification
- Clustering
- Multiple stages of clustering without replacement
- BRR and jackknife replication weights

See [SVY] **svyset** for details.

Subpopulation estimation is available for all estimation commands.

Tabulations and summary statistics are also available, including means, proportions, ratios, and totals over multiple subpopulations, and direct standardization of means, proportions, and ratios.

26.22 Multiple imputation

Multiple imputation (MI) is a statistical technique for estimation in the presence of missing data. If you fit the parameters of y on x_1, x_2, and x_3 using any of the other Stata estimation commands, parameters are fit on the data for which y, x_1, x_2, and x_3 contain no missing values. This process is known as listwise or casewise deletion because observations for which any of y, x_1, x_2, or x_3 contain missing values are ignored or, said differently, deleted from consideration. MI is a technique to recover the information in those ignored observations when the missing values are missing at random (MAR) or missing completely at random (MCAR). Data are MAR if the probability that a value is missing may depend on observed data but not on unobserved data. Data are MCAR if the probability of missingness is not even a function of the observed data.

MI is named for the imputations it produces to replace the missing values in the data. MI does not just form replacement values for the missing data, it produces multiple replacements. The purpose is not to create replacement values as close as possible to the true ones, but to handle missing data in a way resulting in valid statistical inference.

There are three steps in an MI analysis. First, one forms M imputations for each missing value in the data. Second, one fits the model of interest separately on each of the M resulting datasets. Finally, one combines those M estimation results into the desired single result.

The `mi` command does this for you. It can be used with most of Stata's estimation commands, including survey, survival, and panel and multilevel models. See [MI] **intro**.

26.23 Multivariate and cluster analysis

Most of Stata's multivariate capabilities are to be found in the *Multivariate Statistics Reference Manual*, although there are some exceptions.

1. `mvreg` fits multivariate regressions.
2. `manova` fits MANOVA and MANCOVA models, one-way and up—including two-way factorial, three-way factorial, etc.—and it fits nested and mixed-design models. Also see [U] **26.4 ANOVA, ANCOVA, MANOVA, and MANCOVA** above.
3. `canon` estimates canonical correlations and their corresponding loadings. Canonical correlation attempts to describe the relationship between two sets of variables.
4. `pca` extracts principal components and reports eigenvalues and loadings. Some people consider principal components a descriptive tool—in which case standard errors as well as coefficients are relevant—and others look at it as a dimension-reduction technique.
5. `factor` fits factor models and provides principal factors, principal-component factors, iterated principal-component factors, and maximum-likelihood solutions. Factor analysis is concerned with finding few common factors $\widehat{\mathbf{z}}_k$, $k = 1, \dots, q$ that linearly reconstruct the original variables $\mathbf{y}_i$, $i = 1, \dots, L$.
6. `tetrachoric`, in conjunction with `pca` or `factor`, allows you to perform PCA or factor analysis on binary data.
7. `rotate` provides a wide variety of orthogonal and oblique rotations after `factor` and `pca`. Rotations are often used to produce more interpretable results.
8. `procrustes` performs Procrustes analysis, one of the standard methods of multidimensional scaling. It can perform orthogonal or oblique rotations, as well as translation and dilation.

9. mds performs metric and nonmetric multidimensional scaling for dissimilarity between observations with respect to a set of variables. A wide variety of dissimilarity measures are available and, in fact, are the same as those for cluster.
10. ca performs correspondence analysis, an exploratory multivariate technique for analyzing cross-tabulations and the relationship between rows and columns.
11. mca performs multiple correspondence analysis (MCA) and joint correspondence analysis (JCA).
12. mvtest performs multivariate of multivariate normality along with tests of means, covariances, and correlations.
13. cluster provides cluster analysis; both hierarchical and partition clustering methods are available. Strictly speaking, cluster analysis does not fall into the category of statistical estimation. Rather, it is a set of techniques for exploratory data analysis. Stata's cluster environment has many different similarity and dissimilarity measures for continuous and binary data.
14. discrim and candisc perform discriminant analysis. candisc performs linear discriminant analysis (LDA). discrim also performs LDA, and it performs quadratic discriminant analysis (QDA), kth nearest neighbor (KNN), and logistic discriminant analysis. The two commands differ in default output. discrim shows the classification summary, candisc shows the canonical linear discriminant functions, and both will produce either.

26.24 Pharmacokinetic data

There are four estimation commands for analyzing pharmacokinetic data. See [R] **pk** for an overview of the pk system.

1. pkexamine calculates pharmacokinetic measures from time-and-concentration subject-level data. pkexamine computes and displays the maximum measured concentration, the time at the maximum measured concentration, the time of the last measurement, the elimination time, the half-life, and the area under the concentration–time curve (AUC).
2. pksumm obtains the first four moments from the empirical distribution of each pharmacokinetic measurement and tests the null hypothesis that the distribution of that measurement is normally distributed.
3. pkcross analyzes data from a crossover design experiment. When one is analyzing pharmaceutical trial data, if the treatment, carryover, and sequence variables are known, the omnibus test for separability of the treatment and carryover effects is calculated.
4. pkequiv performs bioequivalence testing for two treatments. By default, pkequiv calculates a standard confidence interval symmetric about the difference between the two treatment means. pkequiv also calculates confidence intervals symmetric about zero and intervals based on Fieller's theorem. Also, pkequiv can perform interval hypothesis tests for bioequivalence.

26.25 Specification search tools

There are three other commands that are not really estimation commands but are combined with estimation commands to assist in specification searches: stepwise, fracpoly, and mfp.

stepwise, one of Stata's prefix commands, provides stepwise estimation. You can use the stepwise prefix with some, but not all, estimation commands. In [R] **stepwise** is a table of the estimation commands that are currently supported, but do not take it too literally. It was accurate as of the day that Stata 12 was released, but if you install the official updates, stepwise may now work with other commands, too. If you want to use stepwise with some estimation command, you should try it. Either it will work or you will get the message that the estimation command is not supported by `stepwise`.

`fracpoly` and `mfp` are commands to assist you in performing fractional-polynomial functional specification searches.

26.26 Obtaining new estimation commands

This chapter has discussed all the estimation commands included in Stata 12 the day that it was released; by now, there may be more. To obtain an up-to-date list, type `search estimation`.

And, of course, you can always write your own commands; see [R] **ml**.

26.27 Reference

Gould, W. W. 2000. sg124: Interpreting logistic regression in all its forms. *Stata Technical Bulletin* 53: 19–29. Reprinted in *Stata Technical Bulletin Reprints*, vol. 9, pp. 257–270. College Station, TX: Stata Press.

27 Commands everyone should know

Contents

27.1 42 commands

Putting aside the statistical commands that might particularly interest you, here are 42 commands that everyone should know:

Getting online help	[U] **4 Stata's help and search facilities**
`help`, `hsearch`, `net search`, `search`, `findit`	
Keeping Stata up to date	
`ado`, `net`, `update`	[U] **28 Using the Internet to keep up to date**
`adoupdate`	[R] **adoupdate**
Operating system interface	
`pwd`, `cd`	[D] **cd**
Using and saving data from disk	
`save`	[D] **save**
`use`	[D] **use**
`append`, `merge`	[U] **22 Combining datasets**
`compress`	[D] **compress**
Inputting data into Stata	[U] **21 Inputting and importing data**
`import`	[D] **import**
`edit`	[D] **edit**
Basic data reporting	
`describe`	[D] **describe**
`codebook`	[D] **codebook**
`list`	[D] **list**
`browse`	[D] **edit**
`count`	[D] **count**
`inspect`	[D] **inspect**
`table`	[R] **table**
`tabulate`	[R] **tabulate oneway** and [R] **tabulate twoway**

Data manipulation	[U] **13 Functions and expressions**
`generate, replace`	[D] **generate**
`egen`	[D] **egen**
`rename`	[D] **rename**, [D] **rename group**
`clear`	[D] **clear**
`drop, keep`	[D] **drop**
`sort`	[D] **sort**
`encode, decode`	[D] **encode**
`order`	[D] **order**
`by`	[U] **11.5 by varlist: construct**
`reshape`	[D] **reshape**
Keeping track of your work	
`log`	[U] **15 Saving and printing output—log files**
`notes`	[D] **notes**
Convenience	
`display`	[R] **display**

27.2 The by construct

If you do not understand the `by` *varlist*`:` construct, `_n`, and `_N`, and their interaction, and if you process data where observations are related, you are missing out on something. See

[U] **13.7 Explicit subscripting**
[U] **11.5 by varlist: construct**

Say that you have a dataset with multiple observations per person, and you want the average value of each person's blood pressure (`bp`) for the day. You could

```
. egen avgbp = mean(bp), by(person)
```

but you could also

```
. by person, sort: gen avgbp = sum(bp)/_N
. by person: replace avgbp = avgbp[_N]
```

Yes, typing two commands is more work than typing just one, but understanding the two-command construct is the key to generating more complicated things that no one ever thought about adding to `egen`.

Say that your dataset also contains `time` recording when each observation was made. If you want to add the total time the person is under observation (last time minus first time) to each observation, type

```
. by person (time), sort: gen ttl = time[_N]-time[1]
```

Or, suppose you want to add how long it has been since the person was last observed to each observation:

```
. by person (time), sort: gen howlong = time - time[_n-1]
```

If instead you wanted how long it would be until the next observation, type

```
. by person (time), sort: gen whennext = time[_n+1] - time
```

`by` *varlist*`:`, `_n`, and `_N` are often the solution to difficult calculations.

28 Using the Internet to keep up to date

Contents

28.1 Overview

Stata can read files over the Internet. Just to prove that to yourself, type the following:

```
. use http://www.stata.com/manual/chapter28, clear
```

You have just reached out and gotten a dataset from our website. The dataset is not in HTML format, nor does this have anything to do with your browser. We just copied the Stata data file `chapter28.dta` onto our server, and now people all over the world can use it. If you have a webpage, you can do the same thing. It is a convenient way to share datasets with colleagues.

Now type the following:

```
. update query
```

We promise that nothing bad will happen. `update` will read a short file from www.stata.com that will allow Stata to report whether your copy of Stata is up to date. Is your copy up to date? Now you know. If it is not, we will show you how to update it—it is no harder than typing `update`.

Now type the following:

```
. net from http://www.stata.com
```

That will go to www.stata.com and tell you what is available from our user-download site. The material there is not official, but it is useful. More useful is to type

```
. search kernel regression, net
```

or equivalently,

```
. net search kernel regression
```

That will search the entire web for additions to Stata having to do with kernel regression, whether the additions are from the *Stata Journal*, *Stata Technical Bulletin*, Statalist, archive sites, or private user sites.

To summarize: Stata can read files over the Internet:

1. You can share datasets, do-files, etc., with colleagues all over the world. This requires no special expertise, but you do need to have a webpage.
2. You can update Stata; it is free, easy, and fast.
3. You can find and add new features to Stata; it is also free, easy, and fast.

Finally, you can create a site to distribute new features for Stata.

28.2 Sharing datasets (and other files)

There is just nothing to it: you copy the file as-is (in binary) onto the server and then let your colleagues know the file is there. This works for `.dta` files, `.do` files, `.ado` files, and, in fact, all files.

On the receiving end, you can `use` the file (if it is a `.dta` dataset) or you can copy it:

```
. use http://www.stata.com/manual/chapter28, clear
. copy http://www.stata.com/manual/chapter28.dta mycopy.dta
```

Stata includes a copy-file command and it works over the Internet just as `use` does; see [D] **copy**.

❑ Technical note

If you are concerned about transmission errors, you can create a checksum file before you copy the file onto the server. In placing `chapter28.dta` on our site, we started with `chapter28.dta` in our working directory and typed

```
. checksum chapter28.dta, save
```

This created the new file `chapter28.sum`. We then placed both files on our server. We did not have to create this second file, but because we did, when you `use` the data, Stata will be able to detect transmission errors and warn you if there are problems.

How would Stata know? `chapter28.sum` is a short file containing the result of a mathematical calculation made on the contents of `chapter28.dta`. When your Stata receives `chapter28.dta`, it repeats the calculation and then compares that result with what is recorded in `chapter28.sum`. If the results are different, then there must have been a transmission error.

Creating a checksum file is optional.

See [D] **checksum**.

❑

28.3 Official updates

Although we follow no formal schedule for the release of updates, we typically provide updates to Stata approximately once a month. You do not have to update that often, although we recommend that you do. There are two ways to check whether your copy of Stata is up to date:

type	or select
`. update query`	**Help > Check for Updates**

After that if an update is available, you should

type	or click on
`. update all`	*Install available updates*

After you have updated your Stata, to find out what has changed

type	or select
`. help whatsnew`	**Help > What's New?**

28.3.1 Frequently asked questions about updating

1. Could something go wrong and make my Stata become unusable?

   No. The updates are copied to a temporary place on your computer, Stata examines them to make sure they are complete before copying them to the official place. Thus either the updates are installed or they are not.

2. I do not have access to the Internet from within Stata. Is there a way to update Stata manually?

   Yes. Open your Internet browser to http://www.stata.com/support/updates/ and follow the instructions on that page.

28.4 Downloading and managing additions by users

Try the following:

type

```
. net from http://www.stata.com
```

or select

Help > SJ and User-written Programs

and click on one of the links.

28.4.1 Downloading files

We are not the only ones developing additions to Stata. Stata is supported by a large and highly competent user community. An important part of this is the *Stata Journal* (SJ) and the *Stata Technical Bulletin* (STB). The *Stata Journal* is a refereed, quarterly journal containing articles of interest to Stata users. For more details and subscription information, visit the *Stata Journal* website at http://www.stata-journal.com.

The *Stata Journal* is a printed and electronic journal with corresponding software. If you want the journal, you must subscribe, but the software is available for free from our website at http://www.stata-journal.com.

The predecessor to the *Stata Journal* was the *Stata Technical Bulletin* (STB). The STB was also a printed and electronic journal with corresponding software. Individual STB issues may still be purchased. The STB software is available for free from our website at http://www.stata.com.

Below are instructions for installing the *Stata Journal* and the *Stata Technical Bulletin* software from our website.

Installing the Stata Journal software

1. Select **Help > SJ and User-written Programs**.
2. Click on *Stata Journal*.
3. Click on *sj2-2*.
4. Click on *st0001_1*.
5. Click on *click here to install*.

or

1. Type: `. net from http://www.stata-journal.com/software`
2. Type: `. net cd sj2-2`
3. Type: `. net describe st0001_1`
4. Type: `. net install st0001_1`

The above could be shortened to

```
. net from http://www.stata-journal.com/software/sj2-2
. net describe st0001_1
. net install st0001_1
```

You could also type

```
. net sj 2-2
. net describe st0001_1
. net install st0001_1
```

Installing the STB software

1. Select **Help > SJ and User-written Programs**.
2. Click on *STB*.
3. Click on *stb58*.
4. Click on *sg84_3*.
5. Click on *click here to install*.

or type

1. Type: `. net from http://www.stata.com`
2. Type: `. net cd stb`
3. Type: `. net cd stb58`
4. Type: `. net describe sg84_3`
5. Type: `. net install sg84_3`

The above could be shortened to

```
. net from http://www.stata.com/stb/stb58
. net describe sg84_3
. net install sg84_3
```

28.4.2 Managing files

You now have the `concord` command, because we just downloaded and installed it. Convince yourself of this by typing

```
. help concord
```

and you might try it out, too. Let's now list the additions you have installed—that is probably just `concord`—and then get rid of `concord`.

In command mode, you can type

```
. ado dir
[1] package sg84_3 from http://www.stata.com/stb/stb58
      STB-58 sg84_3.  Concordance correlation coefficient: minor corrections
```

If you had more additions installed, they would be listed. Now knowing that you have *sg84_3* installed, you can obtain a more thorough description by typing

```
. ado describe sg84_3
```

```
[1] package sg84_3 from http://www.stata.com/stb/stb58

TITLE
      STB-58 sg84_3.  Concordance correlation coefficient:  minor corrections
DESCRIPTION/AUTHOR(S)
      STB insert by Thomas J. Steichen, RJRT
                    Nicholas J. Cox, University of Durham, UK
      Support:  steicht@rjrt.com, n.j.cox@durham.ac.uk
      After installation, see help concord
INSTALLATION FILES
      c/concord.ado
      c/concord.sthlp
INSTALLED ON
      5 Oct 2002
```

You can erase *sg84_3* by typing

```
. ado uninstall sg84_3
package sg84_3 from http://www.stata.com/stb/stb58
      STB-58 sg84_3.  Concordance correlation coefficient:  minor corrections
(package uninstalled)
```

You can do all of this from the point-and-click interface, too. Pull down **Help** and select **SJ and User-written Programs** and then click on *List*. From there, you can click on *sg84_3* to see the detailed description of the package and from there you can click on *click here to uninstall* if you want to erase it.

For more information on the ado command and the corresponding menu, see [R] **net**.

28.4.3 Finding files to download

There are two ways to find useful files to download. One is simply to thumb through sites. That is inefficient but entertaining. If you want to do that,

1. Select **Help > SJ and User-written Programs**.
2. Click on *Other Locations*.
3. Click on *links*.

What you are doing is starting at our download site and then working out from there. We maintain a list of other sites and those sites will have more links. You can do this from command mode, too:

```
. net from http://www.stata.com
. net cd links
```

The efficient way to find files—at least if you know what you are looking for—is to search. There are two ways to do that. If you suspect what you are looking for might already be in Stata (or published in the SJ), use Stata's `search` command:

```
. search concordance correlation
```

Equivalently, you could select **Help > Search**. Either way, you will learn about *sg84_3* and you can even click to install it.

If you want to search for additions over the net, that is, the SJ and archive sites and user sites, type

```
. net search concordance correlation
```

or select **Help > Search**, and this time click *Search net resources*, rather than the default "*Search documentation and FAQs*".

28.4.4 Updating additions by users

After you have installed some user-written features, you should periodically check whether any updates or bug fixes area available for those commands. You can do this with the `adoupdate` command. Simply type `adoupdate` to see if any updates are available, and if they are, type `adoupdate, update` to obtain the updates. See [R] **adoupdate** for more details.

28.5 Making your own download site

There are two reasons you may wish to create your own download site:

1. You have datasets and the like, you want to share them with colleagues, and you want to make it easier for colleagues to download the files.
2. You have written Stata programs, etc., that you wish to share with the Stata user community.

Making a download site is easy; the full instructions are found in [R] **net**.

At the beginning of this chapter, we pretended that you had a dataset you wanted to share with colleagues. We said you just had to copy the dataset onto your server and then let your colleagues know the dataset is there.

Let's now pretend that you had two datasets, `ds1.dta` and `ds2.dta`, and you wanted your colleagues to be able to learn about and fetch the datasets by using the `net` command or by pulling down **Help** and selecting **SJ and User-written Programs**.

First, you would copy the datasets to your home page just as before. Then you would create three more files, one to describe your site named `stata.toc` and two more to describe each "package" you want to provide:

begin stata.toc
```
v 3
d My name and affiliation (or whatever other title I choose)
d Datasets for the PAR study
p ds1 The base dataset
p ds2 The detail dataset
```
end stata.toc

begin ds1.pkg
```
v 3
d ds1.  The base dataset
d My name or whatever else I wanted to put
d This dataset contains the baseline values for ...
d Distribution-Date: 26sep2011
p ds1.dta
```
end ds1.pkg

begin ds2.pkg

```
v 3
d ds1.  The detail dataset
d My name or whatever else I wanted to put
d This dataset contains the follow-up information ...
d Distribution-Date: 26sep2011
p ds2.dta
```

end ds2.pkg

The `Distribution-Date` line in the description should be changed whenever you change your package. This line is used by `adoupdate` to determine if a user who has installed your package needs to update it.

Here is what users would see when they went to your site:

```
. net from http://www.myuni.edu/hande/~aparker

http://www.myuni.edu/hande/~aparker
My name and whatever else I wanted to put

Datasets for the PAR study

PACKAGES you could -net describe-:
    ds1               The base dataset
    ds2               The detail dataset

. net describe ds1

package ds1 from http://www.myuni.edu/hande/~aparker

TITLE
      ds1.  The base dataset
DESCRIPTION/AUTHOR(S)
      My name and whatever else I wanted to put
      This dataset contains the baseline values for ...
      Distribution-Date: 26sep2011
ANCILLARY FILES                                    (type net get ds1)
      ds1.dta

. net get ds1
checking ds1 consistency and verifying not already installed...

copying into current directory...
      copying  ds1.dta
ancillary files successfully copied.

. _
```

See [R] **net**.

Subject and author index

This is the subject and author index for the *User's Guide*. You may also want to consult the combined subject index (and the combined author index) in the *Quick Reference and Index*.

Semicolons set off the most important entries from the rest. Sometimes no entry will be set off with semicolons, meaning that all entries are equally important.

D

E

F

G

H

I

J

K

L

P

Q

R

S

W

Y